Advances in Intelligent Systems and Computing

Volume 411

Series editor

Janusz Kacprzyk, Polish Academy of Sciences, Warsaw, Poland
e-mail: kacprzyk@ibspan.waw.pl

About this Series

The series "Advances in Intelligent Systems and Computing" contains publications on theory, applications, and design methods of Intelligent Systems and Intelligent Computing. Virtually all disciplines such as engineering, natural sciences, computer and information science, ICT, economics, business, e-commerce, environment, healthcare, life science are covered. The list of topics spans all the areas of modern intelligent systems and computing.

The publications within "Advances in Intelligent Systems and Computing" are primarily textbooks and proceedings of important conferences, symposia and congresses. They cover significant recent developments in the field, both of a foundational and applicable character. An important characteristic feature of the series is the short publication time and world-wide distribution. This permits a rapid and broad dissemination of research results.

Advisory Board

Chairman

Nikhil R. Pal, Indian Statistical Institute, Kolkata, India
e-mail: nikhil@isical.ac.in

Members

Rafael Bello, Universidad Central "Marta Abreu" de Las Villas, Santa Clara, Cuba
e-mail: rbellop@uclv.edu.cu

Emilio S. Corchado, University of Salamanca, Salamanca, Spain
e-mail: escorchado@usal.es

Hani Hagras, University of Essex, Colchester, UK
e-mail: hani@essex.ac.uk

László T. Kóczy, Széchenyi István University, Győr, Hungary
e-mail: koczy@sze.hu

Vladik Kreinovich, University of Texas at El Paso, El Paso, USA
e-mail: vladik@utep.edu

Chin-Teng Lin, National Chiao Tung University, Hsinchu, Taiwan
e-mail: ctlin@mail.nctu.edu.tw

Jie Lu, University of Technology, Sydney, Australia
e-mail: Jie.Lu@uts.edu.au

Patricia Melin, Tijuana Institute of Technology, Tijuana, Mexico
e-mail: epmelin@hafsamx.org

Nadia Nedjah, State University of Rio de Janeiro, Rio de Janeiro, Brazil
e-mail: nadia@eng.uerj.br

Ngoc Thanh Nguyen, Wroclaw University of Technology, Wroclaw, Poland
e-mail: Ngoc-Thanh.Nguyen@pwr.edu.pl

Jun Wang, The Chinese University of Hong Kong, Shatin, Hong Kong
e-mail: jwang@mae.cuhk.edu.hk

More information about this series at http://www.springer.com/series/11156

Himansu Sekhar Behera
Durga Prasad Mohapatra
Editors

Computational Intelligence in Data Mining—Volume 2

Proceedings of the International Conference on CIDM, 5–6 December 2015

 Springer

Editors
Himansu Sekhar Behera
Department of Computer Science
  Engineering and Information Technology
Veer Surendra Sai University of Technology
Sambalpur, Odisha
India

Durga Prasad Mohapatra
Department of Computer Science
  and Engineering
National Institute of Technology
Rourkela, Odisha
India

ISSN 2194-5357   ISSN 2194-5365   (electronic)
Advances in Intelligent Systems and Computing
ISBN 978-81-322-2729-8   ISBN 978-81-322-2731-1   (eBook)
DOI 10.1007/978-81-322-2731-1

Library of Congress Control Number: 2014956493

Printed on acid-free paper

This Springer imprint is published by SpringerNature
The registered company is Springer (India) Pvt. Ltd.

Preface

The 2nd International Conference on "Computational Intelligence in Data Mining (ICCIDM-2015)" is organized by R.I.T., Berhampur, Odisha, India on 5 and 6 December 2015. ICCIDM is an international forum for representation of research and developments in the fields of Data Mining and Computational Intelligence. More than 300 perspective authors had submitted their research papers to the conference. This time the editors have selected 96 papers after the double-blind peer review process by elegantly experienced subject expert reviewers chosen from the country and abroad. The proceedings of ICCIDM is a mix of papers from some latest findings and research of the authors. It is being a great honour for us to edit the proceedings. We have enjoyed considerably working in cooperation with the International Advisory, Program and Technical Committee to call for papers, review papers and finalize papers to be included in the proceedings.

This International Conference aims at encompassing a new breed of engineers and technologists making it a crest of global success. All the papers are focused on the thematic presentation areas of the conference and they have provided ample opportunity for presentation in different sessions. Research in data mining has its own history. But, there is no doubt about the tips and further advancements in the data mining areas will be the main focus of the conference. This year's program includes exciting collections of contributions resulting from a successful call for papers. The selected papers have been divided into thematic areas including both review and research papers and which highlight the current focus of Computational Intelligence Techniques in Data Mining. The conference aims at creating a forum for further discussion for an integrated information field incorporating a series of technical issues in the frontier analysis and design aspects of different alliances in the related field of intelligent computing and others. Therefore, the call for paper was on three major themes like Methods, Algorithms, and Models in Data Mining and Machine learning, Advance Computing and Applications. Further, the papers discussing the issues and applications related to the theme of the conference were also welcomed to ICCIDM.

The proceedings of ICCIDM are to be released to mark this great day of ICCIDM more special. We hope the author's own research and opinions add value to it. First and foremost are the authors of papers, columns and editorials whose works have made the conference a great success. We had a great time putting together this proceeding. The ICCIDM conference and proceedings are a credit to a large group of people and everyone should be proud of the outcome. We extend our deep sense of gratitude to all for their warm encouragement, inspiration and continuous support for making it possible.

Hope all of us will appreciate the good contributions made and justify our efforts.

Acknowledgments

The 2015 edition of ICCIDM has drawn hundreds of research articles authored by numerous academicians, researchers and practitioners throughout the world. We thank all of them for sharing their knowledge and research findings on an international platform like ICCIDM and thus contributing towards producing such a comprehensive conference proceedings of ICCIDM.

The level of enthusiasm displayed by the Organizing Committee members right from day one is commendable. The extraordinary spirit and dedication shown by the Organizing Committee in every phase throughout the conference deserves sincere thanks from the core of my heart.

It has indeed been an honour for us to edit the proceedings of the conference. We have been fortunate enough to work in cooperation with a brilliant International Advisory, Program and Technical Committee consisting of eminent academicians to call for papers, review papers and finalize papers to be included in the proceedings.

We would like to express our heartfelt gratitude and obligations to the benign reviewers for sparing their valuable time and putting in effort to review the papers in a stipulated time and providing their valuable suggestions and appreciation in improvising the presentation, quality and content of this proceedings. The eminence of these papers is an accolade not only to the authors but also to the reviewers who have guided towards perfection.

Last but not the least, the editorial members of Springer Publishing deserve a special mention and our sincere thanks to them not only for making our dream come true in the shape of this proceedings, but also for its hassle-free and in-time publication in Advances in Intelligent Systems and Computing, Springer.

The ICCIDM conference and proceedings are a credit to a large group of people and everyone should be proud of the outcome.

Conference Committee

Patron

Er. P.K. Patra, Secretary, Roland Group of Institutions, Odisha, India

Convenor

Dr. G. Jena, Principal, RIT, Berhampur, Odisha, India

Organizing Secretary

Mr. Janmenjoy Nayak, DST INSPIRE Fellow, Government of India

Honorary General Chair

Prof. Dr. P.K. Dash, Director, Multi Disciplinary Research Center, S'O'A University, Odisha, India

General Chair

Prof. Dr. A. Abraham, Director, Machine Intelligence Research Labs (MIR Labs) Washington, USA

Honorary Advisory Chair

Prof. Dr. S.K. Pal, Padma Shri Awarded Former Director, J.C. Bose National Fellow and INAE Chair Professor Distinguished Professor, Indian Statistical Institute, Kolkata, India

Prof. Dr. L.M. Patnaik, Ex-Vice Chancellor, DIAT, Pune; Former Professor, IISc, Bangalore, India

Prof. Dr. D. Sharma, University Distinguished Professor, Professor of Computer Science, University of Canberra, Australia

Program Chair

Prof. Dr. B.K. Panigrahi, Ph.D., Associate Professor, Department of Electrical Engineering, IIT Delhi, India

Prof. Dr. H.S. Behera, Ph.D., Associate Professor, Department of Computer Science Engineering and Information Technology, Veer Surendra Sai University of Technology (VSSUT), Burla, Odisha, India

Prof. Dr. R.P. Panda, Ph.D., Professor, Department of ETC, VSSUT, Burla, Odisha, India

Volume Editors

Prof. H.S. Behera, Associate Professor, Department of Computer Science Engineering and Information Technology, Veer Surendra Sai University of Technology (VSSUT), Burla, Odisha, India

Prof. D.P. Mohapatra, Associate Professor, Department of Computer Science and Engineering, NIT, Rourkela, Odisha, India

Technical Committee

Dr. P.N. Suganthan, Ph.D., Associate Professor, School of EEE, NTU, Singapore

Dr. Istvan Erlich, Ph.D., Chair Professor, Head, Department of EE and IT, University of DUISBURG-ESSEN, Germany

Dr. Biju Issac, Ph.D., Professor, Teesside University, Middlesbrough, England, UK

Dr. N.P. Padhy, Ph.D., Professor, Department of EE, IIT, Roorkee, India

Dr. Ch. Satyanarayana, Ph.D., Professor, Department of Computer Science and Engineering, JNTU Kakinada

Dr. M. Murugappan, Ph.D, Senior Lecturer, School of Mechatronic Engineering, Universiti Malaysia Perlis, Perlis, Malaysia

Dr. G. Sahoo, Ph.D., Professor and Head, Department of IT, B.I.T, Meshra, India

Dr. R.H. Lara, Ph.D., Professor, The Electrical Company of Quito (EEQ), Ecuador

Dr. Kashif Munir, Ph.D., Professor, King Fahd University of Petroleum and Minerals, Hafr Al-Batin Campus, Kingdom of Saudi Arabia

Dr. L. Sumalatha, Ph.D., Professor, Department of Computer Science and Engineering, JNTU Kakinada

Dr. R. Boutaba, Ph.D., Professor, University of Waterloo, Canada

Dr. K.N. Rao, Ph.D., Professor, Department of Computer Science and Engineering, Andhra University, Visakhapatnam

Dr. S. Das, Ph.D., Associate Professor, Indian Statistical Institute, Kolkata, India

Dr. H.S. Behera, Ph.D., Associate Professor, Department of Computer Science Engineering and Information Technology, Veer Surendra Sai University of Technology (VSSUT), Burla, Odisha, India

Dr. J.K. Mandal, Ph.D., Professor, Department of CSE, Kalyani University, Kolkata, India

Dr. P.K. Hota, Ph.D., Professor, Department of EE, VSSUT, Burla, Odisha, India

Dr. S. Panda, Ph.D., Professor, Department of EEE, VSSUT, Burla, Odisha, India

Dr. S.C. Satpathy, Professor and Head, Department of Computer Science and Engineering, ANITS, AP, India

Dr. A.K. Turuk, Ph.D., Associate Professor and Head, Department of CSE, NIT, Rourkela, India

Dr. D.P. Mohapatra, Ph.D., Associate Professor, Department of CSE, NIT, Rourkela, India

Dr. R. Behera, Ph.D., Asst. Professor, Department of EE, IIT, Patna, India

Dr. S. Das, Ph.D., Asst. Professor, Department of Computer Science and Engineering, Galgotias University

Dr. M. Patra, Ph.D., Reader, Berhampur University, Odisha, India

Dr. S. Sahana, Ph.D., Asst. Professor, Department of CSE, BIT, Mesra, India

Dr. Asit Das, Ph.D., Associate Professor, Department of CSE, IIEST, WB, India

International Advisory Committee

Prof. G. Panda, IIT, BBSR
Prof. Kenji Suzuki, University of Chicago
Prof. Raj Jain, W.U, USA
Prof. P. Mohapatra, University of California
Prof. S. Naik, University of Waterloo, Canada
Prof. S. Bhattacharjee, NIT, Surat
Prof. BrijeshVerma, C.Q.U, Australia
Prof. Richard Le, Latrob University, AUS
Prof. P. Mitra, IIT, KGP

Prof. Amit Das, IIEST, Kolkata
Prof. Michele Nappi, University of Salerno, Italy
Prof. P. Bhattacharya, NIT, Agaratala
Prof. G. Chakraborty, Iwate Prefectural University

Conference Steering Committee

Publicity Chair
Dr. R.R. Rath, RIT, Berhampur
Prof. S.K. Acharya, RIT, Berhampur
Dr. P.D. Padhy, RIT, Berhampur
Mr. B. Naik, VSSUT, Burla

Logistic Chair
Dr. B.P. Padhi, RIT, Berhampur
Dr. R.N. Kar, RIT, Berhampur
Prof. S.K. Nahak, RIT, Berhampur
Mr. D.P. Kanungo, VSSUT, Burla
Mr. G.T. Chandra Sekhar, VSSUT, Burla

Organizing Committee
Prof. D.P. Tripathy, RIT, Berhampur
Prof. R.R. Polai, RIT, Berhampur
Prof. S.K. Sahu, RIT, Berhampur
Prof. P.M. Sahu, RIT, Berhampur
Prof. P.R. Sahu, RIT, Berhampur
Prof. Rashmta Tripathy, RIT, Berhampur
Prof. S.P. Tripathy, RIT, Berhampur
Prof. S. Kar, RIT, Berhampur
Prof. G. Prem Bihari, RIT, Berhampur
Prof. M. Surendra Prasad Babu, AU, AP
Prof. A. Yugandhara Rao, LCE, Vizianagaram
Prof. S. Halini, SSCE, Srikakulam
Prof. P. Sanyasi Naidu, GITAM University
Prof. K. Karthik, BVRIT, Hyderabad
Prof. M. Srivastava, GGU, Bilaspur
Prof. S. Ratan Kumar, ANITS, Vizag
Prof. Y. Narendra Kumar, LCE, Vizianagaram
Prof. B.G. Laxmi, SSCE, Srikakulam
Prof. P. Pradeep Kumar, SSCE, Srikakulam
Prof. Ch. Ramesh, AITAM
Prof. T. Lokesh, Miracle, Vizianagaram
Prof. A. Ajay Kumar, GVPCE, Vizag
Prof. P. Swadhin Patro, TAT, BBSR
Prof. P. Krishna Rao, AITAM

Prof. R.C. Balabantaray, IIIT, BBSR
Prof. M.M.K. Varma, GNI, Hyderabad
Prof. S.K. Mishra, RIT, Berhampur
Prof. R.K. Choudhury, RIT, Berhampur
Prof. M.C. Pattnaik, RIT, Berhampur
Prof. S.R. Dash, RIT, Berhampur
Prof. Shobhan Patra, RIT, Berhampur
Prof. J. Padhy, RIT, Berhampur
Prof. S.P. Bal, RIT, Berhampur
Prof. A.G. Acharya, RIT, Berhampur
Prof. G. Ramesh Babu, SSIT, Srikakulam
Prof. P. Ramana, GMRIT, AP
Prof. S. Pradhan, UU, BBSR
Prof. S. Chinara, NIT, RKL
Prof. Murthy Sharma, BVC, AP
Prof. D. Cheeranjeevi, SSCE, Srikakulam
Prof. B. Manmadha Kumar, AITAM, Srikakulam
Prof. B.D. Sahu, NIT, RKL
Prof. Ch. Krishna Rao, AITAM
Prof. L.V. Suresh Kumar, GMRIT, AP
Prof. S. Sethi, IGIT, Sarang
Prof. K. Vijetha, GVP College of Engineering, Vizag
Prof. K. Purna Chand, BVRIT, Hyderabad
Prof. H.K. Tripathy, KIIT, BBSR
Prof. Aditya K. Das, KIIT, BBSR
Prof. Lambodar Jena, GEC, BBSR
Prof. A. Khaskalam, GGU, Bilaspur
Prof. D.K. Behera, TAT, BBSR

About the Conference

The International Conference on "Computational Intelligence in Data Mining" (ICCIDM) has become one of the most sought-after International conferences in India amongst researchers across the globe. ICCIDM 2015 aims to facilitate cross-cooperation across diversified regional research communities within India as well as with other International regional research programs and partners. Such active discussions and brainstorming sessions among national and international research communities are the need of the hour as new trends, challenges and applications of Computational Intelligence in the field of Science, Engineering and Technology are cropping up by each passing moment. The 2015 edition of ICCIDM is an opportune platform for researchers, academicians, scientists and practitioners to share their innovative ideas and research findings, which will go a long way in finding solutions to confronting issues in related fields.

The conference aims to:

- Provide a sneak preview into the strengths and weakness of trending applications and research findings in the field of Computational Intelligence and Data Mining.
- Enhance the exchange of ideas and achieve coherence between the various Computational Intelligence Methods.
- Enrich the relevance and exploitation experience in the field of data mining for seasoned and naïve data scientists.
- Bridge the gap between research and academics so as to create a pioneering platform for academicians and practitioners.
- Promote novel high-quality research findings and innovative solutions to the challenging problems in Intelligent Computing.
- Make a fruitful and effective contribution towards the advancements in the field of data mining.
- Provide research recommendations for future assessment reports.

By the end of the conference, we hope the participants will enrich their knowledge by new perspectives and views on current research topics from leading scientists, researchers and academicians around the globe, contribute their own ideas on important research topics like Data Mining and Computational Intelligence, as well as collaborate with their international counterparts.

Contents

Contents

About the Editors

Prof. Himansu Sekhar Behera is working as an Associate Professor in the Department of Computer Science Engineering and Information Technology, Veer Surendra Sai University of Technology (VSSUT)—An Unitary Technical University, Established by Government of Odisha, Burla, Odisha. He has received M.Tech. in Computer Science and Engineering from N.I.T., Rourkela (formerly R.E.C., Rourkela) and Doctor of Philosophy in Engineering (Ph.D.) from Biju Pattnaik University of Technology (BPUT), Rourkela, Government of Odisha, respectively. He has published more than 80 research papers in various international Journals and Conferences, edited 11 books and is acting as a Member of the Editorial/Reviewer Board of various international journals. He is proficient in the field of Computer Science Engineering and served in the capacity of program chair, tutorial chair and acted as advisory member of committees of many national and international conferences. His research interests include Data Mining, Soft Computing, Evolutionary Computation, Machine Intelligence and Distributed System. He is associated with various educational and research societies like OITS, ISTE, IE, ISTD, CSI, OMS, AIAER, SMIAENG, SMCSTA, etc.

Prof. Durga Prasad Mohapatra received his Ph.D. from Indian Institute of Technology Kharagpur and is presently serving as Associate Professor in NIT Rourkela, Odisha. His research interests include software engineering, real-time systems, discrete mathematics and distributed computing. He has published more than 30 research papers in these fields in various international Journals and Conferences. He has received several project grants from DST and UGC, Government of India. He has received the Young Scientist Award for the year 2006 by Orissa Bigyan Academy. He has also received the Prof. K. Arumugam National Award and the Maharashtra State National Award for outstanding research work in Software Engineering for the years 2009 and 2010, respectively, from the Indian Society for Technical Education (ISTE), New Delhi. He is going to receive the Bharat Sikshya Ratan Award for significant contribution in academics awarded by the Global Society for Health and Educational Growth, Delhi.

A Sensor Based Mechanism
for Controlling Mobile Robots with ZigBee

Sreena Narayanan and K.V. Divya

Abstract Mobile robots are now widely used in the daily life. A mobile robot is a one which allows motion in different directions and it can be used as a prototype in certain applications. By controlling the mobile robot using a sensor it can be used as a prototype for wheelchairs and thus they can assist the physically disabled people in their movement. This is very useful for them in their personnel as well as professional life. Thus the mobile robots are entered into the human day-to-day life. The sensor can capture the electrical impulses during the brain activity. And they are converted into commands for the movement of mobile robot. ZigBee is a wireless protocol used for the interaction between the computer and the mobile robot. Brain-computer interface is the communication system that enables the interaction between user and mobile robot. Electroencephalogram signals are used for controlling the mobile robots.

Keywords Mobile robots · ZigBee · Brainwave starter kit · Brain-Computer interface (BCI) · Electroencephalogram (EEG)

1 Introduction

Robot technologies have been introduced for use of robots in environments like industries, nuclear plants, space etc. Now it has been entered into medical fields, home uses etc. Today human-robot interaction takes several forms. Brain computer interface is one of the methods used for their interaction. It provides a simple binary response for the control of a device. So human brain computer interaction is a communication between brain signals and any device in which the brain signals are converted into messages those can be used by a computer or any other external

S. Narayanan (✉) · K.V. Divya
Computer Science and Engineering, Vidya Academy of Science
and Technology, Thalakkotukara, India
e-mail: sri90na@gmail.com

© Springer India 2016

H.S. Behera and D.P. Mohapatra (eds.), *Computational Intelligence in Data Mining—Volume 2*, Advances in Intelligent Systems and Computing 411, DOI 10.1007/978-81-322-2731-1_1

devices in certain applications. These signals have been extracted for controlling the external devices and this process includes the identification of human thoughts and the user's desired action. Desired action may vary depending on different users. Brain Machine Interface is another name for Brain Computer Interface which act as an interface in applying these signals to the devices.

Robots those are supporting the human life in their homes, at public places are called life supporting robots. Suitable environments are prepared for the robots in industrial plants, nuclear facilities etc. In the everyday life some problems will arise due to the requirements of users. In daily life, not only the technologies for motion control of manipulation and mobility to perform the actual task but also human machine interface technologies are important. Mobile robots are creating a growing demand for them because of their support provided to the disabled people. Healthy users can operate these mobile robots by using any conventional input devices. Most commonly used are keyboards, mouse or a joystick. But it was identified that these devices became very difficult to use by people who are physically impatient and elderly individuals. So thus arises the need for controlling the mobile robots using the human brain thoughts. Sensors are used for capturing these brain signals. For humans non-invasive methods are more preferable. Mainly the signal sensing can be done either by using non-invasive method or invasive method. Here non-invasive methods are most commonly used. EEG is an example for this. Human brain signals are captured using the EEG and these signals ate sufficient enough to control the mobile robots in an indoor environment which contains many rooms, corridors and doorways. Sensor is the latest technology used for capturing the brain signals. Specific sensors are used for capturing different signals. So the intention of this project is to develop a mobile robot which can be used as a prototype for wheelchair such that it can assist the physically disabled people in their daily life for their movement. They can do control the wheelchair independent of others.

2 Related Work

Mobile robots are gradually entered into the human life because of their accuracy and efficiency. There are many approaches for developing these mobile robots. The main variation will be in sensing the human brain signals. It can be done either by an EEG cap or using a sensor. This mobile robot can be used as a prototype for wheelchair which can assist the physically disabled people in their daily life. Here both the patient and the family members feel comfort by using this. In [1] Luzheng Bi and Xin-An Fan presented two classes of brain controlled robots to assist the disabilities in people. The classifications are brain controlled manipulators and mobile robots. The neuron impulses generated by various brain activity can be recorded either by invasive or non-invasive method. Invasive method needs surgery to implant the electrodes directly on or inside the cortex. But the non-invasive method doesn't need so. So this one is most commonly used. Non-invasive BCI uses different types of brain signals as inputs. Some pf them are Electroencephalogram (EEG), Magnetoencephalogram (MEG), Blood oxygen level dependent signals (BOLD) and deoxyhemoglobin concentrations. Among these EEG is most widely used because of their low cost and convenient in use. These types of mobile robots don't need any computational intelligence. And also the users are in charge of their movements as possible. According to this the brain controlled mobile robots are classified into two categories. The first one is direct control by BCI. Here BCI system directly translates the brain signals into commands for controlling the mobile robots. The performance of the BCI limits the robots. The second category was developed from the concept of shared control. Here the control over the robot is shared by the user who is using the BCI and an autonomous navigation system which is an intelligent controller. This one uses the intelligence and thus ensures the safety and also accuracy is improved. The cost and complexity is very high. Because this uses an array of sensors.

Arai in [2] proposed an eye based system. But this is not robust against various user races and illumination conditions. The existing input devices are used for interacting with the digital instruments. But these cannot be used by the handicapped persons. They are mainly classified into five categories.

(1) Biopotential method: uses the potential from user's body actions.
(2) Voice based method: utilizes user's voice as input and voice analysis is also done.
(3) Motion based method: it uses normal organ movements
(4) Image analysis method: this one uses a camera to record the user's desire and it is converted into a digital one.
(5) Search coil method: it uses induced voltage.

EWC uses the gaze direction and eye blink. Gaze direction is expressed by the horizontal angle of gaze. This gaze direction and eye blink are used to provide the direction and timing command for the motion. When the user looks at an appropriate angle the computer input system will send the command to the EWC. The working is based on three keys left, right and down. The stop key is not used because the wheelchair will automatically stops when the user changes the gaze direction. This has a problem related to the robustness. Chief BCI systems developed a device for the patients suffering from the amyotrophic lateral sclerosis which is called a spelling device. Here the patient can select letters from an alphabet only through the mental activity. This was proposed by Obermaier in [3]. In this brain computer interface the information transfer rate is given in bits per trial which is used as an evaluation measurement. Here the EEG patterns are classified and the classification is mainly based on the Hidden Markov model and band power estimates. This paper proposed a method which combines the EEG patterns into subsets of two, three, four and five mental tasks which was based on the seperability. A letter can be selected based on duration of 2 min and this is mainly based on the binary decision. First the process will start with a group containing all letters and then finally reaches a group containing the desired letter and another one. The selection of the letters was mainly based on the successive isolation of the desired letter. The reliability of classification affects the aped of spelling device. And this reliability will decrease when the number of tasks will increase.

Philips in [4] presented an Adaptive shared control of brain actuated simulated wheelchair. The shared control is mainly used for providing assistance to the user who is controlling the mobile robot. The assistance is provided in a constant and identical manner. If the user is capable of doing the control by him then lesser the shared control will be needed. In order to provide the shared control at first it should be identified that when or how the assistance is needed. This paper proposed three levels of assistance. They will be activated when the user needs them. The first level of assistance is:

Collision Avoidance: This act as an emergency stop. This will prevent the users from colliding with obstacles. When the user steers the wheelchair very close to an obstacle then its translational velocity will be decreased until the wheelchair comes to a stopping condition. The activation of this behavior has to be determined and for that a laser scanner is used in front of the wheelchair. The activation threshold is set as 0.4 m. The appropriateness of this behavior will be very high when the threshold is within this limit otherwise it is low. The second one is:

Obstacle Avoidance: This is similar to the collision avoidance. This calculates a pair which will steer the wheelchair away from the obstacle. The last one is:

Orientation Recovery: This contains an algorithm which corrects the orientation of the wheelchair when it is misaligned from the actual goal direction. For that first the present direction of the wheelchair has to be calculated. After that a shortest path is chosen based on the current direction and the goal direction. According to that the robot will be moved. A goal map can be used here. The first two levels of assistance are enabled during the whole session.

3 Method

The design architecture of the proposed system is shown in the Fig. 1. This architecture shows the stepwise procedure of how the mobile robots are controlled.

Human brain consists of millions of interconnected neurons. And these neurons will have different patterns. The pattern of interconnection of these neurons are represented as thoughts and emotional states. Different electrical waves are produced according to this pattern. The patterns are mainly depend on the human thoughts. Unique signals are produced for each brain activity. i.e. a muscle contraction will also generate an unique electrical signal. So these electrical signals are captured by the sensor which contains a reference electrode and a sensing tip. And these signals are passed to the bluetooth. The computer will extract the data, identify the brain signals and then processed it. This processed data is given to the robotic module where there is a microprocessor which will move the robot according to the data it received. The serial data transmission and reception is done by using the ZigBee. The robotic module also contains the motor which helps in the motion of robot. The microprocessor data is embedded into the robotic module

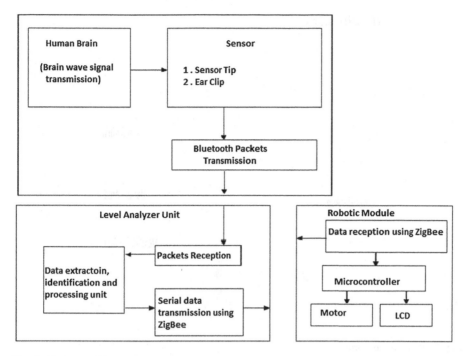

Fig. 1 System architecture

using a pickit2 programmer. With this entire system we can move a mobile robot according to the human brain signal and it can be turned by blink muscle contraction. There are different types of sensors are available. Different sensors are used for sensing different types of human brain signals. Here in this proposed system the electrical waves corresponding to the eye blink and attention are sensed.

a. *Sensor-Brain Wave Starter Kit*

A sensor is a transducer which is used to capture the human brain signals. The name of the sensor used here is Brain Wave Starter Kit. Think Gear is the dry sensor technology used in the sensor. This technology is used to measure, amplify, filter and analysis of EEG signals and the brainwaves. Beyond that eSense algorithms are used inside this. The sensor is like a headset as shown in the Fig. 2. This headset allows the measurement of the wearer's state of mind and make it available to different applications for the processing. The headset mainly consists of a sensing tip or sensor arm, an ear clip and ear loop and an adjustable head band. When these information from the sensors are given to different applications they can respond to different mental activity. The sensing tip is attached to the forehead of the person and this is the real sensor which will detect the electrical waves. The ear clip is fitted on the ear. The electrodes are on the sensor arm. Electrodes are main things for sensing purpose. The ear clip is used as a reference and ground electrode. This is to earth the noisy signals from the human body. The sensor arm is resting on the

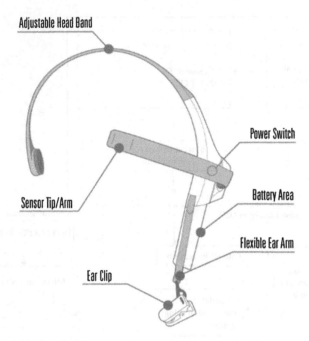

Fig. 2 Brain wave starter kit

forehead above the aye i.e. at the FP1 position. The sensor also contains a single AAA battery whose lifetime is about 8 h.

The ear clip contains the reference pick up voltage. The electrical signal sensed using the sensor arm and this reference voltage are subtracted through the common mode rejection. This is used to serve as a single EEG channel. When a person is doing any activity a neuron will fire in the brain and corresponding to those tiny electrical impulses are released. Brain-computer interface is the technology used to monitor these electrical impulses. So the sensor tip will sense these electrical impulses and these signals are input to the think gear technology.

The sensor arm is placed at the forehead i.e. at the FP1 position. This position is selected because this is an area of minimal hair. So this will offer high EEG clarity for the accurate delivery. Moreover this position is ideal to measure the electrical waves which can be used for processing with the attention and meditation algorithms. This position can also enable blink detection. The sensor will also capture the ambient noise from the body which was generated due to the muscle contraction. So the ear clip will clear out all the noise fro the body and the ambient environment.

b. *Technology Working*

Think gear technology uses a think gear chip. The signals are then interpreted with the eSense algorithms such as attention and meditation algorithms. After subtracting the signals they are amplified to 800× to enhance the faint EEG signal to strong one. Then the signals are passed through the analog and digital low and high pass filters. Then the signals are retained to 1–50 Hz range. And the signals are sampled at 128 or 512 Hz. The signal is analyzed in the time domain at each second. This is to detect and correct the noise artifacts as much as possible. So that maximum original signal will be retained. After the signal is filtered a standard FFT is applied on this signal and it is checked again for the noise which is done in the frequency domain using the same eSense algorithms. Lots of the tiny electrical signals released are combined together to form larger more complex waves. The headset can be considered as a microphone recording an audio. These complex waves are detected using this headset and then it is given to the think gear chip where the signals are processed. The measured signal and the processed interpretations are converted into digital measures as output. And this digital measure is given as input to the computer system.

c. *Algorithm*

Step 1   The brainwave starter kit will capture the signals corresponding to the eye blink and attention.

    (a) Thinkgear chip inside the sensor will process the signals and then convert it into the corresponding digital measures between 0 and 100.

Step 2   Digital measures given as input to system or level analyzer unit
Step 3   Compare the values of eye blink and attention

    (a) If the value of eye blink >70 then move right.
    (b) Else if the value of eye blink <40 then move left.
    (c) For attention if the value >40 then move forward.
    d) Else if the value <30 then move backward.

Step 4   Plot the values in the graph
Step 5   Send these to the microcontroller through the ZigBee unit
Step 6   Microcontroller will compare the values and makes the mobile robot to move

d. *eSense Algorithms*

The sensor uses the attention, meditation and eye blink detection algorithms. The attention meter in this algorithm indicates the intensity of user's level of attention. Its value range is from 0 to 100. When the user focuses on an object or a thought then the attention level of the user will increase. And if he is distracted from this then the value will automatically decreases. The blink detection algorithm will detect the user's eye blink. Its value also ranges between 0 and 100. A higher value indicates a strong blink and a low value indicates a weak blink. Usually values between 40 and 60 are considered as a neutral condition. And the values are classified into ranges of 0–20, 20–40, 60–80, 80–100.

e. *Initial Steps*

(1) Insert the Mind Wave disc to install drivers and desired apps.
(2) Insert the Mind Wave USB adapter (only after installing drivers).
(3) Run the Mind Wave Manager to connect your Mind Wave.

(4) Put on your Mind Wave headset.
(5) Exercise your mind.

The values of different waves are also detected as the output of the sensor.
Delta Waves (deep sleep and unconscious).
Theta Waves (drowsiness and deep relaxation).
Alpha Waves (relaxation and meditation).
Beta Waves (focus and attention).
Gamma Waves (relaxed to some senses and memory).

4 Hardware Description

The hardware part mainly includes the ZigBee and the mobile robot which itself
contains a microcontroller.

A. *ZigBee*

ZigBee and IEEE802.15.4 are standards-based protocols that provide the network
infrastructure required for wireless sensor network applications. ZigBee addresses
the unique needs of most remote monitoring and control sensory network applications. i.e. it defines the network and application layers. For sensor network
applications the major requirements are:

(1) Long batter life.
(2) Low cost.
(3) Small footprint.
(4) Mesh networking.

ZigBee will support these requirements. The main layers in this protocol are
Application layer (APL), Application framework, Application objects, ZigBee
Device Object (ZDO), Application Support Sub layer (APS), Service Security
Provider (SSP), Network layer, Medium Access Control layer (MAC), Physical
layer. The key features of this protocol are (1) high performance, (2) low cost,
(3) advanced networking and security, (4) low power, (5) easy to use. ZigBee is
always used as a pair. One is used for receiving the commands from the computer
and the other is used to transmit these input commands to the robotic module.

ZigBee is the only standards-based technology that addresses the unique needs
of most monitoring and control sensory network applications. It is used as an
interface for providing communication between the level analyzer unit and the
mobile robot. The microcontroller code embedded in the mobile robot allow the
movements of mobile robot in all the four directions.

B. *Mobile Robot*

Mobile robot contains the microcontroller which will receive the commands from
the computer through this ZigBee. Wireless communication is done between the
system and the mobile robot. So the input commands are passed as data packets

through ZigBee module. And this will help the mobile robot to move either I one of the four directions. The four key directions are right, left, forward and backward. Eye blink is use for the left and right direction movement and attention signals are processed for the forward and backward movements. Microcontrollers are quickly replacing computers when it comes to programming robotic device. The program code will be written using the computer system and it is then embedded into the microcontroller using any software which in turn can control the mobile robot. PIC microcontroller is used here. Microchip, the second largest 8-bit microcontroller in the world is the manufacturer of the PIC microcontroller. PIC microcontrollers have a 8-bit data memory bus and a 12, 14 or 16-bit program memory bus depending on the family. The assembler of PIC is known as MPASM and it comes with MPLAB. The pin diagram of PIC16F874A contains the pins RA0-RA6, RB0-RB7, RC0-RC7, RD0-RD7, RE0-RE2.The registers used are TRIS, PORT, ADC, USART and TIMER. The microcontroller will compare the data from the device and then the robot is moved according to that.

5 Results and Discussions

Figure 3 shows the simulation part of the mobile robot. The mobile robot mainly contains the microcontroller in which the code for it's working were embedded. The code includes working of motor for the robotic movement and the lcd display showing the current movement of the mobile robot. The green coloured box in the

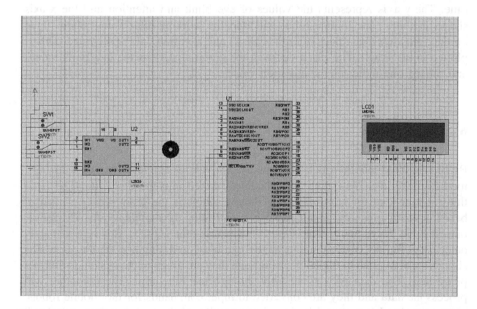

Fig. 3 Simulation result

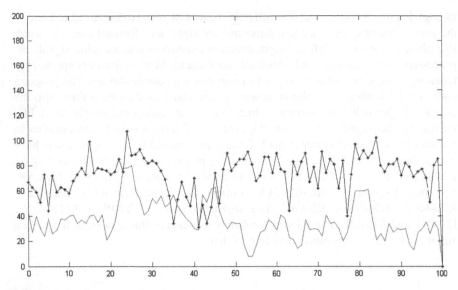

Fig. 4 Graphical representation of eye blink and attention values

figure shows the lcd display. For that it is connected to the data pins of the microcontroller. The microcontroller here used is the PIC16F877A. The left part of the figure contains the motor working.

The eye blink and attention values are in the range of 0–100. Figure 4 shows the graphical representation of the eye blink and attention values plotted against the time. The y-axis represents the values of eye blink and attention and the x-axis shows the time. The mobile robot will start the working after a continuous three blinks. And after that values of further blinks and attention will be recorded. This is plotted in the graph. The black curve in the graph represents the eye blink values and the red one indicates the attention values.

6 Conclusion

Brain controlled mobile robots are now used in various applications like industries, medical fields etc. Now they are gradually entered into the daily life. The main thing that needed for controlling the mobile robots is the brain waves. A sensor based mechanism is used in this paper for controlling the mobile robots. The sensor named Brain Wave Starter Kit is used for capturing the signals. This sensor mainly detects the signals corresponding to the eye blink and attention of the user. The signals are converted to digital measures and they are given to the system for further processing. The data packets are sent through the Bluetooth. The computer will process the data and they are converted to input commands for the mobile robots. These commands are issued in the microcontroller of the mobile robot such that it can move according to those commands. ZigBee is used as an interface between the computer and the mobile robot. The main advantages of this system were less overhead due to the wireless communication through the Bluetooth and ZigBee. This mobile robot can be used as prototype for the wheelchairs. So the user can operate the wheelchair without depending others. This reduces the complexity in data transmission communication. This is highly efficient and easy to model and use. An android application for notifying the patient's relatives through mobile phones can be considered as a future enhancement.

References

1. Bi, L., Member, IEEE, Fan, X.-A., Liu, Y:, EEG-based brain-controlled mobile robots: a survey. IEEE Trans. Human-Mach. Syst. **43**(2) (2013)
2. Arai, K., Mardiyanto, R.: Eyes based electric wheel chair control system. (IJACSA) Int. J. Adv. Comput. Sci. Appl. **2**(12) (2011)
3. Obermaier, B., Neuper, C., Guger, C., Associate Member, IEEE. In: Pfurtscheller, G.: Information transfer rate in a five-classes brain–computer interface. IEEE Trans. Neural Syst. Rehabil. Eng. **9**(3) (2001)
4. Philips, J., del R. Millán, J., Vanacker, G., Lew, E., Galán, F., Ferrez, P.W., Van Brussel, H., Nuttin, M.: Adaptive shared control of a brain-actuated simulated wheelchair. In: Proceedings of the 2007 IEEE 10th International Conference on Rehabilitation Robotics, June 12–15, Noordwijk, The Netherlands
5. del R. Millán, J., Renkens, F., Mouriño, J., Student Member, IEEE, Gerstner, W.: Noninvasive brain-actuated control of a mobile robot by human EEG. IEEE Trans. Biomed. Eng. **51**(6) (2004)
6. Tanaka, K., Matsunaga, K., Wang, H.O.: Electroencephalogram-based control of an electric wheelchair. IEEE Trans. Robot. **21**(4), 762–766 (2005)
7. Gandhi, V., Prasad, G., McGinnity, T.M., Coyle, D.H., Behera, L.: Intelligent adaptive user interfaces for BCI based robotic control. In: Proceedings of the Fifth International Brain-Computer Interface, Meeting 2013
8. Graimann, B., Allison, B., Pfurtscheller, G.: Brain-Computer Interfaces: A Gentle Introduction. Springer, Berlin (2010)
9. Hongbo, W.: Mobile Robot Positioning Based on ZigBee Wireless Sensor Networks and Vision Sensor
10. Gautam, G., Sumanth, G., Karthikeyan K.C., Sundar, S., Venkataraman, D.: Eye movement based electronic wheel chair for physically challenged persons. Int. J. Sci. Technol. Res. **3**(2), 2014
11. Arora, A., Bhattacharyya, S.: An approach towards brain actuated control in the field of robotics using eeg signals: a review. In: International Conference of Advance Research and Innovation (ICARI-2014)
12. Gandhi, V., Prasad, G., Senior Member, IEEE, Coyle, D., Senior Member, IEEE, Behera, L., Senior Member, IEEE, McGinnity, T.M., Senior Member, IEEE: Quantum neural network-based EEG filtering for a brain–computer interface. IEEE Trans. Neural Netw. Learn. Syst. **25**(2) (2014)
13. Chandra Jain, D.: A scenario of brain computer interaction with different types of face recognition techniques. Int. J. Comp. Sci. Eng. Tech. (IJCSET)
14. Jayabhavani, G.N., Raajan, N.R.: Brain enabled mechanized speech synthesizer using brain mobile interface. Int. J. Eng. Tech. (IJET)

15. Guger, C., Ramoser, H., Pfurtscheller, G.: Real-time EEG Analysis with Subject-Specific Spatial Patterns for a Brain-Computer Interface
16. Li, Y., Guan, C., Li, H., Chin, Z.: A self training semi-supervised SVM algorithm and its application in an EEG based brain computer interface speller system. Sci. Direct Pattern Recogn. Lett. **29** (2008)
17. Yanco, H.A., Drury, J.: Classifying Human-robot interaction: an updated taxonomy
18. Kuno, Y., Shimada, N., Shirai, Y.: A robotic wheelchair based on the integration of human and environmental observations. In: IEEE Robotics and Automation Magazine 26, March 2003
19. Yanco, H.A., Drury, J.L.: A Taxonomy for Human-Robot Interaction, AAAI Fall Symposium on Human-Robot Interaction, AAAI Technical Report FS-02-03., pp. 111–119, November 2002
20. Birbaumer, N., Ghanayim, N., Hinterberger, T., Iversen, I., Kotchoubey, B., Kübler, A., Perelmouter, J., Taub§, E., Flor, H.: A spelling device for the paralysed, Macmillan Magazines Ltd, Nature, vol. 398, 25 March 1999

A Two-Warehouse Inventory Model with Exponential Demand Under Permissible Delay in Payment

Trailokyanath Singh and Hadibandhu Pattanayak

Abstract The objective of the proposed paper is to develop an optimal policy of an inventory model that minimizes the total relevant cost per unit time. In this model, a two-warehouse system considers an owned warehouse (OW) with limited storage capacity and a rented warehouse (RW) with unlimited storage capacity. The demand rate is an exponential function of time, the rate of deterioration of OW is more than that of RW and the supplier provides the purchaser a permissible delay of payment. The results have been validated with the help of numerical examples.

Keywords Deterioration · Exponentially increasing demand · Permissible delay in payment · Two-warehouse model

2000 Mathematics Subject Classification 90B05

1 Introduction

In recent years, most of the inventory researchers have been trying to develop the more realistic and practicable inventory models for deteriorating items. In the past few decades, several researchers have studied the inventory model for deteriorating item. Deterioration refers the change, decay, damage, spoilage, vaporization, etc. of the products. Ghare and Schrader [1], the earliest researchers who developed an exponentially decaying inventory model. Their model led the foundation for

T. Singh (✉)
Department of Mathematics, C. V. Raman College of Engineering,
Bhubaneswar 752054, Odisha, India
e-mail: trailokyanaths108@gmail.com

H. Pattanayak
Department of Mathematics, Institute of Mathematics and Applications,
Andharua, Bhubaneswar 751003, Odisha, India
e-mail: h.pattnayak@gmail.com

© Springer India 2016
H.S. Behera and D.P. Mohapatra (eds.), *Computational Intelligence in Data Mining—Volume 2*, Advances in Intelligent Systems and Computing 411, DOI 10.1007/978-81-322-2731-1_2

modeling the inventory items by the differential equation considering demand rate as a function of time. An extensive survey of literature concerning the advances of inventory models for deteriorating items was conducted by Raafat [2], Goyal and Giri [3] and Li et al. [4]. In the traditional EOQ model, it was assumed that the purchaser must pay for items as soon as it is received by the system. But actually now-a-days a supplier grants a certain fixed period to the retailer to increase the demand. During this fixed period, no interest is charged by the supplier. Therefore, this delay period is known as trade credit period. During the trade credit period, customers can sell items, accumulate revenues and finally earn interest. The main purpose of the permissible delay period is to encourage the customers to buy more, to increase market share or to deplete inventories of certain items. For the business scenario, different delay period with different price discounts are offered by the suppliers to encourage the customers to order more quantities. In this respect, Goyal [5], the first researcher who developed an EOQ model under the condition of permissible delay in payment. Aggarwal and Jaggi [6], Khanra et al. [7] and Singh and Pattanayak [8] established their models for deteriorating items under permissible delay in payments.

Traditionally, in the EOQ model, a single warehouse is used to store with inventories. But the single warehouse with unlimited capacity is not always true or the assumption of unlimited capacity of it is unrealistic in real life situations. On the other hand, it is practical to consider another warehouse to store excess items for seasonal production or price discount for bulk purchase etc. called the rented warehouse (RW) while the first is called the own warehouse (OW). Because of better preserving facility and a lower deterioration rate, RW charges a higher holding cost (including the material handling cost and deterioration cost) than OW. Therefore, for the economical point of view, RW is stored after OW and RW is cleared before OW. Generally, the two-warehouse inventory models are developed for the storage of deteriorating inventory. In the real life situations, when some new products are launched to the markets or the seasonal product such as the output of the harvest or alternative price discounts for bulk purchase is available or the demand of the items is very high or the cost of procuring items is higher than the other inventory related cost etc., the management may purchase more items at a time. These items cannot be accommodated in the existing storehouse located at busy market place known as OW and for storing the excess items; an additional warehouse called RW is hired on the rental basis which may be located little away from it. In few decades, several researchers have studied in the field of two-warehouse model. Initially, Hartley [9], considered the effect of a two warehouse model with RW storage policy in his research. Later, Sharma [10] developed a model by assuming a single deteriorating item, constant demand and deterioration rate. Pakkala and Achary [11] studied the two-warehouse inventory model for deteriorating items with finite replenishment rate. In these models they considered the demand as constant. Benkherouf [12] extend Sharma's model and relaxed he assumptions of a fixed length of cycle and a fixed stored item in OW. The demand rate is taken as the function of time in their model. A two-warehouse model with constant demand rate, different deterioration rates and shortages under inflation was studied by Yang [13]. In most of literature of two-warehouse inventory models, it is a customary for the enterprisers to store the goods in OW first and then RW and clears the goods of RW first and then OW. Yang and Chang [14] proposed a two-warehouse inventory model for deteriorating items with partial backlogging and permissible delay in payment under inflation. Singh and Pattnayak [15] studied a two-warehouse inventory model for deteriorating items with linear increasing demand under conditionally permissible delay in payment.

This study proposes a two-warehouse inventory model of deteriorating items with exponentially increasing demand rate is considered under conditionally permissible delay in payment. The demand rate is likely to increase in the case of some new electronic products lunched to the markets like computer chips, modern TV sets etc., harvest items like paddy, wheat etc. and seasonal fruits like mango, oranges etc. For such items the demand is likely to increase very fast, almost exponentially with time. As the demand for such products increases with time, the present model is applicable. Finally an optimal policy is developed for the determination of optimal ordering time and the total relevant cost with exponentially increasing demand rate. Recently, Liang and Zhou [16] established a two-warehouse model for deteriorating items under conditionally permissible delay in payment. As demand pattern is always dynamic, therefore the demand rate is considered an exponentially increasing demand pattern and conditionally delay in payment is permitted in order to make the model relevant and more realistic.

2 Assumptions and Notations

The following assumptions are used to develop the mathematical model:

  (i) The demand rate for the item is deterministic and increasing exponentially with respect to time.
 (ii) The deterioration rate for the item in OW and RW are different rates and RW offers better preserving facility than OW.
(iii) The inventory model deals with a single item.
 (iv) The OW provides a fixed capacity while RW provides unlimited capacity. In order to reduce the inventory costs, it is better to consume the goods of RW than that of OW.
  (v) Shortages are not permitted.
 (vi) The initial inventory level is zero and lead time is taken as zero.
(vii) The replenishment rate is infinite and replenishment is instantaneous.
(viii) There is no replacement or repair of deteriorated units during the cycle.
 (ix) The inventory holding cost of RW is higher than that of OW.
  (x) The unit selling price is greater than unit purchase cost.
 (xi) For the optimal solution, the maximum deteriorating quantity for the items in the OW is less than exponentially increasing demand rate.

(xii) The supplier offers the retailer a delay period in paying for purchasing cost
and the retailer can accumulate revenues by selling items and by earning
interests. Total relevant costs include cost of placing orders, cost of carrying
inventory, costs of deterioration, interest payable opportunity cost and
opportunity interest earned.

The model is developed with the following notations:

   (i) $D(t)$: the time-dependent demand rate, $D(t) = A_1 e^{\lambda t}$, where $A_1 > 1$ is the
       initial demand and $0 < \lambda < 1, (A_1 > \lambda)$ are constants.
  (ii) A_o: the ordering cost of inventory per order where $A_o > 0$.
 (iii) P: the unit selling price of the item where $P > 0$.
  (iv) C: the unit purchase cost of the item where $C > 0$ and $C < P$.
   (v) w: the fixed capacity of OW.
  (vi) $I_o(t)$: the level of inventory at any instant of time t, in the interval $[0, T]$ in
       the OW.
 (vii) $I_r(t)$: the level of inventory at any instant of time t, in the interval $[0, t_w]$ in
       the RW.
(viii) h_o: the inventory carrying cost per unit per unit time in OW (excluding
       interest charges).
  (ix) h_r: the inventory carrying cost per unit per unit time in RW (excluding
       interest charges) where $h_r > h_o$.
   (x) γ: the constant rate of deterioration in OW where $0 < \gamma \ll 1$.
  (xi) β: the constant rate of deterioration in RW where $0 < \beta \ll 1, \gamma > \beta$ and
       $(h_r - h_o) > C(\gamma - \beta)$.
 (xii) M: the permissible delay period (fraction of the year) in settling the
       accounts with the suppliers.
(xiii) I_c: interest charged per rupee in stock per cycle by the supplier.
 (xiv) I_e: interest that can be earned per rupee per cycle.
  (xv) t_w: the time at which the inventory level reaches to w.
 (xvi) T: the length of the replenishment cycle.
(xvii) $Z_i, i = 1, 2, 3$: the total relevant costs.

3   Mathematical Model

During the interval $[0, t_w]$, the instantaneous inventory level in RW and OW at any t
are governed by the following differential equations:

$$\frac{dI_r(t)}{dt} + \beta I_r(t) = -D(t), \qquad 0 \le t \le t_w \tag{1}$$

where $D(t) = A_1 e^{\lambda t}$, $A_1 > 1$ is the initial demand and $0 < \lambda < 1(A_1 > \lambda)$ with the boundary condition $I_r(t_w) = 0$ and

$$\frac{dI_0(t)}{dt} + \gamma I_0(t) = 0, \qquad 0 \le t \le t_w \tag{2}$$

with initial condition $I_0(0) = w$, respectively.

Using the conditions above, the solutions of Eqs. (1) and (2) are given by

$$I_r(t) = \frac{A_1}{\beta + \lambda} \left[e^{(\beta + \lambda)t_w - \beta t} - e^{\lambda t} \right], \qquad 0 \le t \le t_w \tag{3}$$

$$\text{and } I_0(t) = w e^{-\gamma t}, \qquad 0 \le t \le t_w. \tag{4}$$

Further, when $t \in [t_w, T]$, the inventory level in OW is governed by the following differential equation:

$$\frac{dI_0(t)}{dt} + \gamma I_0(t) = -D(t), \qquad t_w \le t \le T \tag{5}$$

with the boundary condition $I_0(T) = 0$.

The solution of Eq. (5) is given by

$$I_0(t) = \frac{A_1}{\gamma + \lambda} \left[e^{(\gamma + \lambda)T - \gamma t} - e^{\lambda t} \right], \qquad t_w \le t \le T. \tag{6}$$

Now, the total annual relevant cost Z consists of the following elements:

(i) Cost of placing orders: $(CO) = \frac{A_0}{T}$.

(ii) Annual cost of carrying inventory: (CC)

The annual cost of carrying inventory in RW during the interval $[0, t_w]$ and the annual cost of carrying inventory in OW during $[0, T]$ are

$$\frac{h_r}{T} \int_0^{t_w} I_r(t) dt = \frac{h_r}{T} \int_0^{t_w} \left[\frac{A_1}{\beta + \lambda} \left\{ e^{(\beta + \lambda)t_w - \beta t} - e^{\lambda t} \right\} \right] dt$$

$$= \frac{A_1 h_r}{T(\beta + \lambda)} \left[\frac{1}{\beta} \left\{ e^{(\beta + \lambda)t_w} - e^{\lambda t_w} \right\} + \frac{1}{\lambda} \left(1 - e^{\lambda t_w} \right) \right] \tag{7}$$

and $\frac{h_0}{T} \int_0^T I_0(t) dt = \frac{h_0}{T} \left[\int_0^{t_w} I_0(t) dt + \int_{t_w}^T I_0(t) dt \right]$

$$= \frac{A_1 h_0}{(\gamma + \lambda)T} \left[\frac{1}{\gamma} \left(e^{(\gamma + \lambda)T - \gamma t_w} - e^{\lambda T} \right) + \frac{1}{\lambda} \left(e^{\lambda t_w} - e^{\lambda T} \right) \right] + \frac{w h_0}{\gamma T} \left(1 - e^{-\gamma t_w} \right) \tag{8}$$

respectively.

(i) Annual costs of deteriorating units: CD

The total annual cost of deteriorated units CD is C times the sum of the amounts of the deteriorated items in both RW and OW during the interval $[0, T]$, i.e.,

$$
\begin{aligned}
CD &= \frac{C}{T}\left[\beta \int_0^{t_w} I_r(t)dt + \gamma \int_0^T I_o(t)dt\right]\\
&= \frac{\beta A_1 C}{(\beta+\lambda)T}\left[\frac{1}{\beta}\left(e^{(\beta+\lambda)t_w} - e^{\lambda t_w}\right) + \frac{1}{\lambda}\left(1 - e^{\lambda t_w}\right)\right]\\
&\quad + \frac{\gamma A_1 C}{(\gamma+\lambda)T}\left[\frac{1}{\gamma}\left(e^{(\gamma+\lambda)T-\gamma t_w} - e^{\lambda T} + \frac{1}{\lambda}\left(1 - e^{\lambda t_w}\right)\right)\right] + \frac{wC}{T}\left(1 - e^{-\gamma t_w}\right).
\end{aligned}
$$

$$(9)$$

(ii) The annual interest chargeable cost:

In this respect, there arise three possibilities for the annual interest payable opportunity costs.

Case (I): $M \le t_w < T$.

The annual interest chargeable cost (IC_1)

$$
\begin{aligned}
&= \frac{CI_c}{T}\left[\int_M^{t_w} I_r(t)dt + \int_M^T I_o(t)dt + \int_{t_w}^T I_o(t)dt\right]\\
&= \frac{A_1 CI_c}{(\beta+\lambda)}\left[\frac{1}{\beta}\left(e^{(\beta+\lambda)t_w} - e^{\lambda t_w}\right) + \frac{1}{\lambda}\left(e^{\lambda M} - e^{\lambda t_w}\right)\right]\\
&\quad + \frac{A_1 CI_c}{(\gamma+\lambda)T}\left[\frac{1}{\gamma}\left(e^{(\gamma+\lambda)T-\gamma t_w} - e^{\lambda T} + \frac{1}{\lambda}\left(e^{\lambda T} - e^{\lambda t_w}\right)\right)\right] + \frac{wCI_c}{\gamma T}\left(e^{-\gamma M} - e^{-\gamma t_w}\right).
\end{aligned}
$$

$$(10)$$

Case (II): $t_w < M \le T$.

The annual interest chargeable cost (IC_2)

$$
= \frac{CI_c}{T}\int_M^T I_o(t)dt = \frac{CI_c A_1}{T(\gamma+\lambda)}\left[\frac{1}{\gamma}\left\{e^{(\gamma+\lambda)T-\gamma M} - e^{\lambda T}\right\} + \frac{1}{\lambda}\left(e^{\lambda M} - e^{\lambda T}\right)\right]. \quad (11)
$$

Case (III): $M > T$.

No interests are charged for the items.

(v) The annual opportunity interest

Case (I): $M \le T$.

The annual interest earned is (IE_1)

$$= \frac{PI_e}{T} \int_0^M tD(t)dt = \frac{PI_e A_1}{T\lambda} \left[Me^{\lambda M} + \frac{1}{\lambda}\left(1 - e^{\lambda M}\right) \right].$$ (12)

Case (II): $M > T$.
The annual interest earned is (IE_2)

$$
\begin{aligned}
&= \frac{PI_e}{T} \left[\int_0^T tD(t)dt + (M - T) \int_0^M D(t)dt \right] \\
&= \frac{PI_e A_1}{T\lambda} \left[Te^{\lambda T} + \left(M - T - \frac{1}{\lambda}\right)\left(e^{\lambda T} - 1\right) \right].
\end{aligned}
$$ (13)

According to the assumptions, the annual relevant cost $Z(t_w, T)$ for the retailers = cost of placing orders + inventory carrying cost in RW + inventory holding cost in OW + cost of deteriorating items + interest payable opportunity cost − opportunity interest earned.

$$\text{i.e., } Z(t_w, T) = \begin{cases} Z_1, & M \leq t_w \leq T, \\ Z_2, & t_w < M \leq T, \\ Z_3, & M > T, \end{cases}$$ (14)

where

$$
\begin{aligned}
Z_1 = {}& \frac{A_o}{T} + \frac{(h_r + \beta C)A_1}{T(\beta + \lambda)} \left[\frac{1}{\beta}\left(e^{(\beta + \lambda)t_w} - e^{\lambda t_w}\right) + \frac{1}{\lambda}\left(1 - e^{\lambda t_w}\right) \right] \\
& + \frac{(h_o + \gamma C)A_1}{T(\gamma + \lambda)} \left[\frac{1}{\gamma}\left(e^{(\gamma + \lambda)T - \gamma t_w} - e^{\lambda T}\right) + \frac{1}{\lambda}\left(e^{\lambda t_w} - e^{\lambda T}\right) \right] \\
& + \frac{CI_c A_1}{(\beta + \lambda)T} \left[\frac{1}{\beta}\left(e^{(\beta + \lambda)t_w - \beta M} - e^{\lambda t_w}\right) + \frac{1}{\lambda}\left(e^{\lambda M} - e^{\lambda t_w}\right) \right] \\
& + \frac{CI_c A_1}{(\gamma + \lambda)T} \left[\frac{1}{\gamma}\left(e^{(\gamma + \lambda)T - \gamma t_w} - e^{\lambda T}\right) + \frac{1}{\lambda}\left(e^{\lambda t_w} - e^{\lambda T}\right) \right] \\
& + \frac{w}{\gamma T} \left[(h_o + \gamma C)(1 - e^{-\gamma t_w}) + CI_c\left(e^{-\gamma M} - e^{-\gamma t_w}\right) \right] \\
& - \frac{PI_e A_1}{\lambda T} \left[Me^{\lambda M} + \frac{1}{\lambda}\left(1 - e^{\lambda M}\right) \right],
\end{aligned}
$$ (15)

$$
\begin{aligned}
Z_2 =\ & \frac{A_o}{T} + \frac{(h_r + \beta C)A_1}{T(\beta + \lambda)} \left[\frac{1}{\beta} \left(e^{(\beta + \lambda)t_w} - e^{\lambda t_w} \right) + \frac{1}{\lambda} \left(1 - e^{\lambda t_w} \right) \right] \\
& + \frac{(h_o + \gamma C)A_1}{T(\gamma + \lambda)} \left[\frac{1}{\gamma} \left(e^{(\gamma + \lambda)T - \gamma t_w} - e^{\lambda T} \right) + \frac{1}{\lambda} \left(e^{\lambda t_w} - e^{\lambda T} \right) \right] \\
& + \frac{CI_c A_1}{T(\gamma + \lambda)} \left[\frac{1}{\gamma} \left(e^{(\gamma + \lambda)T - \gamma M} - e^{\lambda T} \right) + \frac{1}{\lambda} \left(e^{\lambda M} - e^{\lambda T} \right) \right] \\
& + \frac{w}{\gamma T} [(h_o + \gamma C)(1 - e^{-\gamma t_w})] - \frac{PI_e A_1}{\lambda T} \left[Me^{\lambda M} + \frac{1}{\lambda} \left(1 - e^{\lambda M} \right) \right],
\end{aligned}
\tag{16}
$$

and

$$
\begin{aligned}
Z_3 =\ & \frac{A_o}{T} + \frac{(h_r + \beta C)A_1}{T(\beta + \lambda)} \left[\frac{1}{\beta} \left(e^{(\beta + \lambda)t_w} - e^{\lambda t_w} \right) + \frac{1}{\lambda} \left(1 - e^{\lambda t_w} \right) \right] \\
& + \frac{(h_o + \gamma C)A_1}{T(\gamma + \lambda)} \left[\frac{1}{\gamma} \left(e^{(\gamma + \lambda)T - \gamma t_w} - e^{\lambda T} \right) + \frac{1}{\lambda} \left(e^{\lambda t_w} - e^{\lambda T} \right) \right] \\
& + \frac{w}{\gamma T} [(h_o + \gamma C)(1 - e^{-\gamma t_w})] - \frac{PI_e A_1}{\lambda T} \left[Te^{\lambda T} + \left(M - T - \frac{1}{\lambda} \right) (e^{\lambda T} - 1) \right].
\end{aligned}
\tag{17}
$$

The objective of the model is to find the optimal values of t_w^* and T^* in order to minimize the total relevant cost $Z(t_w, T)$.

For the minimization of Z_1, the necessary conditions are

$$
\begin{aligned}
\frac{\partial(Z_1)}{\partial t_w} =\ & \frac{(h_r + \beta C)A_1}{T(\beta + \lambda)} \left[\frac{1}{\beta} \left\{ (\beta + \lambda)e^{(\beta + \lambda)t_w} - \lambda e^{\lambda t_w} \right\} - e^{\lambda t_w} \right] \\
& + \frac{(h_o + \gamma C)A_1}{T(\gamma + \lambda)} \left[e^{\lambda t_w} - e^{(\gamma + \lambda)T - \gamma t_w} \right] + \frac{w}{T}(h_o + \gamma C + CI_c)e^{-\gamma t_w} \\
& + \frac{CI_c A_1}{(\beta + \lambda)T} \left[\frac{1}{\beta} \left((\beta + \lambda)e^{(\beta + \lambda)t_w - \beta M} - \lambda e^{\lambda t_w} \right) - e^{\lambda t_w} \right] \\
& + \frac{CI_c A_1}{(\gamma + \lambda)T} \left[e^{\lambda t_w} - e^{(\gamma + \lambda)T - \gamma t_w} \right] = 0.
\end{aligned}
\tag{18}
$$

and

$$
\frac{\partial(Z_1)}{\partial T} = \frac{(h_o + \gamma C + CI_c)A_1}{(\gamma + \lambda)T} \left[\frac{1}{\gamma} \left((\gamma + \lambda)e^{(\gamma + \lambda)T - \gamma t_w} - \lambda e^{\lambda T} \right) - e^{\lambda T} \right] - \frac{Z_1}{T} = 0.
\tag{19}
$$

Let t_w^* and T_1^* be the optimal solutions of Eqs. (18) and (19) and the solution set (t_w^*, T_1^*) will be optimal solution if its Hessian matrix $Z_1(t_w^*, T_1^*)$ is positive definite provided $\left[\dfrac{\partial^2(Z_1)}{\partial t_w^2}\right]_{(t_w^*,T_1^*)} > 0$, $\left[\dfrac{\partial^2(Z_1)}{\partial T^2}\right]_{(t_w^*,T_1^*)} > 0$ and $\left[\dfrac{\partial^2(Z_1)}{\partial t_w^2}\cdot\dfrac{\partial^2(Z_1)}{\partial T^2}-\right.$

$\left.\dfrac{\partial^2(Z_1)}{\partial t_w\partial T}\cdot\dfrac{\partial^2(Z_1)}{\partial T\partial t_w}\right]_{(t_w^*,T_1^*)} > 0.$

Similarly, for the minimization of Z_2, the necessary conditions are

$$
\begin{aligned}
\frac{\partial(Z_2)}{\partial t_w} &= \frac{(h_r+\beta C)A_1}{T(\beta+\lambda)}\left[\frac{1}{\beta}\left\{(\beta+\lambda)e^{(\beta+\lambda)t_w}-\lambda e^{\lambda t_w}\right\}-e^{\lambda t_w}\right] \\
&+ \frac{(h_o+\gamma C)A_1}{T(\gamma+\lambda)}\left[e^{\lambda t_w}-e^{(\gamma+\lambda)T-\gamma t_w}\right]+\frac{w}{T}(h_o+\gamma C)e^{-\gamma t_w}=0.
\end{aligned}
\tag{20}
$$

and

$$
\begin{aligned}
\frac{\partial(Z_2)}{\partial T} &= \frac{(h_o+\gamma C)A_1}{(\gamma+\lambda)T}\left\{\frac{1}{\gamma}\left((\gamma+\lambda)e^{(\gamma+\lambda)T-\gamma t_w}-\lambda e^{\lambda T}\right)-e^{\lambda T}\right\} \\
&+ \frac{CI_cA_1}{(\gamma+\lambda)T}\left\{\frac{1}{\gamma}\left((\gamma+\lambda)e^{(\gamma+\lambda)T-\gamma M}-\lambda e^{\lambda T}\right)-e^{\lambda T}\right\}-\frac{Z_2}{T}=0.
\end{aligned}
\tag{21}
$$

Let t_w^* and T_2^* be the optimal solutions of Eqs. (20) and (21) and the solution set (t_w^*,T_2^*) will be optimal solution if its Hessian matrix $Z_2(t_w^*,T_2^*)$ is positive definite provided $\left[\dfrac{\partial^2(Z_2)}{\partial t_w^2}\right]_{(t_w^*,T_2^*)} > 0$, $\left[\dfrac{\partial^2(Z_2)}{\partial T^2}\right]_{(t_w^*,T_2^*)} > 0$ and $\left[\dfrac{\partial^2(Z_2)}{\partial t_w^2}\cdot\dfrac{\partial^2(Z_2)}{\partial T^2}-\right.$

$\left.\dfrac{\partial^2(Z_2)}{\partial t_w\partial T}\cdot\dfrac{\partial^2(Z_2)}{\partial T\partial t_w}\right]_{(t_w^*,T_2^*)} > 0.$

The minimization of Z_3, the necessary conditions are

$$
\begin{aligned}
\frac{\partial(Z_3)}{\partial t_w} &= \frac{(h_r+\beta C)A_1}{T(\beta+\lambda)}\left[\frac{1}{\beta}\left\{(\beta+\lambda)e^{(\beta+\lambda)t_w}-\lambda e^{\lambda t_w}\right\}-e^{\lambda t_w}\right] \\
&+ \frac{(h_o+\gamma C)A_1}{T(\gamma+\lambda)}\left[e^{\lambda t_w}-e^{(\gamma+\lambda)T-\gamma t_w}\right]+\frac{w}{T}(h_o+\gamma C)e^{-\gamma t_w}=0.
\end{aligned}
\tag{22}
$$

and

$$
\begin{aligned}
\frac{\partial(Z_3)}{\partial T} &= \frac{(h_o+\gamma C)A_1}{(\gamma+\lambda)T}\left[\frac{1}{\gamma}\left((\gamma+\lambda)e^{(\gamma+\lambda)T-\gamma t_w}-\lambda e^{\lambda T}\right)-e^{\lambda T}\right] \\
&- \frac{PI_eA_1}{\lambda T}\left\{\lambda e^{\lambda T}-e^{\lambda T}+1\right\}-\frac{Z_3}{T}=0.
\end{aligned}
\tag{23}
$$

Let t_w^* and T_3^* be the optimal solutions of Eqs. (22) and (23) and the solution set (t_w^*,T_2^*) will be optimal solution if its Hessian matrix $Z_3(t_w^*,T_3^*)$ is positive definite provided $\left[\dfrac{\partial^2(Z_3)}{\partial t_w^2}\right]_{(t_w^*,T_3^*)} > 0$, $\left[\dfrac{\partial^2(Z_3)}{\partial T^2}\right]_{(t_w^*,T_3^*)} > 0$ and $\left[\dfrac{\partial^2(Z_3)}{\partial t_w^2}\cdot\dfrac{\partial^2(Z_3)}{\partial T^2}-\right.$

$\left.\dfrac{\partial^2(Z_3)}{\partial t_w\partial T}\cdot\dfrac{\partial^2(Z_3)}{\partial T\partial t_w}\right]_{(t_w^*,T_3^*)} > 0.$

4 Algorithms for Finding the Optimal Solution

The Newton-Rapson's method is applied to find the optimal solution of the model. The following steps are as follows:

Step 1: Determine t_w^{1*} and T_1^* from Eqs. (18) and (19) where $t_w^* = t_w^{1*}$ and $T^* = T_1^*$. If the condition $M \le t_w^{1*} < T_1^*$ is satisfied, then determine $Z_1^* = Z_1^*(t_w^{1*}, T_1^*)$ from Eq. (15); otherwise go to Step 2.

Step 2: Determine t_w^{2*} and T_2^* from Eqs. (20) and (21) where $t_w^* = t_w^{2*}$ and $T^* = T_2^*$. If the condition $t_w^{2*} < M \le T_2^*$ is satisfied, then determine $Z_2^* = Z_2^*(t_w^{2*}, T_2^*)$ from Eq. (16); otherwise go to Step 3.

Step 3: Determine t_w^{3*} and T_3^* from Eqs. (22) and (23) where $t_w^* = t_w^{3*}$ and $T^* = T_3^*$. If the condition $t_w^{3*} < M \le T_3^*$ is satisfied, then determine $Z_3^* = Z_3^*(t_w^{3*}, T_3^*)$ from Eq. (17); otherwise go to Step 4.

Step 4: If $(t_w^*, T^*) = \arg\min\{Z_1^* = Z_1^*(t_w^{1*}, T_1^*), Z_2^* = Z_2^*(t_w^{2*}, T_2^*), Z_3^* = Z_3^* (t_w^{3*}, T_3^*)\}$, then determine t_w^*, T^* and Z^*.

5 Numerical Examples

Example 1: Let us consider the parameters of the two-warehouse inventory model as $A_1 = 2000$ units per year, $\lambda = 0.4$ units per year, $A_o = $ Rs. 1600 per order, $h_r = $ Rs. 4 per unit per order, $h_o = $ Rs. 1 per unit per order, $w = 120$ units, $C = $ Rs. 10 per unit, $P = $ Rs. 16 per unit per year, $I_c = $ Rs. 0.16 per rupee per year, $I_e = $ Rs. 0.12 per rupee per year, $M = 0.25$ year, $\gamma = 0.2$ and $\beta = 0.08$:

Using the step-by-step procedure, the optimal solutions are $t_w^* = 0.300324$ year and $T^* = 0.646559$ year and the corresponding $Z^* = $ Rs. 4271.69.

Therefore, the both warehouses will be empty after 0.346235 year as RW vanishes at 0.300324 year.

Example 2: Let us consider the parameters of the two-warehouse inventory model as $A_1 = 2000$ units per year, $\lambda = 0.3$ units per year, $A_o = $ Rs. 1550 per order, $h_r = $ Rs. 3 per unit per order, $h_o = $ Rs. 1 per unit per order, $w = 120$ units, $C = $ Rs. 10 per unit, $P = $ Rs. 15 per unit per year, $I_c = $ Rs. 0.15 per rupee per year, $I_e = $ Rs. 0.12 per rupee per year, $M = 0.25$ year, $\gamma = 0.1$ and $\beta = 0.05$:

Using the step-by-step procedure, the optimal solutions are $t_w^* = 0.422787$ year and $T^* = 0.933838$ year and the corresponding $Z^* = $ Rs. 3764.23.

Therefore, the both warehouses will be empty after 0.511051 year as RW vanishes at 0422787 year.

6 Conclusions

To be precise, an approach is made to associate costs and to determine the inventory control policy which minimizes the total relevant costs. In this paper, a two-warehouse inventory model of deteriorating items with exponentially increasing demand rate is considered under conditionally permissible delay in payment. The demand rate is likely to increase in the case of some new electronic products lunched to the markets like computer chips, modern TV sets etc., harvest items like paddy, wheat etc. and seasonal fruits like mango, oranges etc. For such items the demand is likely to increase very fast, almost exponentially with time. For selling more items, suppliers offer delay periods. Finally an optimal policy is developed for the determination of optimal ordering time and the total relevant cost with exponentially increasing demand rate.

Further extensions of this two-warehouse inventory model can be done for generalized demand pattern, stock-dependent demand, quantity discount, a bulk release pattern etc. Another possible future direction of research is to consider the inflation and time value of money.

References

1. Ghare, P.N., Schrader, G.F.: A model for exponentially decaying inventories. J. Ind. Eng. 238–243 (1963)
2. Raafat, F.: Survey of literature on continuously deteriorating inventory model. J. Oper. Res. Soc. **42**, 27–37 (1991)
3. Goyal, S.K., Giri, B.C.: Recent trends in modeling of deteriorating inventory. Eur. J. Oper. Res. **134**, 1–16 (2001)
4. Li, R., Lan, H., Mawhinney, J.R.: A review on deteriorating inventory study. J. Serv. Sci. Manag. **3**(1), 117–129 (2010)
5. Goyal, S.K.: Economic order quantity under conditions of permissible delay in payments. J. Oper. Res. Soc. **36**, 35–38 (1985)
6. Aggarwal, S.P., Jaggi, C.K.: Ordering policies of deteriorating items under permissible delay in payments. J. Oper. Res. Soc. **46**, 658–662 (1995)
7. Khanra, S., Ghosh, S.K., Chaudhuri, K.S.: An EOQ model for a deteriorating item with time dependent quadratic demand under permissible delay in payment. Appl. Math. Comput. **218**(1), 1–9 (2011)
8. Singh, T., Pattanayak, H.: An ordering policy with time-proportional deterioration, linear demand and permissible delay in payment. Smart Innovation, Syst. Technol. **33**(3), 649–658 (2015)
9. Hartley, R.V.: Operations Research—A Managerial Emphasis, pp. 315–317. Good Year Publishing Company, California (1976)
10. Sarma, K.V.S.: A deterministic order level inventory model for deteriorating items with two storage facilities. Eur. J. Oper. Res. **29**, 70–73 (1987)
11. Pakkala, T.P.M., Achary, K.K.: A deterministic inventory model for deteriorating items with two-warehouses and finite rate. Int. J. Prod. Econ. **32**, 291–299 (1992)
12. Benkherouf, L.: A deterministic order level inventory model for deteriorating items with two storage facilities. Int. J. Prod. Econ. **48**, 167–175 (1997)

13. Yang, H.L.: Two-warehouse inventory models for deteriorating items with shortages under inflation. Eur. J. Oper. Res. **157**, 344–356 (2004)
14. Yang, H.L., Chang, C.T.: A two-warehouse partial backlogging inventory model for deteriorating items with permissible delay in payment under inflation. Appl. Math. Model. **37**, 2717–2726 (2013)
15. Singh, T., Pattnayak, H.: A two-warehouse inventory model for deteriorating items with linear demand under conditionally permissible delay in payment. Int. J. Manag. Sci. Eng. Manag. **9** (2), 104–113 (2014)
16. Liang, Y., Zhou, F.: A two-warehouse inventory model for deteriorating items under conditionally permissible delay in payment. Appl. Math. Model. **35**, 2221–2231 (2011)

Effect of Outlier Detection on Clustering Accuracy and Computation Time of CHB K-Means Algorithm

K. Aparna and Mydhili K. Nair

Abstract Data clustering is one of the major areas of research in data mining. Of late, high dimensionality dataset is becoming popular because of the generation of huge volumes of data. Among the traditional partitional clustering algorithms, the bisecting K-Means is one of the most widely used for high dimensional dataset. But the performance degrades as the dimensionality increases. Also, the task of selection of cluster for further bisection is a challenging one. To overcome these drawbacks, we incorporate two constraints namely, stability-based measure and Mean Square Error (MSE) on the novel partitional clustering method, CHB-K-Means algorithm. In the experimental analysis, the performance is analyzed with respect to computation time and clustering accuracy as the number of outliers detected varies. We infer that an average clustering accuracy of 75 % has been achieved and the computation time taken for cluster formation also decreases as more number of outliers is detected.

Keywords Outlier detection · Clustering · CHB K-Means algorithm

1 Introduction

Clustering is one of the most important concepts in data mining. Its importance can be seen in diverse fields of science. The clustering process involves determining clusters of similar objects from a huge volume of dataset. Traditional clustering algorithms become computationally complex as the size of the dataset grow in size. A comprehensive study and analysis of various clustering algorithms is discussed in

K. Aparna (✉)
BMS Institue of Technology & Management, Bengaluru 560 064, Karnataka, India
e-mail: aparnak.bmsit@gmail.com

M.K. Nair
MS Ramaiah Institue of Technology, Bengaluru 560 054, Karnataka, India
e-mail: mydhili.nair@gmail.com

© Springer India 2016
H.S. Behera and D.P. Mohapatra (eds.), *Computational Intelligence in Data Mining—Volume 2*, Advances in Intelligent Systems and Computing 411, DOI 10.1007/978-81-322-2731-1_3

[1]. Because of the huge accumulation of data in the recent years, most of the current datasets are of high dimensions. As a result, the similarity between objects is not very valid leading to inaccurate results [2]. Though high dimensional datasets provide an insight into useful patterns, they also come with lot of computational challenges [3]. This has given rise to feature selection/extraction problem [4] which is a very significant aspect of knowledge discovery and data mining. Even though Bisecting K-Means [5] is an efficient method for the clustering of high dimensional data, there are some problems associated with it. The two main drawbacks of the Bisecting K-Means taken for solving this paper are, (1) Handling the presence of outliers, and (2) dependency of similarity measure when handling high dimensional data.

In this paper, we have developed a novel partitional clustering algorithm called CHB-K Means (Constraint based High dimensional Bisecting K-Means) to form clusters out of high dimensional data. This algorithm executes in two major phases namely final data matrix formation [6] and constraint based Bisecting K-Means algorithm. In the initial process, three matrices are formed namely, weighted attribute matrix, binary matrix and final data matrix [6]. The weighted attribute matrix helps to detect outlier data points and then the clustering process is done by making use of constraint based Bisecting K-Means algorithm. The constraints used in this approach are stability of the clusters and Mean Square Error (MSE). Based on these parameters the clustering process is executed and the desired number of clusters is generated.

The rest of the paper is organized as follows. Section 2 describes the related research in the field of high dimensional data clustering. Section 3 explains in details the various steps adopted in this approach and Sect. 4 illustrates the experimentations with results and discussions of the proposed approach using different datasets. Finally the conclusion is given in Sect. 5.

2 Review of Related Work

A number of related works are available for clustering of data records using partitional clustering. Recently, the clustering of high dimensional dataset is gaining attention among the data mining researchers. A brief review of some of the recent work on clustering of data by using partitional methods is presented here.

The authors in [7] have come out with a new approach for formation of clusters. The paper deals with the continuous data sets. The dataset is initially sorted in a particular order. All the datasets that were closest to the given centroid was considered as part of a single cluster. Their experimental results have shown that the new approach performs better in terms of consuming less computational time. The authors also suggest that the approach can be extended for discrete datasets.

In [8], H.S. Behera et al. have proposed a new model for clustering called IHKMCA (Improved Hybridized K-Means Clustering Algorithm) in which the Canonical Variate Analysis is applied to the high dimensional dataset in order to reduce its dimensionality without affecting the original data. To this reduced low dimensional data set, the authors have applied the Genetic Algorithm in order to obtain the initial centroid values. K-Means algorithm is then applied to this modified reduced data set. The experiments have shown better results compared to the traditional algorithms in terms of time complexity.

In [9] the authors have used a new algorithm in order to initialize the clusters before applying the K-Means algorithm. PCA technique is used initially for dimensionality reduction. From the reduced dataset, the initial seed values are selected which is a pre-requisite for the K-Means algorithm. These initial centroid values are then normalized using Z-score before giving them as input to the K-Means algorithm. The new algorithm is tested using some of the benchmark datasets and the results show better performance in terms of efficiency.

Jun Gu et al. [10] have proposed a new semisupervised spectral clustering method, i.e., SSNCut, with two types of constraints: must-link (ML) constraints on document pairs with high MS (or GC) similarities and cannot-link (CL) constraints on those with low similarities. They have empirically demonstrated the performance of SSNCut on MEDLINE document clustering, by using 100 data sets of MEDLINE records. Experimental results show that SSNCut outperformed a linear combination method and several well-known semisupervised clustering methods, being statistically significant.

In [11], the authors have made an attempt to improve the fitness of cluster centers by hybridizing the K-Means algorithm with the improved PSO. The proposed method has been compared with other traditional methods and the results obtained are very effective, steady and stable.

3 Constrained HB-K Means: An Algorithm for High Dimensional Data Clustering Using Constrained Bisecting K-Means

In this step, the clustering process is done with the help of the data matrix achieved in [6]. Here, at first, two clusters are mined from the final data matrix and then, the cluster that has more data points is given to the clustering procedure again for mining two clusters from it. This procedure is repeated until we obtain the desired number of clusters. The cluster centroid has to be optimized for getting better results, which is an add-on in constrained HB K-Means algorithm. There are two constraints added in the proposed methodology—the stability and mean square error (MSE). The stability is a measure to improve the quality of cluster. The proposed method deals with improving the stability of K-Means clustering by generating stable clusters according to K value. For instance, if the K value is set to 2, there will be two possibilities for cluster formation, i.e. either horizontal partitioning or vertical partitioning. So it can be accounted that either of the clusters will be with relevant number of data. In other cases, if the K value is set to 5, the number of data points in each cluster can vary i.e. it can be either too low or too high, which will result in instability. So in order to rectify the problem, we initiate stability score for each cluster centroid. The cluster centroid with highest stability will be considered and the rest is discarded.

The next major constraint considered for the proposed approach is the Mean Square Error (MSE). The MSE value is calculated in order to efficiently identify the most related clusters in the data group. The MSE is a measure of closeness of the data points. The distance between the clusters should be minimized for stable clusters. So the aim of introducing MSE to the method is to get the best clusters from the high dimensional data. The MSE is calculated as the difference between the squares of the data points, which is concerned as a better value than the simple difference. In the proposed method, the better cluster that can be utilized for the further clustering process is identified using MSE. The cluster with minimum MSE is considered as the relevant cluster for further processing. Basically, any amount of the center of a distribution should be associated with some measure of error. If we say that the centroid c_i is a good measure of cluster, then presumably we are saying that c_i represents the entire distribution in a better way than the other centroids. Hence it can be inferred that the MSE (c), a centroid, is the measure of quality of that centroid, which can be qualified for an efficient clustering process. The MSE (c) can be calculated as follows:

$$MSE(c) = E(c - c_i)^2$$

Here c is the cluster centroid which is subjected for minimization and c_i is the other cluster centroids in total. So the value of c_i that minimizes the MSE is selected for further calculations. Now we move on to the total process of the proposed constraint based clustering process. Randomly choose a data point, say $m_L \in R^P$ from the final data matrix D_F. This data point is used to calculate the mean value of the data points for splitting. We are considering this data point as a randomly selected point for calculating the Euclidean distance between the value and each and every data point. Calculate the mean $m_R \in R^P$ as $m_R = 2M - m_L$. The m_R value is also calculated and is considered as a point for calculating the distance of with data points. The M value needed for calculating the mean is as follows:

$$M = \frac{1}{n-L} \sum_{i=1}^{n-L} D_{Fi}$$

Split the data matrix $D_F = [D_{F1}, D_{F2}, \ldots D_{F(n-L)}]$ into two sub-clusters D_L and D_R, in accordance with the following condition:

$$\begin{cases} D_{Fi} \in D_L & if \quad E_M(D_{Fi}, m_L) \leq E_M(D_{Fi}, m_R) \\ D_{Fi} \in D_R & if \quad E_M(D_{Fi}, m_L) > E_M(D_{Fi}, m_R) \end{cases}$$

where

$$E_M(D_{Fi}, m_L) = \sqrt{\begin{array}{c} D_{Fi}{}^{(1)}, m_L^{(1)})^2 * bi^{(1)} + (D_{Fi}^{(2)}, m_L^{(2)})^2 * bi^{(2)} + \\ \ldots\ldots + (D_{Fi}{}^{(n-L)}, m_L^{(n-L)})^2 * bi^{(n-L)} \end{array}}$$

$$E_M(D_{Fi}, mR) = \sqrt{\sum_{j=1}^{m} (D_{Fi}^j - mR^j)^2 * bi^j}$$

Using the distance calculated using the above formula, we partition the data matrix into two clusters D_L and D_R.

$$C_L = \frac{1}{N_L} \sum_{j=1}^{N_L} D_{Lj} \qquad C_R = \frac{1}{N_R} \sum_{j=1}^{N_R} D_{Rj}$$

Now, the calculated centroid values of the resulting clusters are used for determining the stopping criteria for the proposed algorithm. The generated clusters have to be revamped for stability. This cluster centroid can give either best solution or normal solution. The stability of the cluster has to be calculated for the best solutions. The stability of the cluster centroids gives more clear solution in generating clusters. We have now calculated the C_L and C_R. For processing the clusters in terms of stability, another set of data points are selected and considered as centroids. Now calculate C_L' and C_R' for the newly selected points. Then, calculate the distance between the centroids so as to find the instability of clusters. The cluster with low instability has to be considered for the further process.

$$instable(C_L) = dist(C_L, C_L')$$
$$instable(C_R) = dist(C_R, C_R')$$

The respective C value with less instability is considered for the further processing of the proposed method. The second constraint considered for the proposed approach is MSE. The MSE of each cluster should be minimum for the clustering results. The MSE is calculated as the mean of squared distance between each data point to the cluster centroid.

$$MSE(C_L) = mean(dist(d_i, C_L))^2$$
$$MSE(C_R) = mean(dist(d_i, C_R))^2$$

The cluster centroid for further processing is selected based on the following condition:

$$if\,(instable(C_L)\,and\,MSE(C_L) < C_L)$$
$$C_L\,is\,selected\,for\,cluster\,formation$$
$$if\,(instable(C_R)\,and\,MSE(C_R) < C_R)$$
$$C_R\,is\,selected\,for\,cluster\,formation$$

The centroid values are calculated as the mean value of the points included in the respective clusters. Thus the clusters are formed based on the constraints, stability and MSE, which will provide better clustering accuracy as compared to the conventional bisecting algorithms.

Algorithm CHB K-Means Clustering

```
Input: D, k
Output: k clusters
Step 1. Select the input data, D
Step 2. Preprocess the data D
Step 3. Calculate mean of values of data matrix using
```

$$M_j = \frac{1}{n}\sum_{i=1}^{n} D_{ij}$$

```
Step 4. Calculate standard deviation using,
```

$$S_j = \sqrt{\frac{1}{n}\sum_{i=1}^{n}(D_{ij} - M_j)^2}$$

```
Step 5. Set the range r_ij for the data matrix and
construct the weighted attribute matrix,
```

$$WAM_{ij} = \frac{f_{ij}(f_{ij} - 1)}{n(n-1)}$$

```
Step 6. Calculate mean of values of attribute frequency
matrix using,
```

$$MA_j = \frac{1}{n}\sum_{i=1}^{n} WAM_{ij}$$

```
Step 7. Find the value of K using,
```

$$K = MA_j * \varphi$$

```
Step 8. Generate binary matrix as,
```

$$b_{ij} = \begin{cases} 1 & ; \ if \ WAM_{ij} > K \\ 0 & ; \ if \ WAM_{ij} \le K \end{cases}$$

Step 9. Set the value of L as outlier threshold
Step 10. Find column wise mean CT$_j$ of the WAM$_{i,j}$ using

$$CT_j = \frac{1}{m}\sum_{j=1}^{m} WAM_{ij}$$

and sort in descending order
Step11. Discard the top L values from the sorted list.
Step12. Select the binary matrix BM and outlier removed matrix D$_F$
Step13. Select a data point, eg: m$_L$ ∈ R^P from the final data matrix D$_F$
Step14. Find mean m$_R$ ∈ R^P as m$_R$ = 2M - m$_L$ using

$$M = \frac{1}{n-L}\sum_{i=1}^{n-L} D_{Fi}$$

Step15. Calculate the distance of the data point with respect to

$$E_M(D_{Fi}, m_L) = \sqrt{\sum_{j=1}^{m}(D_{Fi}{}^{j} - m_L{}^{j})^2 * bi^{j}} \quad .$$

Step16. Calculate C$_L$ and C$_R$,
Step17. Calculate instable(C$_L$) and instable(C$_R$)
Step18. Calculate MSE(C$_L$) and MSE(C$_R$)
Step19.

$$if\,(instable(C_L)\,\&\,MSE(C_L) < C_L)$$

$$C_L\ is\ selected\ for\ cluster\ formation$$

$$if\,(instable(C_R)\,\&\,MSE(C_R) < C_R)$$

$$C_R\ is\ selected\ for\ cluster\ formation$$

Step 20. Apply HBK-Means clustering [6]
Step 21. Stop

4 Results and Discussions

This section presents the experimental results of the proposed algorithm and the detailed discussion of the results obtained. The proposed approach is implemented in MATLAB. Here, we have tested our proposed approach using the Spambase (dataset 1) and Pen-Based Recognition of Handwritten Digits (dataset 2) Datasets. The testing was done using a computer with Intel Core 2 Duo CPU with clock speed 2.2 and 4 GB RAM. A number of metrics for comparing high dimensional data clustering algorithms were recently proposed in the literature. The performance metrics used in our paper are clustering accuracy and computation time.

4.1   Performance Analysis Based on Clustering Accuracy

In this section, we plot the performance analysis of the proposed Constraint based Bisecting K-Means algorithm. The performance of the proposed approach is evaluated to the clustering accuracy of the approach. The process is conducted by varying the outlier threshold. We have selected two datasets and the experiment is conducted by varying the total number of clusters.

Figures 1 and 2 present the clustering accuracy based on dataset 1 and dataset 2 by the proposed CHB K-Means algorithm. The analysis is conducted by varying the outlier threshold from 0 to 250. Both the graphs shows varied accuracy as the values of outlier threshold varies. This happens because of the change in data after the removal of outliers. For the dataset 1, the highest accuracy is obtained when the number of clusters is 7 and the rate is 0.789. The dataset 2 has achieved highest clustering accuracy of 0.77 when the number of clusters is 7. It can be inferred that as the number of clusters and the number of outliers detected increases, the clustering accuracy rate also increases.

4.2   Performance Analysis Based on Computation Time

In this section, we discuss about the performance analysis of the proposed approach with respect to the computation time. The computation time of the proposed approach is evaluated by varying the outlier threshold from 0 to 250. The number of clusters is varied from 2 to 7 for the ease of clustering and mapping the results.

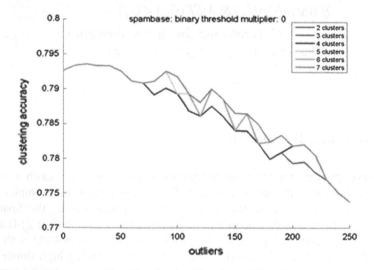

Fig. 1 Clustering accuracy: effect of outlier detection threshold in dataset 1

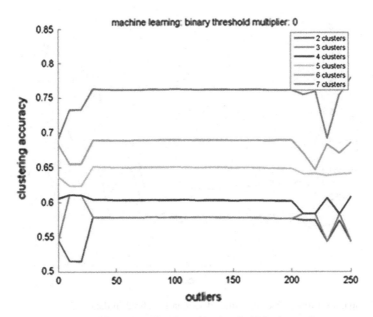

Fig. 2 Clustering accuracy: effect of outlier detection threshold in dataset 2

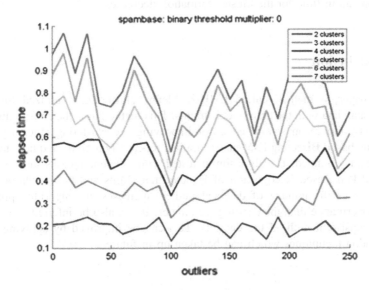

Fig. 3 Computation time: effect of outlier detection threshold in dataset 1

For the dataset 1, the highest time consumed is 1.01 ms when the number of clusters is 7 as shown in Fig. 3 and the dataset 2 has achieved the highest computation time of 0.7 ms for the total number of clusters being 7 as shown in Fig. 4.

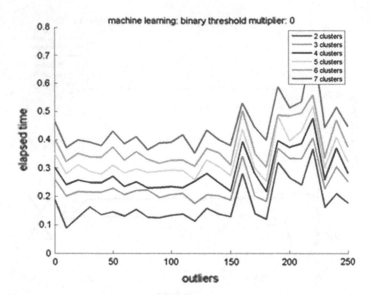

Fig. 4 Computation time: effect of outlier detection threshold in dataset 2

It can be inferred from the graphs that as the number of detected outliers increases, the computation time for the cluster formation decreases.

5   Conclusion

In this paper, a method called CHB-K Means clustering is utilized for high dimensional data with the help of constraints. This method uses the constraints such as stability and mean square error for improving the clustering efficiency. The Constraint based Bisecting K-Means is applied repeatedly until the required number of clusters is obtained. The algorithm is tested with different datasets such as Spam Base and Pen-Based Recognition of Handwritten Digits and the results obtained shows that as the number of detected outliers increases, the algorithm provides better performance than the existing techniques. It can also be inferred that results are not consistent with all the datasets. This can be improved by applying some optimization techniques which can be taken up in future.

References

1. Aparna, K., Nair, M.K.: Comprehensive study and analysis of partitional data clustering techniques. Int. J. Bus. Anal. **2**(1), 23–38 (2015)
2. Bouguessa, M., Wang, S.: Mining projected clusters in high-dimensional spaces. IEEE Trans. Knowl. Data Eng. **21**(4), 507–522 (2008)

3. John Stone, I.M.: Non-linear dimensionality reduction by LLE (2009)
4. Jain, A.K., Murty, M.N., Flynn, P.J:, Data clustering: a review. ACM Comput. Surv. **31**(3), 264–323 (1999)
5. Savaresi, S.M., Boley, D.L.: On the performance of bisecting K-means and PDDP. In: Proceedings of the First SIAM International Conference on Data Mining, pp. 1–14 (2001)
6. Aparna, K., Nair, M.K.: HB-K means: an algorithm for high dimensional data clustering using bisecting k-means. In: Accepted for publication in International Journal of Applied Engineering Research, Scopus Indexed Journal
7. Prasanna, K., Sankara Prasanna Kumar, M., Surya Narayana, G.: A novel benchmark K-means clustering on continuous data. Int. J. Comput. Sci. Eng. (IJCSE) **3**(8), 2974–2977 (2011)
8. Behera, H.S., Lingdoh, R.B., Kodamasingh, D.: An improved hybridized K means clustering algorithm (IHKMCA) for high dimensional dataset and it's performance analysis. Int. J. Comput. Sci. Eng. (IJCSE) **3**(3), 1183–1190 (2011)
9. Napoleon, D., Pavalakodi, S.: New method for dimensionality reduction using K-means clustering algorithm for high dimensional data set. Int. J. Comput. Appl. **13**(7), 41–46 (2011)
10. Gu, J., Feng, W., Zeng, J., Mamitsuka, H.: Efficient semisupervised MEDLINE document clustering with MeSH-semantic and global-content constraints. IEEE Trans. Cybern. **43**(4), 1265–1276 (2013)
11. Nayak, J., Naik, B., Kanungo, D.P., Behera, H.S.: An improved swarm based hybrid k-means clustering for optimal cluster centers. In: Information Systems Design and Intelligent Applications, Springer, India, pp. 545–553 (2015)

A Pragmatic Delineation on Cache Bypass Algorithm in Last-Level Cache (LLC)

Banchhanidhi Dash, Debabala Swain and Debabrata Swain

Abstract Last-level cache is a high level cache memory shared by all core in a multi-core chip to improve the overall efficiency of modern processors. Most of the cache management policies are based on the residency of the cache block, access time, frequency tour, recency and reuse distance to lower down the total miss count. However, in a contemporary multi level cache processor a bypass replacement reduces both miss penalty and processor stall cycles to accelerate the overall performance. This paper has analyzed different bypass replacement approaches. It also gives motivation behind different bypass replacement techniques, used in the last-level cache with inclusive and non-inclusive behavior with improved placement and promotion.

Keywords Bypass algorithm · Last level cache · Cache optimization

1 Introduction

A cache memory is a small and fast-fleeting storage designed to speed up the data in the CPU chip. It stores the data/instructions accessed most frequently and recently by the processor. The processor can't access a RAM directly. So, a newly accessed data/instruction block first brought into cache then processor accesses it. For further transactions of the block, the processor refers only cache memory. It ultimately improves the speed up and bandwidth of the memory and processor. In a multi level hierarchy system, the memory levels are designed as per their increasing size and distance from the processor. The cache layer closest to the processor is L1 which is faster than all other levels. The outer most cache level is known as last level cache

B. Dash · D. Swain (✉)
School of Computer Engineering, KIIT University, Bhubaneswar, India
e-mail: debabala.swain@gmail.com

D. Swain
Department of Computer Engineering, VIT, Pune, India

© Springer India 2016   37
H.S. Behera and D.P. Mohapatra (eds.), *Computational Intelligence in Data Mining—Volume 2*, Advances in Intelligent Systems and Computing 411, DOI 10.1007/978-81-322-2731-1_4

or LLC. The Level 1 or L1 is the inner most cache located in the processor chip. The next level is level 2 or L2. It is faster than all higher levels of cache and RAM as well. In a three level hierarchies L3 is the last-level cache or LLC, which is slower than both L1 and L2 and faster than RAM. It comes in a size of MBs generally 8-16 MB.

This paper is mostly intended to give a descriptive analysis on different bypass techniques in the last level cache in a three level hierarchies with their desired and drawback features.

1.1 Exclusive and Inclusive Cache

The memory system is a traditional optimization target for enhancing the overall performance of a computational architecture. Inclusive and exclusive are two ways to represent the residing of the block in the multi level hierarchy. Enhancing the size of cache memory is not only a better solution to improve the performance and the utility of cache memory a different level. In inclusive cache, data present at different level concurrently and the duplication of data to decrease the effective cache capacity, but simplify cache coherence mechanism in a multiprocessor based system. Back invalidation results the inclusion of victim and bandwidth need is the main limitation in inclusive cache system performance is degraded and incurs loss of valuable data and duplicate data cannot be referenced again. With exclusive cache there is no coherency between levels. It implies that the design of an exclusive cache perfectly manages the complete cache region and no back invalidation and inclusion victim involved. The system shows better efficiency with exclusive cache even if not supporting the coherency with respect to a different level (Fig. 1).

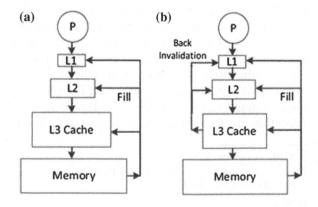

Fig. 1 a Exclusive model.
b Inclusive model [1]

2 Literature Survey

2.1 Replacement Using Inclusion and Evolution of a Block of LLC

This paper [2] proposes two models in a multi level hierarchy using the inclusion characteristics. In the first model L1, L2 are exclusively and L3 is inclusive and the second model the first two levels are inclusive and third level is exclusive with the second level. These two models compared with different cache memory performance evaluation parameter and found that the second model gives better results through superior hit analysis and lower miss penalty.

2.2 Replacement in Inclusive LLC Blocks Using Bypass Buffer

In this paper [1] the LLC hierarchy consists of a LLC and a bypass buffer (BB). The bypass buffer keeps tags of the data blocks; the working set present in the LLC is not evicted to make room for less useful data. But an evicted block from the LLC or Bypass Buffer overthrows a block from L1 and L2 to convince inclusion property. As compared to the updated LLC design the inclusion of bypass buffer (BB) adds extra hardware which is cost effective in hardware and its implementation (Fig. 2).

Fig. 2 Bypass buffer representation in inclusive LLC [1]

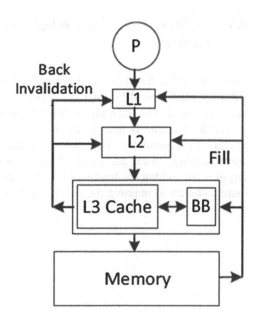

This paper carries out an important observation on the lifetime of bypassed blocks to motivate low overhead BB idea. The BB also aids the device of bypassing and reduces the hardware cost as compared to other cache bypassing algorithms. In this algorithm, the buffer produced high gain and arrived at an overall better performance result.

2.3 Bypass and Insertion Algorithm in LLC

In this paper [3] three different algorithms are used to find the age of a block and insertion into cache with each memory access by the processor. They are 1. TC-AGE PLOICY 2. TC-UC-AGE policy and 3. TC-UC-RANK policy. The initial insertion algorithm TC-AGE PLOICY sets the age counter of all TC1 blocks to three and sets an age counter 1 to all TC0 blocks. The second insertion algorithm TC-UC-AGE policy preserves graded ages to the TC0 blocks. Third insertion algorithm TC-UC-RANK policy rates them by considering the age counter of live blocks from one to three. From the above observation it has been found that the LRU has no significance in exclusive LLC. So a number of new proposals are given for the implementation of selective bypassing in a multi level hierarchy. The age counter is updated by two characteristics. 1. By considering trip counter of a block in L2 and LLC from the time of insertion to expulsion. 2. The hit counters of a block in L2 till its stay So it combines the trip counter and hit counter to improve the overall performance and throughput.

2.4 Replacement Using Global Priority Factor in Last-Level Caches

This paper [4] is based on the design of a dead block predictor that compares with RRIP (Re Referenced Interval Prediction) and provides better performance in the last level cache. The main idea behind this method is the insertion of a cache block in last level cache is determined by using the priority of a block using global priority table predictor. The table is updated by finding the hash value of the corresponding block addresses. This method provides a better design which requires less additional hardware to store the predict information and provides a better solution to improve the overall performance of the system.

Fig. 3 Illustration of
MRUTour for block A [5]

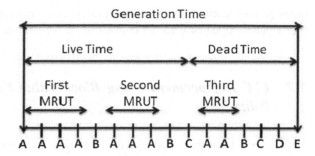

2.5 Replacement in LLC Using MRUT Based Prediction

This paper [5] proposes an algorithm known as MRUT replacement algorithm applied in the last level cache in a multi level cache hierarchy design. The idea behind this algorithm is that it selects the block who has only one MRUT. The MRUT (Most Recently Used Tour) of a block is the number of time the block holds the MRU position when it is stored in the last level cache. This algorithm uses the MRU bit that represents the block which has a single or multiple bit and one more MRU bit to indicate which block in MRU. Block with single MRUT ignoring the MRU block can be selected randomly as a victim block and the block with multiple MRUT non MRU block is used to select the victim. This algorithm shows better performance over LRU and recently proposed adaptive algorithm by reducing the MPKI (Miss Per Kilo Instruction) and improves IPC as compared to LRU and other proposals (Fig. 3).

2.6 Algorithm for Selective Bypassing and Cache Block Inclusion at LLC

This paper [6] presents a selective bypassing algorithm at LLC. This algorithm considers the past access format of a cache line in response to a cache miss and each cache line holds its own tag value in the array which is identified by a single bit. The tag value becomes set or reset based on accessing the retained block and for, a miss with the inclusion of a new block. The selected bit would decide the retention of cache block in last level cache. This model maintains and stores the access and bypasses the historical value in a table. A saturating counter for 2-bits is used to store for each cache block. On an LLC miss on any block, the block's counter in the bypass buffer is referred. If the counter value becomes less than 3, then it predicts that block not to re-access till it's stay in the cache and subsequently bypass. The counter of that block gets increments. However, if the counter value is 3, then the block is inserted into the cache. The fundamental replacement algorithms choose a victim block to be replaced. But SCIP is self-determining which can work with any other primary cache replacement algorithm. After the replacement of a victimized block buffers updates with the counter value for next accesses.

2.7 LLC Replacement Using Block Value Based Insertion Policy

This paper [7] has proposed a method of using insertion of cache blocks using an evaluated value for each block. Here miss penalty, hit benefit is two parameters considered to calculate a value called block value. It preserves the blocks with higher value and a few number of the targeted block pair to know about the association between replacing block and targeted block. In case of a miss the value of the new block will be compared with the targeted block. If its value is less, then this policy will try to remove the new block rather the old targeted block. VIP design provides better cache performance in uni-core and multi-core processors with less storage requirement. The value of incoming block and victim block plays an important role while performing insertion decision in this algorithm (Table 1).

Comparative representation of Conferred Bypass algorithm in LLC

Table 1 Comparative study of different proposed bypass algorithm

S. no.	Algorithm/model	Parameters considered	Performance measure	Advantage	Disadvantage
1	In model-1 L1 and L2 are exclusive cache with L3 as inclusive cache [2]	Size of main memory in blocks, access time of L1, L2 and L3 cache in cycles/s	CPU access time, hit rate, and global miss rate	Improve hit rate with reduced cache latency	This model does not follow coherency property in a multi level cache memory system strictly
2	In model-2 L2 inclusive and L3 exclusive with L2 [2]	Size of main memory in blocks, access time of L1, L2 and L3 cache in cycles/s	No. of hits, hit analysis, CPU execution time, global miss count	Improve hit count with reduced cache latency as compared to model-1	This model does not follow coherency property in a multi level cache memory system strictly
3	Inclusive LLC blocks using bypass buffer (BB) [1]	LLC miss rate, L1 cache, L2 cache, L3 cache	Efficient tracking using bypass buffer, speed up, % of bypass buffer hit, normalized weighted speed up, energy	1. Small BB is sufficient to reap most of the performance gains bypassing	1. Easy for exclusive cache, but find difficulties to maintain coherency strictly

(continued)

Table 1 (continued)

S. no.	Algorithm/model	Parameters considered	Performance measure	Advantage	Disadvantage
			consumption, memory bandwidth reduction	2. Reduces hardware cost significantly	2. Not cost effective due to the design requirement
4	Global priority table based model [4]	The storage overhead of different predictor, power cost, normalized LLC misses, speed up	Measured in a single core, multicore shared LLC	Enhanced performance with fewer hardware requirements and power consumption	Does not represent the performance for dead block predictor
5	Replacement algorithm based on MRU-tour in LLC [5]	Frequency, recency, and total number of live count and dead count for any block	Weighted value of a block, % of miss rate, weighted speed up normalized with respect to LRU	The hardware design is simply due to random assortment of victim block	Design cost becomes high as compared to other replacement policies
6	Algorithm for selective bypassing and cache block inclusion at LLC [6]	L1 instruction cache, and data cache L2 cache, L3 cache	1. On uniprocessor speeds up to 75 and 18 % average relative to NRU 2. On quad core speed up to 35 and 9 % on average relative to NRU	Robust performance and ability to improve the cache performance and efficiency	Hardware overhead
7	Value based insertion policy (VIP) algorithm [7]	Miss penalty, hit benefit, valid bit, prefetch bit, victim block tag	1. Measures the miss latency 2. Estimated the new block value 3. Normalises weighted speed up	Estimated value of a block can be easily computed with negligible changes to the current cache design	VIP does not work for heterogeneous processor and its needs more changes on hardware cache design

3 Conclusion

This paper, represents a universal evaluation of all foremost replacement techniques in the last-level cache using bypass techniques. Also the precise comparison is bestowed through the working principles, potency and shortcoming. Lastly, we have highlighted some aggressive key points for cache-bypassing and discovered directions for future investigation. All cache replacement policies are intended to preserve the blocks which are previleged by processor access. Also an attempt to reduce the miss penalty so that all the cache reference can be fulfilled with minimized processor latency. This goal of this paper is expressed by following vital reviews:

1. We have illustrated the memory system with its hierarchical significance along with a last level cache from the context of processor performance.
2. We have demonstrated the bypassing concept its types and advantage over coherency and inclusion cache property.
3. We have discussed various ways of replacement concepts in the last level cache using bypass algorithms. Also we are aimed to propose an inventive bypass algorithm to evaluate a weight in LLC using multiple cache aspects.

The above study will help the readers and researchers to develop innovative contributions towards developing new replacement techniques for cache optimization.

References

1. Gupta, S., Gao, H., Zhou, H.: Adaptive cache bypassing for inclusive last-level caches. In: 2013 IEEE 27th International Symposium on Parallel & Distributed Processing. http://dl.acm.org/citation.cfm?id=2511367
2. Soni, M., Nath, R.: Proposed multi level cache models based on inclusion & their evaluation. Int. J. Eng. Trends Technol. (IJETT) **4**(7), (2013). www.ijettjournal.org/volume-4/issue-7/IJETT-V4I7P134.pdf
3. Chaudhuri, M., Gaur, J., Subramoney, S.: Bypass and insertion algorithms for exclusive last-level caches. In: IEEE, Computer Architecture (ISCA), 2011 38th Annual International Symposium on 4–8 June 2011, pp. 81–92. ISSN:1063-6897. www.cse.iitk.ac.in/users/mainakc/pub/isca2011.pdf
4. Yu, B., Ma, J., Chen, T., Wu, M.: Global priority table for last-level caches. In: 2011 Ninth IEEE International Conference on Dependable Autonomic and Secure Computing. www.Computer.org/csdl/proceeding/dasc/2011/4612/00/4612a279.pdf
5. Valero, A., Sahuquillo, J.: MRU-tour-based replacement algorithms for last-level caches. In: 23rd International Symposium on Computer Architecture and High Performance Computing. IEEE (2011). www.computer.org/csdl/proceedings/sbac-pad/2011/4573a112.pdf

6. Kharbutli, M., Jarrah, M., Jararweh, Y.: SCIP: selective cache insertion and bypassing to improve the performance of last-level caches. In: 2013 IEEE Jordan Conference on Applied Electrical Engineering and Computing Technologies (AEECT). www.just.edu.jo/~yijararweh/publications.html
7. Li, L., Lu, J., Cheng, X.: Block value based insertion policy for high performance last-level caches. In: Proceeding ICS '14 Proceedings of the 28th ACM international conference on Supercomputing, pp. 63–72. ACM New York, NY, USA ©2014. ISBN:978-1-4503-2642-1. http://dl.acm.org/citation.cfm?id=2597653

Effect of Physiochemical Properties on Classification Algorithms for Kinases Family

Priyanka Purkayastha, Srikar Varanasi, Aruna Malapathi, Perumal Yogeeswari and Dharmarajan Sriram

Abstract Kinase phosphorylates specific substrates by transferring phosphate from ATP. These are important targets for the treatment of various neurological disorders, drug addiction and cancer. To organize the kinases diversity and to compare distantly related sequences it is important to classify kinases with high precision. In this study we made an attempt to classify kinases using four different classification algorithms with three different physiochemical features. Our results suggest that Random Forest gives an average precision of 0.99 for classification of kinases; and when amphiphilic pseudo amino acid composition was used as feature, the precision of the classifier was much higher than compared to amino acid composition and dipeptide composition. Hence, Random forest with amphiphilic pseudo amino acid composition is the best combination to achieve classification of kinases with high precision. Further the same can be extended for subfamilies, which can give more insight into the predominant features specific to kinases subfamilies.

Keywords Kinases · Physiochemical properties · Biological data analysis

1 Introduction

Protein kinases are the important families of protein which maintains the regulation of biological progressions by phosphorylation at posttranslational level of serine, threonine and tyrosine amino acid residues [1]. Protein kinases have become the

P. Purkayastha · P. Yogeeswari (✉) · D. Sriram (✉)
Computer-Aided Drug Design Lab, Department of Pharmacy, Birla Institute
of Technology & Science-Pilani Hyderabad Campus, Hyderabad 500078, A.P, India
e-mail: pyogee@hyderabad.bits-pilani.ac.in

S. Varanasi · A. Malapathi (✉)
Department of Computer Science and Information Systems, Birla Institute
of Technology & Science-Pilani Hyderabad Campus, Hyderabad 500078, A.P, India
e-mail: arunam@hyderabad.bits-pilani.ac.in

© Springer India 2016
H.S. Behera and D.P. Mohapatra (eds.), *Computational Intelligence in Data Mining—Volume 2*, Advances in Intelligent Systems and Computing 411, DOI 10.1007/978-81-322-2731-1_5

most researched families for protein and most of the kinase targets are being studied for the treatment of various neurological disorders, neuropathic pain, drug addiction and cancer. This group of proteins shares the conserved catalytic and regulatory domain, which regulates the catalytic activity of the kinases [2–7]. Therefore, doing a classification of these kinases will be helpful for understanding similar protein kinase family, which could be used for studying similar kinases for designing specific kinase inhibitors. Looking further on protein classification, there are many databases on the kinases like Homokinase et al. [8–10]. Nevertheless, the classification offers less accuracy due to their fundamental classification algorithm [10]. So, in this work we made an attempt to classify the various families of kinases based on amino acid, dipeptide and amphiphilic pseudo amino acid composition. The pre-classified kinases are retrieved from KinBase, where sequences are classified based on sequence similarity and is grouped into 10 families [11]. The families are (i) Tyrosine Kinases (TK); (ii) Tyrosine Kinase-Like (TKL); (iii) Protein kinases with the families with A, G and C (AGC); (iv) Calmodulin/Calcium regulated kinases (CAMK); (v) Casein kinase 1 (CK1); (vi) Cyclin-dependent kinases (CDKs), mitogen-activated protein kinases (MAP kinases), glycogen synthase kinases (GSK) and CDK-like kinases (CMGC); (vii) Other kinases; (viii) STE kinases; (ix) RGC kinases; and (x) Atypical kinases. We found several similar works on classification using amino acids (AC), dipeptide (DC) and amphiphilic pseudo amino acid composition (APAAC), taking that into account we extracted amino acid, dipeptide and amphiphilic pseudo amino acid composition features from kinase sequences [12]. In earlier studies, amino acid composition (AAC) has been used to predict the structural class and localization of proteins using fixed pattern length of 20 [13–15]. The dipeptide composition (DC) is also found to be essential for classification, prediction of protein's subcellular localization and fold recognition using fixed pattern length of 400 [15, 16]. Chou et al., used amphiphilic pseudo amino acid composition (APAAC) for protein structural classification. An improved accuracy of the SVM classifier was found using amphiphilic pseudo amino acid composition for protein classification [17]. Therefore, in our study, we have made an attempt to do a comparative study to classify the kinases and to benchmark the algorithms for classification of kinases. We attempted to use naïve Bayes, Logistic, random forest and SVM (radial basis function) classifiers and validated the performance of the classifiers by calculating precision, recall and ROC.

2 Materials and Methods

2.1 Datasets

The dataset used for classification of kinases was generated using 10 classified families of human and mouse kinases sequences from KinBase excluding all pseudogenes and a few sequences which contains different amino acid residues.

Total of 1065 sequences for human and mouse were obtained as shown in Table 1. The amino acid, dipeptide and amphiphilic pseudo amino acid composition for each sequences were extracted using R. The structure (S) of the dataset was formulated as Eq. (1), the $\bigcup$ symbol expresses union operator.

$$S = AGC_1 \cup Atypical_2 \cup CAMK_3 \cup CK1_4 \ldots \ldots \ldots \cup TKL_{10} \qquad (1)$$

Representing kinases using amino acid composition (AAC). Protein sequence composition has been expressed in 20 dimensional features with AAC. In the recent past, researchers have used AAC for classification of protein and also for predicting sub cellular localization [18]. The AAC is the fraction of each amino acid type in a protein. The fractions of all 20 natural amino acids were calculated by using Eqs. (2) and (3).

$$P_{AAC} = \{p_1 \ldots \ldots p_{20}\} \qquad (2)$$

$$\text{Fraction of each amino acid } (p_1) = \frac{\text{Total number of each amino acid}}{\text{Total number of all possible amino acids present}} \qquad (3)$$

Representing Kinases using Dipeptide composition (DC). Dipeptide composition (DC) has been used earlier by Bhasin and Raghava for prediction of families and subfamilies of G-protein coupled receptors (GPCRs) [19]. DC was used totransform the variable length of proteins to fixed length feature vectors. The DC has been expressed in 400 dimensional features. DC encapsulates information about the fraction of amino acids as well as their local order. In our experiment for classification, we have adopted the same DC-based along with the two other physiochemical properties of protein which include previously defined ACC and

Table 1 The dataset consists of 1065 sequences and are classified into ten classes

Human and mouse kinases families	Number of human sequences	Number of mouse sequences
AGC	63	60
Atypical	44	43
CAMK	74	96
CK1	12	11
CMGC	64	62
OTHERS	81	83
RGC	5	7
STE	47	47
TK	90	90
TKL	43	42
Dataset	523	541
Total	1064	

amphiphilic pseudo amino acid composition (APAAC). The DC was calculated using Eqs. (4) and (5).

$$P_{DC} = \{p_1 \cdots \cdots p_{400}\} \tag{4}$$

$$\text{Fraction of each dipeptide } (p_1) = \frac{\text{Total number of each dipeptide}}{\text{Total number of all possible dipeptides present}} \tag{5}$$

Representing Amphiphilic pseudo amino acid composition (APAAC). APAAC is originally offered by Chou et al. in 2005 [17]. Along with AAC and DC for classification of kinases, none of the classification used APAAC features for classification of kinases. It is found to be an operational descriptor of protein classification and has been used in several studies for analysis of various biological sequences [20–23]. APAAC is consists of 80 dimensional features. The first 20 components of APAAC dimensional vectors consist of the features of naïve amino acid residues. The first 20 vector is represented as in the given Eq. (2) and the added feature components in APAAC consists of hydrophobic and hydrophilic features of kinases sequences which plays important role in protein interaction protein folding. The APAAC has been expressed in $20 + 2\lambda$ components and the vector is represented by using Eq. (6).

$$P_{APAAC} = \{p1 \ldots p20, p20 + 1 \ldots p20 + \lambda, p20 + \lambda + 1 \ldots p20 + 2\lambda\} \tag{6}$$

Where the 20 components of the vector represents the amino acid composition as discussed in Eq. 2 and the rest vectors represents the set of correlation factors 2λ, for hydrophobic and hydrophilic properties of protein. The 2λ values were calculated as described in the paper [20] and the number of weight and correlation factors used, are the default parameters present in ProtR. Therefore, the 80 features were extracted.

2.2 Classification Algorithms

We tested four classifiers using amino acid composition, dipeptide composition and amphiphilic pseudo amino acid composition features. We summarize key properties of those classifiers in the following

Naive bayes classifier (NBC). The probability model for a classifier is a conditional model and is given by the Eq. (7).

$$p(C|x) = p(C|x_1, \ldots \ldots, x_n) \tag{7}$$

For x vector which is consists of the features 1 to n. The problem arises when the number of features is large and model on large number of features with large number of values are taken, such model is not feasible enough. Therefore using Bayes theorem, the probability model can be rewritten as given in Eq. (8).

$$p(C|x) = \frac{p(C)p(x/C)}{p(x)} \tag{8}$$

In practice, we are interested in the numerator of the fraction, since the denominator does not depend on C. The values of the features x_i are given so that the denominator stays effectively constant. So, $p(C|x_1, \ldots, x_n)$ can be rewritten as

$$p(C)p(x_1, \ldots, x_n|C). \tag{9}$$

$$p(C)p(x_1|C)p(x_2, \ldots, x_n|C, x_1) \tag{10}$$

$$p(C)p(x_1|C)p(x_2|C)p(x_3, \ldots, x_n|C, x_1, x_2) \tag{11}$$

$$p(C)p(x_1|C)p(x_2|C)\ldots p(x_n|C, x_1, x_2, \ldots, x_{n-1}) \tag{12}$$

According to naïve conditional independence, each feature x_i is conditionally independent of every other features x_j for $j \neq i$.

$$p(x_i|C, x_j) = p(x_i|C) \tag{13}$$

For $i \neq j$, the joint model can be expressed as

$$p(C, x_1, ..x_n) = p(C)p(x_1|C)p(x_2|C) \tag{14}$$

Therefore, the independence assumptions and the conditional distribution can be written as

$$p(C|x_1, \ldots x_n) = \frac{1}{Z}p(C)\Pi_{i=1}^{n}p(x_i|C) \tag{15}$$

The naïve Bayes classifier combines this naïve Bayes probability model with decision rule. Common rule is to choose the hypothesis which is most probable. This is also known as the maximum a posteriori (MAP) decision rule and is given by the following equation.

$$\text{Classify}(s_1, .., s_n) = \text{argmax} p(C = c)\Pi_{i=1}^{n}p(x_i = x_i|C = c) \tag{16}$$

The naïve Bayes classifier is chosen because it is particularly for high dimensionality data. Despite of its simplicity, Naive Bayes can often outperform other sophisticated classification methods [24].

Logistic Regression. Logistic model is a binary classification model which is based on calculation of success probability. The probability is calculated based on two possible categories 0 and 1. Based on the data available, the probability for both the values of the given input classes can be calculated. The logistic regression is based on logistic function, which can be defined as $P = e^t/(1 + e^t)$ [25]. The logistic regression is a simple but yet powerful classification tool in data mining applications. Therefore the results of logistic regression can be compared with other data mining algorithms for classification of kinases.

Random forest classification. Random forests are the generalization of recursive partitioning which combines a collection of trees called an ensemble. Random forests are a collection of identically distributed trees whose class value is obtained by a variant on majority vote. The classifier consists of a collection of tree like classifiers which uses a large number of decision trees, all of which are trained to tackle the same problem [26]. There are three factors that govern the individuality of the trees:

- Each tree is trained using a random subset of trained samples.
- When the tree is growing the best split on each node in the tree is found by searching through n randomly selected features. For a data set with N features, n is selected and kept smaller than that of N.
- Each tree is made to grow to the fullest so that there is no pruning.

Random forests are tree classifiers that are trained in randomly choosing the subset of input data where the final classification is based on the majority vote by the trees in the forest.

Support vector machine (SVM). SVM classifier seeks to find optimal separating hyperplane by concentrating on the training set data that lies on the class boundaries and are necessary for discriminating the classes. Since SVM is designed for binary classification and can be extended for multiple classification problems. The adopted approach for multiclass classification using SVM is one-against-all approach. A set of binary classifiers are used to reduce multi class problem, where each classifier is trained to separate a class from the rest of the classes. With a SVM classifier, a training data set of n number of classes is represented by $\{x_i, y_i\}$, $i = 1... R$, $y_i \in \{1, -1\}$ in a dimensional space [27]. The optimal separating hyperplane is defined by $\omega \cdot x + b = 0$, where x is the data point on the hyperplane, ω is the hyperplane and b is bias. For linearly separable case, a hyperplane can be defined as $\omega \cdot x_i + b \geq +1$ and $\omega \cdot x_i + b \geq -1$ for $y_i \in \{1, -1\}$, respectively.

For non-linearly separable case, hyperplane equation is introduced with a slack variables called $\{\xi i\}_i^r$ and is represented as given in Eq. (17).

$$y(w \cdot x_i + b) > 1 - \xi i \tag{17}$$

To fit the data in non-linearly separable case, the training dataset is mapped into high dimensional space, this has the spreading effect of data distribution in a manner that alleviates the fitting of linear hyperplane. Since, specifically the training data is projected into a high dimensional space, the cost effective computation in a high dimensional space is reduced using definite kernels. On such majorly used kernel is radial basis function kernel and is given by using Eq. (18), where γ controls the width of the kernel function.

$$k(x, x_i) = \exp(-\gamma||x - x_i||^2) \tag{18}$$

The classification of multiple kinases families is a multi-class problem in machine learning. In this case, the number of families of kinases is 10. SVM is designed to be a binary classifier and hence 10 binary SVMs were adapted to address our problem of classification.

Using the three kinase physiochemical properties, we trained four classification models for all ten classes of kinases which include naïve Bayes, logistic, random forest and SVM classifiers using 10 fold cross validation. The models were built using 2/3rd training data and testing the models with remaining 1/3rd data. The division of training and test data is done randomly.

3   Results and Discussion

The precision, recall and ROC values for four algorithms to classify all ten classes were calculated. The precision and recall measures were calculated for providing the accuracy that the specific classes were predicted. The precision and recall is defined as the following Eqs. (19) and (20).

Precision provides the accuracy measure that a specific class has been predicted. Recall is also known as sensitivity of a classifier and is a measure of the ability of the prediction model to select the instances of a certain class from a dataset. Considering the importance of Precision and Recall, we calculated both the measures for all the classifiers for all ten classes of kinases. The results are shown in three Figs. 1, 2 and 3 (Fig. 1 shows the Precision, recall and ROC values).

Another term called ROC (receiver operating characteristic) was calculated. ROC is an illustrative detail for evaluating the performance of classifier in binary classification. ROC is a graphical plot which can be created by plotting the true positive rate against false positive. Thus, ROC gives the details of the function of fall-out.

Along with precision and recall, the ROC is also calculated for all the classifiers. In case of amino acid composition, Random Forest classifier provides the good precision values than compared to other classifiers. As shown in Table 1, the classifiers performance for amino acid composition for all the ten families AGC, atypical, CAMK, CK1, CMGC, OTHER, RGC, STE, TK and TKL families of kinases. The random forest gave the best precision, recall and ROC value than other classifiers. This implies the fact that when amino acid composition is taken for consideration, random forest classifiers performs the best which could be due to small number of feature set. But out of all 4 classifier, 3 classifiers namely logistic, random forest and SVM classifier outperforms, which could be due to the small number of sequences (instances) present in the families. However, the number of

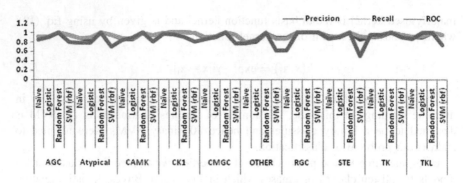

Fig. 1 Performance representation of all four classifiers (naïve bayes, logistic, random forest and SVM) for ten classes of kinases using amino acid composition

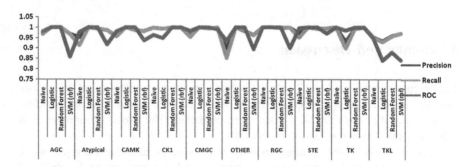

Fig. 2 Performance representation of all four classifiers (naïve bayes, logistic, random forest and SVM) for ten classes of kinases using dipeptide composition

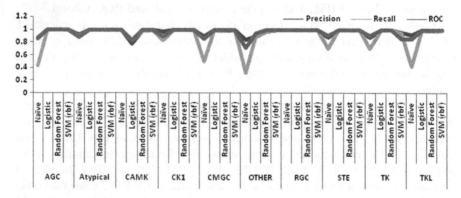

Fig. 3 Performance representation of all four classifiers (naïve bayes, logistic, random forest and SVM) for ten classes of kinases using amphiphilic pseudo amino acid composition

classifier predicting the test data with high precision is less than compared to DC features set and APAAC features set.

$$Precision = \frac{True\ Positives}{(True\ Positives + False\ Positives)} \tag{19}$$

$$Recall = \frac{True\ Positives}{(True\ Positives + False\ Negatives)} \tag{20}$$

Similarly for dipeptide composition (the representation is shown in Fig. 2). The result varies with different families, in case of AGC, random forest and logistic outperforms other classifier with a precision, recall and ROC value of 1. In case of atypical and CAMK family logistic showed good precision, recall and ROC value of 1. For the CK1 family, random forest performed well whereas in case of CMGC, OTHER, RGC and STE families, logistic showed good precision score than other classifiers. Naïve Bayes classifier showed good model validation with the RGC and TK test data. And SVM classifier performs well for TKL family of data using dipeptide composition. The variability in performance of the dataset could be due to the different number of sequences present in each family.

Using APAAC, in most of the family SVM and logistic performed well than compared with random forest and naïve Bayes classifier. As shown in Fig. 3, Logistic and SVM classifier performs well for the families AGC, atypical, CK1, CMGC, RGC, STE and TKL whereas for the family CAMK, SVM performs showed good precision, recall and ROC score. In case of RGC, three classifiers (naïve Bayes, Logistic and SVM classifiers) out of four outperformed with a precision, recall and ROC of 1. And in case of TK, logistic showed good precision value.

Looking at the performance of all the classifiers using AAC, DC and APAAC features with ten different family, all ten families showed good average precision, recall and ROC score using random forest. This suggests that random forest could be the best possible classifier for classification of kinases. And from these set of experiments described in the above figures, Random forest gave an average precision of 0.99; and when APAAC was used as feature for classification, the precision of the classifier was much higher than compared to AAC and DC. Hence, Random forest with APAAC is the best combination to achieve classification of kinases with high precision. Therefore the same can be extended and studied for subfamilies to acquire insight into the predominant features specific to each subfamily of kinases.

4 Conclusion

In this paper an attempt to find the effect of feature selection on different classification algorithms have been studied and experimented exhaustively. We have focused on three different physiochemical properties namely AAC, DC and

APAAC; and four classifiers namely Naïve Bayes, Logistic regression, Random forest and SVM (rbf). Our results suggest that feature selection is an important step for any classifier to give a classification with high precision. From these set of experiments described in results and discussion section, we conclude two observations: Random Forest gives an average precision of 0.99; and when APAAC was used as feature, the precision of the classifier was much higher than compared to AAC and DC. Hence, Random forest with APAAC is the best combination to achieve classification of kinases with high precision. Further the same can be extended and studied for subfamilies, which can give more insight into the predominant features specific to each subfamily of kinases.

Acknowledgments The authors acknowledge CSIR, India, for the financial support.

References

1. Hernandez, A.I., Blace, N., Crary, J.F., Serrano, P.A., Leitges, M., Libien, J.M., Weinstein, G., Tcherapanov, A., Sactor, T.C.: Protein kinase M zeta synthesis from a brain mRNA encoding an independent protein kinase C zeta catalytic domain. Implications for the molecular mechanism of memory. J. Biol. Chem. **278**, 40305–40316 (2008)
2. Cohen, P.: Protein kinases—the major drug targets of the twenty-first century? Nat. Rev. Drug Discov. **1**, 309–315 (2002)
3. Sacktor, T.C., Crary, J.F., Hernandez, A.I., Mirra, S., Shao, C.: Atypical protein kinase C isoforms in disorders of the nervous system and cancer. US7790854 B2 (2010)
4. Laferriere, A., Pitcher, M.H., Haldane, A., Huang, Y., Cornea, V., Kumar, N., Sacktor, T.C., Cervero, F., Coderre, T.J.: PKMzeta is essential for spinal plasticity underlying the maintenance of persistent pain. Mol. Pain **7**, 99 (2011)
5. Li, X.Y., Ko, H.G., Chen, T., Descalzi, G., Koga, K., Wang, H., Kim, S.S., Shang, Y., Kwak, C., Park, S.W., Shim, J., Lee, K., Collingridge, G.L., Kaang, B.K., Zhuo, M.: Alleviating neuropathic pain hypersensitivity by inhibiting PKMζ in the anterior cingulate cortex. Science **330**, 1400–1404 (2010)
6. Li, Y.Q., Xue, Y.X., He, Y.Y., Li, F.Q., Xue, L.F., Xu, C.M., Sacktor, T.C., Shaham, Y., Lu, L.: Inhibition of PKMzeta in nucleus accumbens core abolishes long-term drug reward memory. J. Neurosci. **31**, 5436–5446 (2011)
7. Hartsink-Segers, S.A., Beaudoin, J.J., Luijendijk, M.W., Exalto, C., Pieters, R., Den Boer, M. L.: PKCζ and PKMζ are overexpressed in TCF3-rearranged paediatric acute lymphoblastic leukaemia and are associated with increased thiopurine sensitivity. Leukemia **29**, 304–311 (2015)
8. Milanesi, L., Petrillo, M., Sepe, L., Boccia, A., D'Agostino, N., Passamano, M., Paolella, G.: Systematic analysis of human kinase genes: a large number of genes and alternative splicing events result in functional and structural diversity. BMC Bioinformatics **6**, S20 (2005)
9. Krupa, A., Abhinandan, K.R., Srinivasan, N.: KinG: a database of protein kinases in genomes. Nucleic Acids Res. **32**, D513–D515 (2004)
10. Suresh S, Saranya J, Raja K, Jeyakumar N, HomoKinase: A Curated Database of Human Protein Kinases, ISRN Computational Biology, 5, 2013
11. Manning, G., Whyte, D.B., Martinez, R., Hunter, T., Sudarsanam, S.: The protein kinase complement of the human genome. Science **298**, 1912–1934 (2002)

12. Shepherd, A.J., Gorse, D., Thornton, J.M.: A novel approach to the recognition of protein architecture from sequence using Fourier analysis and neural networks. Proteins **50**, 290–302 (2003)
13. Hua, S., Sun, Z.: Support vector machine approach for protein subcellular localization prediction. Bioinformatics **17**, 721–728 (2001)
14. Chou, K.C., Cai, Y.D.: Using functional domain composition and support vector machines for prediction of protein subcellular location. J. Biol. Chem. **277**, 45765–45769 (2002)
15. Bhasin, M., Raghava, G.P.: Classification of nuclear receptors based on amino acid composition and dipeptide composition. J. Biol. Chem. **279**, 23262–23266 (2004)
16. Shamim, M.T., Anwaruddin, M., Nagarajaram, H.A.: Support vector machine-based classification of protein folds using the structural properties of amino acid residues and amino acid residue pairs. Bioinformatics **23**, 3320–3327 (2007)
17. Zhou, X.B., Chen, C., Li, Z.C., Zou, X.Y.: Using Chou's amphiphilic pseudo-amino acid composition and support vector machine for prediction of enzyme subfamily classes. J. TheorBiol. **3**, 546–551 (2007)
18. Krajewski, Z., Tkacz, E.: Protein structural classification based on pseudo amino acid composition using SVM classifier. Biocybernetics Biomed. Eng. **33**, 77–87 (2013)
19. Bhasin, M., Raghava, G.P.: GPCRpred: an SVM-based method for prediction of families and subfamilies of G-protein coupled receptors. Nucleic Acids Res. **32**, W383–W389 (2004)
20. Chou, K.C.: Using amphiphilic pseudo amino acid composition to predict enzyme subfamily classes. Bioinformatics **21**, 10–19 (2005)
21. Chou, K.C., Cai, Y.D.: Prediction of membrane protein types by incorporating amphipathic effects. J. Chem. Inf. Model **45**, 407–413 (2005)
22. Chou, K.C., Shen, H.B.: Predicting protein subcellular location by fusing multiple classifiers. J. Cell. Biochem. **99**, 517–527 (2006)
23. Chou, K.C., Shen, H.B.: Predicting eukaryotic protein subcellular location by fusing optimized evidence-theoretic K-nearest neighbor classifiers. J. Proteome Res. **5**, 1888–1897 (2006)
24. McCallum, A., Nigam, K.: A comparison of event models for naive bayes text classification. In: AAAI-98 Workshop on Learning for Text Categorization. **752**, 41–48 (1998)
25. Hosmer, Jr D.W., Lemeshow, S.: Applied Logistic Regression. Wiley (2004)
26. Breiman, L.: Random forests. Mach. Learn. **45**, 5–32 (2001)
27. Tong, S., Koller, D.: Support vector machine active learning with applications to text classification. J. Mach. Learn. Res. **2**, 45–66 (2002)

A Fuzzy Knowledge Based Sensor Node Appraisal Technique for Fault Tolerant Data Aggregation in Wireless Sensor Networks

Sasmita Acharya and C.R. Tripathy

Abstract Wireless Sensor Networks (WSNs) have a wide range of applications in the real world. They consist of a large number of small, inexpensive, limited energy and low-cost sensor nodes which are deployed in vast geographical areas for remote sensing and monitoring operations. The sensor nodes are mostly deployed in harsh environments and unattended setups. They are prone to failure due to battery depletion, low-cost design or malfunctioning of some components. The paper proposes a two-level fuzzy knowledge based sensor node appraisal technique (NAT) in which the cluster head (CH) assesses the health status of each non-cluster head (NCH) node by the application of fuzzy rules and challenge-response technique. The CH then aggregates data from only the healthy NCHs and forwards it to the base station. It is a pro-active approach which prevents faulty data from reaching the base station. The simulation is carried out with injected NCH faults at a specified rate. The simulation results show that the proposed NAT technique can significantly improve the throughput, network lifetime and quality of service (QoS) provided by WSNs.

Keywords Wireless sensor networks · Throughput · Network lifetime · Quality of service

1 Introduction

Wireless Sensor Networks (WSNs) are used in many mission critical applications in the real world. They consist of many sensor nodes which are randomly deployed in an area of interest for remote sensing of a phenomenon of interest [1]. All the sensor

S. Acharya (✉) · C.R. Tripathy
Department of Computer Applications, VSS University of Technology, Burla, India
e-mail: talktosas@gmail.com

C.R. Tripathy
e-mail: crt.vssut@yahoo.com

© Springer India 2016   59
H.S. Behera and D.P. Mohapatra (eds.), *Computational Intelligence
in Data Mining—Volume 2*, Advances in Intelligent Systems
and Computing 411, DOI 10.1007/978-81-322-2731-1_6

nodes in a cluster that are not acting as the cluster head are called as the Non-Cluster Head (NCH) nodes. The NCH nodes sense the environment and send the data to the nearby Cluster Head (CH) node. The CH acts as the data fusion agent in the cluster. It aggregates the data received from all the NCHs in a cluster and then forwards it to the base station or sink node. Sometimes, the NCH node may become faulty due to hardware or software malfunctioning and may send its faulty data to the CH. If the number of faulty NCH nodes in a cluster is high, then it may result in faulty data being transmitted to the base station. This in turn will degrade the quality of service (QoS) provided by the WSN and will render the network unreliable for future data transmission.

Fault Tolerance refers to the ability of a system to deliver at a desired level in the presence of faults. The paper proposes a two-level fuzzy knowledge based sensor node appraisal technique (NAT) for fault-tolerant data aggregation in WSNs. The proposed NAT algorithm identifies the faulty NCH nodes in the network through its Level 1 Fuzzy knowledge based Fault Prediction Phase and then Level 2 Challenge-Response Phase. As a result, the CH identifies the faulty sensor nodes in the cluster and can isolate them during data aggregation.

The organization of the paper is as follows. The related work is discussed in Sect. 2. The system model is presented in Sect. 3. The NAT algorithm is presented in Sect. 4. The simulation results are discussed in Sects. 5 and 6 presents the conclusion and future work.

2 Related Work

This section briefly reviews the related work on fault detection in WSNs. A survey on WSNs is presented in [1]. A novel approach for faulty node detection in WSNs using fuzzy logic and majority voting technique is discussed in [2]. An improved distributed fault detection (DFD) scheme to check out the failed nodes in the network is discussed in [3]. A fault tolerance fuzzy knowledge based control algorithm (FTFK) is presented in [4]. The work in [5] addresses the sensor fault detection scheme by using the majority voting technique. A fuzzy approach to data aggregation to reduce power consumption in WSNs is discussed in [6]. An intelligent sleeping mechanism for WSNs is discussed in [7]. Various inter-actor connectivity restoration techniques for wireless sensor and actor networks are discussed in [8]. An Artificial Neural Network (ANN) approach for fault tolerant WSNs is discussed in [9]. A Fuzzy Logic based Joint Intra-cluster and Inter-cluster Multi-hop Data Dissemination Approach in Large Scale WSNs is presented in [10].

3 System Model

This section presents the network topology model and the fault model for the application of NAT algorithm to predict faulty nodes in WSNs.

3.1 The Network Topology Model

A typical grid network model is shown in Fig. 1. The network has many sensor nodes with varying energy levels which are grouped into four grids. The small sensor nodes with low energy form the NCH nodes. The NCH node with the highest residual energy forms the CH. The CHs are rotated between different NCHs in a cluster as per the TDMA schedule for efficient load balancing and enhancing the network lifetime. The NCH nodes sense the environment, collect data and forward it to their respective CHs. The CHs perform in-network data aggregation and compression and forward the data to the base station. The base station aggregates the data received from all the CHs. The maximum number of hops is two.

3.2 The Fault Model

In the fault model, the fault tolerance mechanism of WSN is assessed by observing the network's reaction to NCH faults. The NCH faults are injected into the network at different rates to simulate the fault scenario. For fault simulation, a Poisson distribution is used to schedule a time 'tf' to inject NCH faults into the network. A fault rate of 0.2 means that at time 'tf', 20 % of NCHs in the network fail.

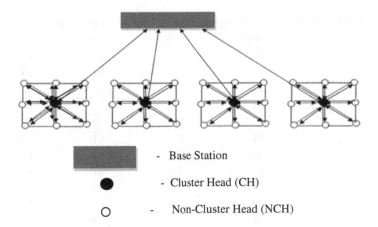

▬	- Base Station
●	- Cluster Head (CH)
○	- Non-Cluster Head (NCH)

Fig. 1 A typical grid network model

4 NAT Algorithm

This section discusses the NAT mechanism and presents the NAT algorithm. The proposed sensor node appraisal technique (NAT) monitors the status of each NCH node in a cluster and assigns a health status (HS) to each node by the application of fuzzy rules. The fuzzy rules are generated using a combination of three inputs—residual node energy (RNE), packet delivery ratio (PDR) and fault ratio (FR). The RNE for an NCH node is given by the remaining energy after each simulation round. The PDR for an NCH node is given by the number of data packets successfully sent to the CH/Total number of data packets. The FR for an NCH node is given by the number of simulation rounds in which the NCH is found to be faulty/Total number of simulation rounds.

The RNE values range from Very Low (VL), Low (L) and Medium (M) to High (H) with a trapezoidal membership function assigned. A membership function (MF) is used to map a point in input space to its membership value in output space. For example, if the residual energy of a NCH node is 0.1 J, it will be assigned Very Low (VL) level. A trapezoidal MF is specified by four parameters. The PDR and FR values range from Very Poor (VP), Poor (P) and Average (A) to Good (G) with a trapezoidal membership function assigned. These values are updated by the CH at the end of each simulation round and are kept in a record table.

The fuzzy output is the Health Status (HS) which can take values—Faulty (F), Healthy (H) or Unhealthy (UH) according to the fuzzy rules and has a triangular membership function assigned. The fuzzy output HS is updated periodically to the CH. This NAT technique is applied pro-actively at two levels—level 1 (Fault Prediction Phase) and level 2 (Challenge-Response Phase). At level 1, the CH applies Fuzzy Logic to predict the health status of each NCH. At level 2, the CH applies the Challenge-Response technique discussed in [4] to all the NCHs from level 1. The CH aggregates data from only the healthy NCH nodes identified by the two-level NAT algorithm.

Table 1 Sample of fuzzy rules

Fuzzy inputs		Fuzzy output	
RNE	PDR	FR	HS
VL	P	P	F
L	P	P	UH
H	G	G	H
M	A	A	H
H	VP	VP	UH
M	P	VP	UH
VL	VP	VP	F
L	VP	VP	F
VL	P	VP	F
VL	VP	P	F
M	VP	VP	F

A sample of fuzzy rules is presented in Table 1. It gives the fuzzy output (HS) corresponding to three fuzzy inputs (RNE, PDR and FR) generated by the application of different fuzzy rules.

4.1 NAT Algorithm

Input: Network configuration, number of simulation rounds 'R'.
Output: A NAT fault identification scheme.

a) Initialization Phase
    Initialize the network topology based on network configuration.
    Initialize energy E_i of all sensor nodes.
    Initialize node 0 as BS.

b) Exploration Phase
    for each simulation round R
        Record the RNE, PDR and FR values for each NCH in a record table to be maintained by the CH of that grid

c) Fault Prediction Phase (Level 1 Checking)
    for each NCH in the grid
        Apply fuzzy rules to predict its HS
        Go to the Challenge-Response Phase (Level 2)

d) Challenge-Response Phase (Level 2 Checking)
    The CH generates challenge data and broadcasts it to the NCH nodes in the grid
    for each NCH node in the grid
        Run the computational checking algorithm (CCA) taking challenge data as input
        Generate a response message
        Send the response message to the initiator CH in the grid
    The CH marks the healthy NCHs (based on the correct response); monitors the unhealthy NCHs and sets the faulty NCHs to sleep mode (by switching off their transceivers).

e) Exploitation Phase (Routing)
    for each simulation round R
        Forward data only from healthy NCH nodes (with HS = 'H') to respective CH by the help of routing table
        Perform in-network data aggregation and compression at CH
        Forward data from CH to node 0 (BS)

Thus, by the application of intelligent NAT algorithm, the faulty NCHs are identified; isolated and only data from healthy NCHs are forwarded to the base station through the respective CHs in the network.

5 Simulation Results and Discussion

This section presents the simulation model, performance metrics and compares the performance of the proposed NAT algorithm with that of existing FTFK and DFD algorithms through simulation. The simulation is done using MATLAB 7.5.0.

5.1 Simulation Model

In each round of simulation, the CHs are selected and grid clusters are formed. Then, each CH runs the NAT algorithm to identify the healthy NCHs. The NCH faults are injected at rate 0.2 into the network. Each CH aggregates the data received only from the healthy NCHs and forwards it to the base station. The CH performs data aggregation at a ratio which is limited to 10 % in the simulation.

5.2 Performance Metrics

The performance of different fault detection mechanisms—FTFK and DFD are compared with that of the proposed NAT through simulation with respect to the following four metrics:

(a) **Network Lifetime (in seconds)**: It is the time until the network is completely partitioned because of the failure of the CHs in the network.
(b) **Residual Energy (in Joules)**: It gives the remaining energy after each simulation round for different algorithms.
(c) **Loss Probability**: It is defined as the ratio of the number of data packets dropped (n1) to the sum of the number of data packets received at the base station (n2) and the number of data packets dropped until the end of simulation. That is, Loss Probability = $n1/(n1 + n2)$.
(d) **Throughput (in bytes)**: It is given by the number of data bytes successfully received at the base station in each round of simulation for each algorithm.

5.3 Simulation Results and Discussion

This section presents and compares the simulation results for FTFK, DFD and the proposed NAT algorithm with respect to the performance metrics outlined in Sect. 5.2.

Table 2 Configuration parameters

Parameter	Value
Network size	1200 m × 1200 m
Number of sensors	450–500
Number of cluster heads	45–50
Simulation time	500 s
Radio range	150 m
Energy transmitted	0.66 W
Energy received	0.395 W
Aggregation ratio	10 %

5.3.1 Simulation Setup

The configuration details of the simulation are presented in Table 2. It specifies the values for different simulation parameters like network size, number of sensor nodes, simulation time, energy and other relevant information.

5.3.2 Results and Discussion

The simulation results for different performance metrics are presented in this section.

Network Lifetime Versus Data Arrival Rate

Figure 2 depicts the network lifetime (in seconds) for different data arrival rates (in Kb/s). It is observed that the network lifetime is the longest for the proposed NAT algorithm and the shortest for DFD algorithm while FTFK performs close to NAT but better than DFD. The better performance of the NAT algorithm is due to its intelligent two-level faulty node identification scheme which reduces the energy consumption in the network and thus enhances the network lifetime.

Residual Energy Versus Number of Rounds

Figure 3 depicts the residual energy (in Joules) in each round for each of the simulated algorithms with NCH faults injected at rate 0.2 into the network. It clearly shows that the residual energy is the maximum for the proposed NAT algorithm and the minimum for DFD. The performance of FTFK lies in between. The better performance of the NAT algorithm is due to its filtering out of the faulty and unhealthy NCH nodes through its two-level checking strategy.

Fig. 2 Network lifetime
versus data arrival rate

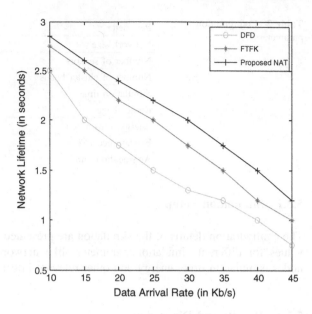

Fig. 3 Residual energy
versus simulation rounds

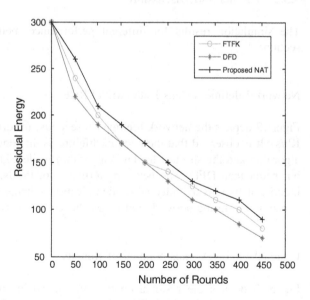

Loss Probability Versus Data Arrival Rate

Figure 4 shows the loss probability versus data arrival rate for the different simulated algorithms. It is observed that the proposed NAT algorithm exhibits the lowest loss probability while DFD has the highest loss probability. The performance of FTFK algorithm lies in between the two extremes. The highest loss probability in

Fig. 4 Loss probability
versus data arrival rate

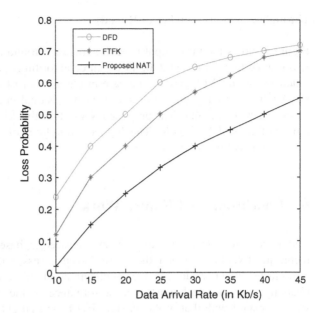

DFD indicates that more number of data packets is dropped during data transmission due to the early death of some sensor nodes. The optimal performance of NAT is due to its intelligent two-level checking strategy which weeds out the data from faulty nodes.

Fig. 5 Throughput versus
simulation rounds

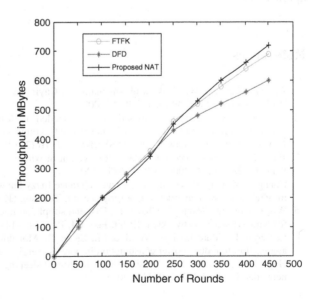

Throughput Versus Number of Rounds

Figure 5 shows the throughput in each simulation round for different algorithms. It is observed that till round 200, all the three algorithms perform uniformly without much variation. With gradual increase in the number of rounds, the throughput of DFD falls drastically due to decrease in the residual energy of sensor nodes and shorter network lifetime. The proposed NAT algorithm gives the highest throughput closely followed by FTFK. The higher throughput of NAT owes to its unique two-level checking strategy.

6 Conclusion and Future Work

The paper presents a two-level fuzzy knowledge based sensor node appraisal technique (NAT) in which the cluster head assesses the health status of each non-cluster head node by the application of a two-level fuzzy logic and challenge-response technique. The performance of the proposed NAT algorithm was compared with that of the existing FTFK and DFD algorithms with respect to various metrics through simulation. It was concluded that the proposed NAT algorithm gives better performance than the other simulated algorithms thus resulting in a more reliable and fault-tolerant WSN as validated through the simulation results. The proposed NAT technique maybe extended in future to improve the network lifetime and fault tolerance capability in different multi-sensor WSN applications.

References

1. Akyildiz, I.F., Su, W., Sankarasubramaniam, Y., Cayirci, E.: A survey on sensor networks. IEEE Commun. Mag. **40**(8), 102–114 (2002)
2. Javanmardi, S., Barati, A., Dastgheib, S.J., Attarzadeh, I.: A novel approach for faulty node detection with the aid of fuzzy theory and majority voting in wireless sensor networks. Int. J. Adv. Smart Sens. Netw. Syst. (IJASSN) **2**(4), 1–10 (2012)
3. Jiang, P.: A new method for node fault detection in wireless sensor networks. Open Access Sens. J. 1282–1294 (2009). ISSN:1424-8220
4. Chang, S., Chung, P., Huang, T.: A fault tolerance fuzzy knowledge based control algorithm in wireless sensor networks. J. Convergence Inf. Technol. (JCIT) **8**(2), (2013)
5. Wu, J., Duh, D., Wang, T., Chang, L.: Fast and simple on-line sensor fault detection scheme for wireless sensor networks. IFIP Int. Fed. Inf. Process. 444–455 (2007)
6. Lazzerini, B., Marcelloni, F., Vecchio, M., Croce, S., Monaldi, E.: A fuzzy approach to data aggregation to reduce power consumption in wireless sensor networks. IEEE 457–462 (2006)
7. Hady, A.A., El-kader, S.M.A., Eissa, H.S.: Intelligent sleeping mechanism for wireless sensor networks. Egypt. Inf. J. 109–115 (2013)

8. Acharya, S., Tripathy, C.R.: Inter-actor connectivity restoration in wireless sensor actor networks: an overview. In: Proceedings of the 48th Annual Convention of CSI, vol. 1, Advances in Intelligent Systems and Computing, vol. 248, Springer International Publishing Switzerland, pp. 351–360. (2014)
9. Acharya, S., Tripathy, C.R.: An ANN approach for fault-tolerant wireless sensor networks. In: Proceedings of the 49th Annual Convention of CSI, vol. 2, Hyderabad, Advances in Intelligent Systems and Computing, vol. 338, Springer International Publishing Switzerland, pp. 475–483. (2015)
10. Ranga, V., Dave, M., Verma, A.K.: A fuzzy logic based joint intra-cluster and inter-cluster multi-hop data dissemination approach in large scale WSNs. In: Proceedings of 2015 International Conference on Future Computational Technologies (ICFCT 2015), vol. 1. Singapore (2015)

A Novel Recognition of Indian Bank Cheques Using Feed Forward Neural Network

S.P. Raghavendra and Ajit Danti

Abstract This paper presents the results from a research to design and develop a bank cheque recognition system for Indian banks. Geometrical features of the bank cheque is the one which is the outcome of the type or category of the cheque which belongs to a specific bank. In this research, an attempt is made to design a model to classify Indian bank cheques according to the features extracted scanned cheque images by applying classification level decision using Feed forward artificial Neural Network (NN). The proposed paper contains a feed forward propagation Neural networks system designed for classification of bank cheque images. Six groups of bank cheque images including SBI, Canara, Axis, Corporation, SBM and Union bank cheques are used in the classification system. The accuracy of the system is analyzed by the variation on the range of the cheque image with different orientation and locations of bank logo and trained. The efficiency of the system is demonstrated through the experimental results extensively.

Keywords Feature extraction · Bank cheque · Neural network · Geometrical invariant features · Recognition

1 Introduction

Cheque refers to the document of official and authenticated financial transaction carried out in a bank, a message about transaction of money in an internal processing of the bank with reference to the account holder. In the context of the cheque image and the classification of banks they belongs usually refers to the change in graphical and visual pattern with respect to different banks.

S.P. Raghavendra (✉) · A. Danti
Department of MCA, JNNCE, Shivamogga, Karnataka, India
e-mail: raghusp_bdvt@yahoo.com

A. Danti
e-mail: ajitdanti@yahoo.com

© Springer India 2016
H.S. Behera and D.P. Mohapatra (eds.), *Computational Intelligence in Data Mining—Volume 2*, Advances in Intelligent Systems and Computing 411, DOI 10.1007/978-81-322-2731-1_7

The concept of geometrical shapes of a logo in a given cheque image includes:

1. The changes in the geometrical structures of the logo represent the different visual patterns of the category of bank the query cheque belongs.
2. The physical characteristics of the cheque image are identified using unique features.
3. By interpreting the identified invariant features of the cheque image, logos of the bank can be recognized.

Bank cheques processing involve one of the highly recommended and authenticated financial transactions, containing some of the potential fields regarding authenticity and transaction processing. Large amount of bank cheques will be validated every day and the task of automatically recognizing and detecting the different bank cheques in user-machine environment is useful.

Bank cheques database will be created manually by scanning standard set of nationalized bank cheques. For instances, many of the researchers used the manual database. In this approach, database is constructed to recognize six nationalized Indian bank chques: SBI, Canara, Axis, Corporation, SBM and Union bank of India. Along with the standard database, hundreds of other cheque images downloaded from the internet are also considered. Second, most of the systems conduct two stages.

- *Feature extraction*
- *Classification and recognition.*

In this research, the Feature extraction and bank logo classification is done by feed forward Artificial Neural Network. Figure 1 shows a typical structure of a Neuron.

This paper proposes an effective method for bank logo recognition from cheque images. Here, the classification of the cheque images is done.

In the first stage of feature extraction, invariant geometrical features re extracted and training vector is constructed. The second stage processing includes, the classification is based on the outcome of the Primary level to recognize and classify into one of six categories of groups of bank viz SBI, Canara, Axis, Corporation, SBM and Union bank of India using Feed-Forward ANN.

Fig. 1 Structure of a neuron

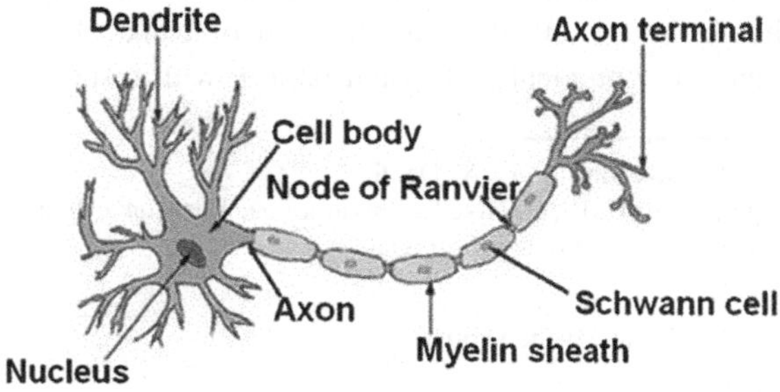

1.1 Literature Survey

Dileep et al. [1] presents a novel technique to design a model to classify human facial expressions according to the features extracted from human facial images by applying 3 Sigma limits in Second level decision using Neural Network (NN). Jayadevan [2] proposed many techniques as a survey paper revealing valuable information for researchers. Wahap [3] presents the results from a research to design and develop a bank cheque recognition system for malayasian banks, the system concentrates on recognizing the courtesy amount and date only. Samanth [4] in his paper proposed a technique for proposes new techniques for automatic processing of CTS bank cheques. Dilip and Danti [5] presents a novel approach for extraction of geometrical features of the face like triangular features, Orientation, Perimeter, and Distance for face recognition and person's internal emotional states which is very much useful in proposed technique also. Jasmine Pemeena Priyadarsini et al. [6] presents a technique which validates the authenticity of the signatures.

Here, six Indian bank cheques viz SBI, Canara, Axis, Corporation, SBM and Union bank are considered. The probability of finding these six bank cheques in any of the image is more compared to the above methodologies proposed by different researchers. This work is not limited to a single database, but also can be applied to different databases and also the images downloaded from the internet.

In this paper, the algorithm has been proposed to recognize the Indian bank cheques using Neural Network technique. Systematic comparison of classification methods is done using geometrical features of the segmented bank logo.

The paper is being developed as follows. Section 1 tells about the corpus of cheques for experiment. Section 2 gives proposed methodology. Section 3 highlights proposed algorithm. Section 4 gives the experimental results. Finally, conclusions are given in Sect. 5.

2 Proposed Methodology

This paper proposes an effective method for classification of Indian banks using logo. Here, the classification of the cheque is done systematically, the experimental results of bank cheque recognition are shown in the Fig. 2.

2.1 Feature Extraction

An algorithm bank logo recognition is proposed to classify input images into one of six groups viz SBI, Canara, Axis, Corporation, SBM and Union bank using Feed-Forward ANN. In the first stage, geometrically invariant features viz centroid,

Fig. 2 Sample experimental results with recognition of bank

eccentricity, convex area, euler number, solidity are determined using the following Eqs. (1)–(5).

$$centroid\left(C_{xy}\right) = \left\{ \begin{array}{l} C_x = \frac{1}{N}\sum_{i=1}^{N} x_i \\ C_y = \frac{1}{N}\sum_{i=1}^{N} y_i \end{array} \right\} \qquad (1)$$

where x and y are horizontal and vertical coordinate

$$eccentricity(c) = \left(\frac{c}{a}\right) \qquad (2)$$

where c is the distance from the center to a focus, a is the distance from that focus to a vertex.

$$Euler\ number = (S - N) \qquad (3)$$

$$solidity = \left(\frac{A_S}{H}\right) \qquad (4)$$

$$elongation = \left(\frac{m_j}{m_n}\right) \qquad (5)$$

where m_j is major axis and m_n is minor axis of logo region.

These features are geometrically invariant of size, scale and orientation. The logos are searched randomly from the image based on elongation constraint which is well under the bounds threshold value considered as the potential logo candidate. Later they are compared with knowledgebase by means of feed forward artificial neural network. Using Eq. (6).

2.2 Classification and Recognition

The second stage processing includes, the classification based on the outcome of the first stage to improve the classification rate effectively. Instead of considering all Neurons (64 × 64) into the Neural Network, the six invariant features of logo of each of the image will be given as input to the Neural Network. Logo of each image is represented by 6 standard feature set. This feature set is fed to the Neural network as training set for the purpose of classification at second stage.

A knowledgebase consist of six invariant features set of each class of bank for all bank cheque images of size 30 × 6 is constructed. Sample experimental results of processing of the training of cheque logo images using Neural Network is shown in the Fig. 3.

Figure 4 gives the diagrammatic representation of "sample neural network architecture"

The output of the NN classifier predicts the cheque type of a bank in the classification level. This can be represented by Eq. (6),

$$y = \sum_{i=1}^{H} v_i \sigma(z_i) \tag{6}$$

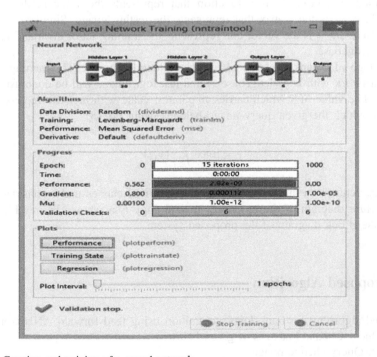

Fig. 3 Creation and training of a neural network

Fig. 4 Simple neural
network architecture

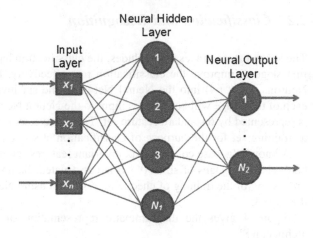

where

$$z_i = \sum\nolimits_{j=1}^{n} w_{ij}x_j + b_i \tag{7}$$

$w_{i\,j}$ indicates the cost, y represents the predicted cheque image of a bank in the recognition. Equation (6) is a function that represents the value in the Neural Network, $\sigma(z_i)$ is the matrix that represents the testing image. Simulation of the Neural Network v_i is the vector that represents the training set of invariant features as knowledgebase.

The logo of a bank is classified by using Neural Network. The training of the neural network with knowledgebase v_i and test logo image will be done using the Eq. (6). The value of y which carries maximum value with index to represent the class for which the given query image belongs using Eq. (8)

$$y = \max \int_{i=1}^{c_n} (y, i) \tag{8}$$

where c_n denotes the number of different supervised classes under consideration. Architecture diagram of the proposed methodology is as shown below. Figure 5 shows the Block diagram of the proposed system.

3 Proposed Algorithm

Proposed algorithm for bank logo recognition using feed forward ANN from the given Query Cheque image is as given below:

Input: Query cheque image

Fig. 5 Block diagram of the proposed system

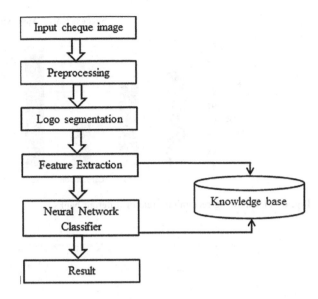

Output: Determine bank class viz SBI, Canara, Axis, Corporation, SBM or Union bank

Step 1: To Train, Input all cheque logo images to the Neural Network and create a knowledgebase with six invariant features

Step 2: Segment the potential logo region by selecting randomly, and extract features using Eqs. (1)–(5)

Step 3: Set the Target for the classification of six categories of bank

Step 4: Create & Train the Neural Network

Step 5: Create knowledgebase, for all cheques using Eq. (6)

Step 6: Determine the index for maximum value using the Eq. (8) and this value will constitutes the type of class for which the query cheque image belongs

4 Experimental Results

The different bank cheques has been detected successfully. In empirical analysis and experiment around, 200 cheques recognized out of 213 bank cheques, the success rate of detection with respect to all six bank cheques is Axis bank is 93.3 %, Canara bank cheques is 92.5 %, Corporation bank cheques is 90 %, SBI bank cheques is 95 %, SBM bank cheques is 90 % and finally Union bank cheques is 93.33 % leading to success rate of 92.35 % (approximately, calculated as the average success rate of above mentioned bank cheques). The average time taken to

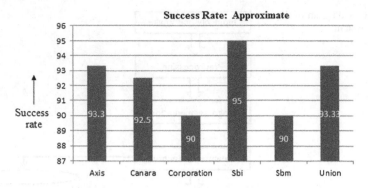

Fig. 6 Plot showing recognition rate of different banks

Fig. 7 Sample misdetecting results

detect is 1 s calculated by dividing the time duration of processing all cheques by total number of all cheques considered. However, proposed method fails to detect the damaged or distorted, watershed and occluded cheque images. Comparative analysis is given in the below given Fig. (6). Sample misdetection results are shown in the Fig. (7).

5 Conclusion

An attempt has been made by designing a novel algorithm in order to classify and detect typical class of the bank based on geometrically invariant features and neural network is presented. The proposed technique is better with respect to speed and accuracy. Single bank cheque image with different geometrical shape or logo were recognized with accuracy of 92.85 %. Poor results will be obtained for distorted or damaged cheques images and are not recognized properly like damaged view of cheques having one segment missing or crossed line or cancelled cheques with oblique line. These issues are considered in our future work.

References

1. Dhanalakshmi, S., Kaviya, J.: Cheque Clearance System Using Various Validation Techniques. 4(12), 1412 (2013). ISSN:2229-3345
2. Shridhar, M., Houle, G.F., Kimura, F.: Document recognition strategies for bank cheques. In: 978-1-4244-3355-1/09/$25.00©2009 IEEE Processing & TCP-IP Protocol. Int. J. Adv. Res. Electr. Electron. Instrum. Eng. (An ISO 3297: 2007 Certified Organization) 3(3), (2014)
3. Wahap, A.R., Khalid, M., Ahmad, A.R., Yusof, R.: A Neural Network Based Bank Cheque Recognition system for Malaysian Cheques
4. Dileep, M.R., Danti, A.: Lines of connectivity-face model for recognition of the human facial expressions. Int. J. Artif. Intell. Mech. 2(2), (2013). ISSN:2320–5121
5. Dileep, M.R, Danti, A.: Two level decision for recognition of human facial expressions using neural network. In: International Conference on Computing and Intelligence Systems, vol. 4, pp. 1368–1373. (2015). ISSN:2278-2397
6. Samant, H., Gaikwad, A., Ingale, V., Sarode, H.: Cheque Deposition System Using Image

Resource Management and Performance Analysis of Model-Based Control System Software Engineering Using AADL

K.S. Kushal, Manju Nanda and J. Jayanthi

Abstract The key principles involved in abstraction, encapsulation, design and development phases of the software structures of a system is, management of their complexities. These structures consist of the necessary components defining the concept of architecture of the system. Complex embedded systems, evolving with time, comprise of complex software and hardware units for its execution. This requires effective and efficient software model-based engineering practices. The complex systems are evolving in-terms of its resources and contemplating the operational dynamics. In this paper we emphasize on the formal foundations of Architecture Analysis and Design Language (AADL) for model-based engineering practices. This engineering process involves models as the centralized and the indispensible artifacts in a product's development life-cycle. The outcome of the approach features the techniques along with the core capabilities of AADL and managing the evolving resources considering impact analysis. A suitable case study, Power Boat Autopilot is considered. The details on the use of AADL capabilities for architectural modeling and analysis are briefly presented in this paper.

Keywords Society of automotive engineers-architecture analysis and design language (SAE AADL) · Model-based engineering (MBE) · Architecture-driven designing

K.S. Kushal (✉) · M. Nanda · J. Jayanthi
Aerospace Electronics & Systems Division, CSIR-National Aerospace
Laboratories, Bengaluru, India
e-mail: ksk261188@gmail.com

M. Nanda
e-mail: manjun@nal.res.in

J. Jayanthi
e-mail: jayanthi@nal.res.in

© Springer India 2016
H.S. Behera and D.P. Mohapatra (eds.), *Computational Intelligence in Data Mining—Volume 2*, Advances in Intelligent Systems and Computing 411, DOI 10.1007/978-81-322-2731-1_8

81

1 Introduction

The application of Model-Based Engineering (MBE) [1, 2] has been indispensible in developing advanced computing systems. Model-Driven Architectures (MDA) [1, 2], Model-Driven Development (MDD) [1, 2], Model-Based Development (MBD) [1, 2] and Model-Centred Development (MCD) [1, 2], are the emerging model-based engineering approaches.

The embedded control systems have become increasingly intensive in software. The software is dependent on commercially available computing hardware modules, challenging their integration. Along the developmental life-cycle, MBE proves to provide great confidence, in the form of predictive analysis at an early stage of abstraction. Integration processes, with great assurance for the developed system will conform to the system requirements. Architecture Analysis and Design Language (AADL), an SAE International (formerly known as Society of Automotive Engineers) standard [3], a model-based software systems engineering is used to capture the static modular software architecture, run-time architecture deployment architectures and its physical environments. AADL is a component-based modeling language. AADL combined with formal foundations and rigorous analyses [3, 4], critical real-time computational factors such as performance, dependability, resource management, security and data integrity are evaluated [5]. A unified system architectural model can be developed and analyzed using AADL, supported with its extensible constructs. Real-time complex embedded systems are resource constrained. Implementation of such constraint-based resources with the proper allocation of processor, memory and the bus resources are suitably provided by AADL.

In this paper we have considered an embedded control system such as Power-Boat Autopilot (PBA) as case study. PBA is a distributed embedded real-time system for speed, navigational and guidance control of a maritime vessel. While PBA is a maritime application, it represents the key elements of vehicle control for a wide range of similar applications involved in aircrafts, spacecrafts and automotive applications. In this system, autopilot is the controlling unit of the system, controlling the trail of the vehicle, without any human intervention. These systems evolve over time and the development of such safety-critical embedded control systems [6] requires less error prone environments in their course of development. We illustrate the developmental procedure involved in the implementation of the speed control unit and its functionality, an integral part of the PBA system. This exemplifies the use of AADL for similar control applications in various applicative domains such as aeronautical, automotive or land vehicle speed control systems. High-level representations such as systems, processes, device components, run-time compositions of all these elements of AADL are discussed in this paper. Allocation of dynamic resources [7] and assignment of values for the analysis of their instances are also discussed in this paper.

The paper is divided into five major sections; Sect. 2 provides an insight about the AADL. Section 3 brief about the implementation characteristics using AADL. The architecture of speed control unit of PBA system and various analyses are briefed in this section. Section 4 discusses various metrics and parameters involved in analysing and evaluating the speed control unit architecture of PBA system. Section 5 concludes and gives an overview of future scope of the work.

2 Literature Survey

2.1 Model-Based Engineering (MBE) for Embedded Real-Time Systems with AADL

Models of embedded real-time systems are the computational systems in an operational environment. They represent critical functional units for these applicative systems. Some of the vitally important operational quality attributes such as performance, safety and dependability [8] are also evaluated. Thus there happens to be three key elements for an embedded real-time control system software architecture, as shown in Fig. 1;

1. Application run-time architecture in terms of communication application tasks.
2. Execution platform expressed in terms of its interconnected networks.
3. Physical deployment platform/environment with which the application interacts.

Along with these three key elements, as shown in Fig. 1, some of the essential interconnection interactions in the form of;

1. Logical Connectivity: between embedded application software and the physical components (e.g. behaviour, control, commands)

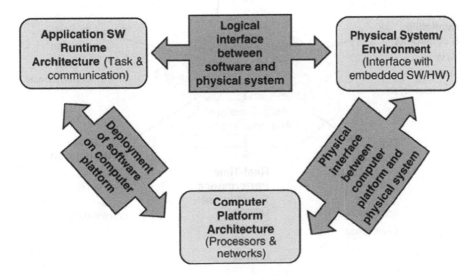

Fig. 1 Elements of embedded software system architectures [1]

2. Physical Connectivity: between the embedded system and the actual physical system
3. Deployment Features: computation of the resources and its management, for the execution of an embedded application software.

These key elements are provided to support the MBE of an embedded real-time control system software architecture.

Their logical connectivity helps us in making the right assumptions about the physical system, its behaviour with faults and its exceptional behaviour. The physical connectivity ensures that systems interact physically and the necessary engineering requirements are met. With the deployment features, we ensure that the resources and their evolution are dynamically allocated to the system, impacting their performance.

These resources are managed effectively. Safety in coherence with performance criteria is also ensured. There have been certain drawbacks at different abstracts in the development life-cycle with evolution of the system architecture. This can be overcome by the architecture-centric model-based engineering approach as shown in Fig. 2.

This approach having two major benefits that will address the evolving design requirements specifications suitably, being;

1. Any changes being made with the architecture of a system is automatically reflected in the analytical models as and when they are regenerated for the revalidation of the system, after the change in the system has been made successfully.
2. Impact of the change in the system model addressing a specific quality attribute over the other can be easily determined by re-evaluating these models with different quality dimensions.

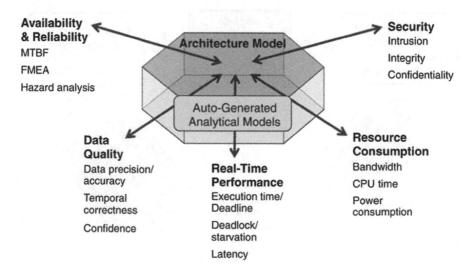

Fig. 2 Architecture-centric model-based engineering [1]

Such model-based approaches were addressed by the introduction of Architecture Analysis and Design Language (AADL) [3]. AADL includes both textual as well as graphical language support. AADL provides an extensible support [9] for the annotation of the models with user-defined analytical properties.

2.2 Working with SAE AADL

Using AADL, the architecture of an embedded control system can be captured as architectural models, hierarchically. The descriptions comprise identifying and detailing software and hardware components, along with their interactions in the form of interface specifications. These implementations are provided as packages. The packages represent the libraries of component specifications that can be used in multiple architectural models. Packages can be made as public or private versions.

- Public Package: contains specifications available to other packages.
- Private Packages: used to protect the details of the component implementations if we do not want to expose the details to others.

In AADL, we define run-time characteristics for the description of software components, such as threads, processes, and data. This includes execution times, scheduling and deadlines. It also includes key elements representing code related information [10]. We define the execution hardware platform components such as processors, memory and buses and characterize them with relevant execution and configuration properties. The resource properties are defined for these components. The resource configuration metrics such as processor speed, memory word size, power, etc. are defined. Additional system's operational environments are defined via devices that represent the system interfaces. A fully specified system implementation is done by integrating all of system's composite elements into a single hierarchical component. In this hierarchy we would establish the major details such as;

- Interactions among the components
- Architectural structure required to define an executable system

Data event exchanges that happen between the defined software and hardware components in terms of physical connections are also defined. An AADL instance model is obtained when we instantiate the top-level system implementation. Various analyses are performed for this top-level system implementation.

This model-based approach assures the quality of the software architecture modeled. This is supported with thorough verification and validation of the architectural model [11]. The resource allocation and management is performed with the evaluation of critical quality attributes generated by AADL. Quality is assured with respect to resource attributes such as performance [5], flow latency, etc. This approach was proposed by Weiss et al. [11]. A better resource utilization and management indicates the capability of the system to execute a large number of involved tasks continuing in meeting the deadlines. This was proposed by Hudak et al. [12]. This work also included the concept of design re-use. Use of pre-emptive fixed-priority scheduling is recommended to achieve better resource utilization and flexible design. The analysis of the scheduling policy leads to the evaluation of gain in resource utilization. They also proposed that analyzing the sensitivity to variations in evolving delay scenarios yields an effective resource management approach.

3   Modeling the Speed Control Unit of PBA Using AADL

In this paper we propose a novel approach to exhibit a high-level system representation of the speed control unit of PBA, developed using AADL system, process, package, and device components. This initial representation of the Speed Control Unit of PBA is as given in the block diagram Fig. 3, with the inclusion of run-time composition details of all the elements. The model elements are associated with necessary properties, required for analysis and generate a model instance. We define the components of the system, with their specifications, as a package using the open-source tool suite for AADL representation, OSATE (Open-Source AADL Tool Environment) v2.0.0 [13], as shown in Fig. 4.

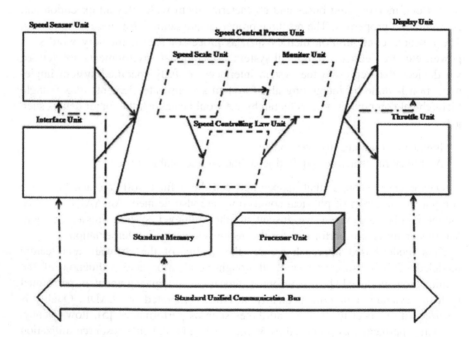

Fig. 3 Block diagram of speed control unit of PBA

Fig. 4 OSATE project environment [13]

The process of defining the components in AADL is similar to identifying objects in an object-oriented methodology. These components may include abstract encapsulations of functionality as well as representations of certain tangible notations of the system and its deployment environment. The description of the speed controller of PBA is reviewed based on the details provided in the block diagram, Fig. 3. A pilot interface unit as an input, for processing relevant PBA information is required. A speed sensor unit that sends the speed data to PBA, a PBA controller to control the functionality and a throttle actuator that responds to PBA commands along with a display unit is required. Each component is identified and a type definition for each of the component, with component name, run-time category and interfaces, are defined. The speed sensor, pilot interface unit, throttle actuator unit and the display unit are modeled as devices and the functions of PBA control unit as processes. The device components providing interconnect through interfaces to the external environment may not be needed to be decomposed extensively. Bus is defined as a subcomponent. The syntax for representing the various devices of the system, are as shown in Fig. 5.

The process component, reflecting the control function has to be implemented as core of the control processing of PBA software architecture. The software run-time components will be contained within its implementation. The interfaces to the components of PBA are defined as ports. These ports are declared within the component type. Their definitions are reflected in each implementation including all the properties and subcomponent declarations. The top-level system implementation requires integration of all the defined components, both software and hardware instances. They are implemented by defining the connections for each of these ports, components and subcomponents. The internal structure of the

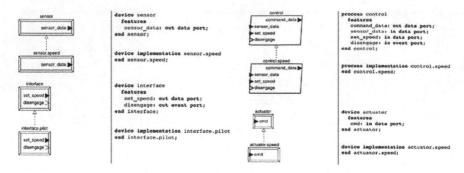

Fig. 5 Component type and their implementations

implementation can be accessed hierarchically in the OSATE tool plug-in environment on Eclipse IDE.

The control software is defined with the composition of the process *Speed_Control*. The speed control unit is autonomous. The speed control process is divided into two subcomponents, defined as threads. A thread which receives input from speed sensor. It scales and filters the data before delivering the processed data to second thread. The second thread executes the speed control laws and outputs commands to the throttle actuator. The interfaces involved in the two threads are different. Thus we define a type and implementation each specific for each thread. Property associations are also used to assign execution characteristics to these threads. We have also assumed that the threads are periodically assigned to the processor. This is done using the *Dispatch_Protocol* property association, each with a period of 50 ms, using the *Period* property association. This implementation representation is as shown in Fig. 6. In AADL, modeling the hardware instances and binding the modeled control software to that hardware instance with execution time and scheduling policies, provide an effective resource management of the system. The representation of such system is as shown in Figs. 7 and 8.

Fig. 6 Process
implementation

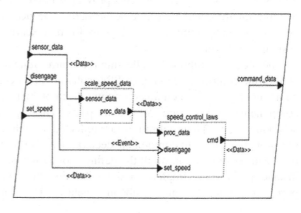

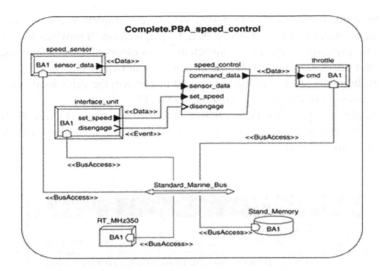

Fig. 7 Integrated software and hardware system

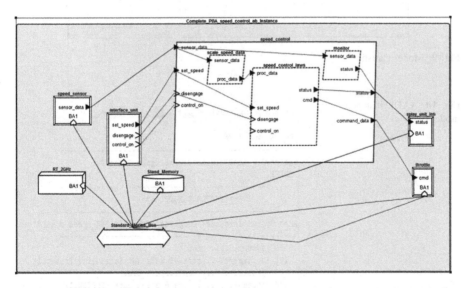

Fig. 8 AADL representation of the instance of the integrated PBA speed control unit architecture

4 Validating the Instance Model: Metrics and Parameters

Various analyses for execution time, latency, scheduling policies are performed on the system. These analyses are performed as the architecture model instance is instantiated without any errors. Upon instantiation, model statistics is computed. It gives the number of component types, component implementations, and flow specifications declared in the instance AADL specification. It provides the number of threads, processes, semantic connections, processors, buses, memories, and devices that are instantiated in AADL instance, as shown in Fig. 9.

AADL Inspector, from Ellidiss is required to perform the behavioral, static and dynamic analyses of the instance created in OSATE.

The Static Analysis generates metrics in two categories; Declarative Model metrics and Instance Model metrics, as shown in Figs. 10 and 11. AADL Declarative Model Metrics.

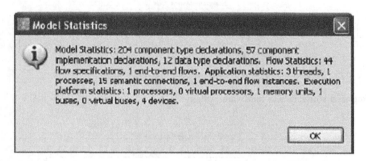

Fig. 9 Model statistics

Fig. 10 AADL instance model metrics

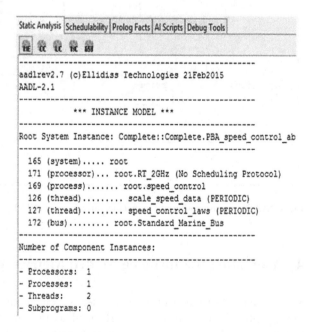

Fig. 11 AADL declarative
model metrics

```
-----------------------------------------------------
            *** DECLARATIVE MODEL ***
-----------------------------------------------------
Number of Packages:...........................1
Number of Component Types:...................12
Number of Component Implementations:.........12
Number of Subcomponents:.....................11
Number of Call Sequences:.....................0
Number of Subprogram Calls:...................0
Number of Features:..........................28
Number of Connections:.......................20
Number of Property Associations:.............15
---- Prototypes -------------------------------------
Number of Prototypes:.........................0
Number of Prototype Bindings:.................0
---- Flows ------------------------------------------
Number of Flow Specifications:................4
Number of Flow Implementations:...............1
---- Modes ------------------------------------------
Number of Modes:..............................2
Number of Mode Transitions:...................2
---- Properties -------------------------------------
Number of Property Sets:......................0
Number of Property Types:.....................0
Number of Property Definitions:...............0
Number of Property Constants:.................0
---- Annexes ----------------------------------------
Number of Annexes:............................2
---- Behavior Annex Items ---------------------------
Number of Behavior Annex Variables:...........0
Number of Behavior Annex States:..............0
Number of Behavior Annex Transitions:.........0
Number of Behavior Annex Conditions:..........0
Number of Behavior Annex Actions:.............0
---- Error Annex v1 Items ---------------------------
Number of Error Annex Properties:.............0
---- Error Annex v2 Items ---------------------------
Number of Error Type Definitions:.............0
Number of Error Type Set Definitions:.........0
Number of Error Type Mappings:................0
Number of Error Type Transformations:.........0
Number of Error Behavior Definitions:.........0
```

The Static Analysis tool encompasses a set of independent rule checkers that verify for semantic correctness. Each rule checker is implemented as a service of the static analysis tool, which are as follows;

- CC—for AADL Consistency Checks as in Fig. 12.
- LC—for AADL Legality rules Checks as in Fig. 13.
- NC—for AADL Naming rules Checker as in Fig. 14.
- 653—for ARINC 653 rules checker

Fig. 12 Consistency checks

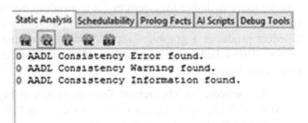

```
Static Analysis  Schedulability  Prolog Facts  AI Scripts  Debug Tools

0 AADL Consistency Error found.
0 AADL Consistency Warning found.
0 AADL Consistency Information found.
```

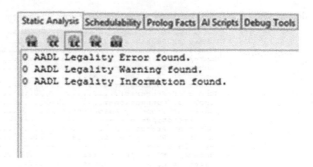

Fig. 13 Legality checks

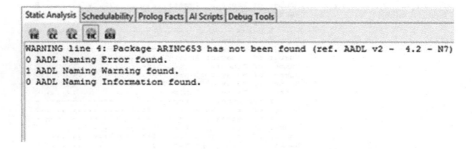

Fig. 14 Naming checks

In the Fig. 14, there is a warning with the naming rules checks; with the message Package ARINC653 has not been found, meaning that the ARINC 653 constraints are not included in the package.

4.1 Schedulability Analysis

The entire process including threads and its implementations are all correlated and executed, as soon as Schedulability tab is selected in the AADL Inspector. This automatically activates the services for scheduling the processes, threads, provided by the third party plug-in called as Cheddar, as shown in Fig. 16. Schedule Table. Cheddar produces a graphical representation of timing behavior of the real-time system. This schedule table is a result of the static simulation. It basically provides three launches, as;

- THE—Scheduling Theoretical Tests as shown in Fig. 15
- SIM—Scheduling Simulation Test as shown in Fig. 18
- Schedule Table as shown in Fig. 16

Fig. 15 Theoretical test

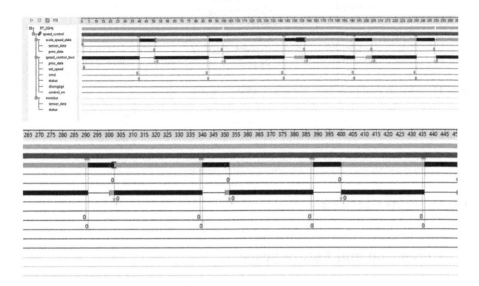

Fig. 16 Schedule table

In the Schedule Table, time lines are displayed for each processor, process, thread and shared data subcomponent of the top-level system instantiated. The time scale and color scheme shared within dynamic simulator as per Fig. 17. Theoretical analysis computes the processor utilization factor and threads response time, as and when the corresponding conditions are met.

The simulation test provides the number of pre-emptions and the context switches as well as thread response time. The scheduling policy of the system is assumed to be Rate Monotonic (RMS). Time period for processes and threads being 50 ms, dynamic simulation was done using Marzhin, an AADL Inspector plug-in. This simulator is event-driven and can analyze wide variety of real-time systems (Fig. 18).

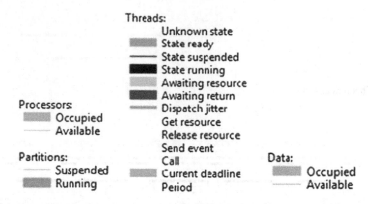

Fig. 17 Color code of time lines in the simulation schedule tests

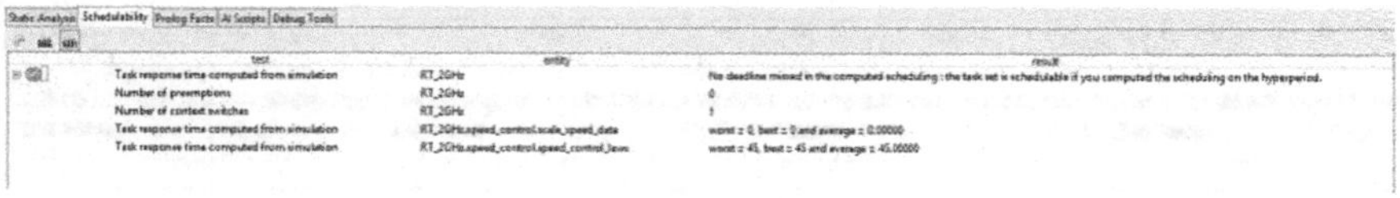

Fig. 18 Simulation schedule test

5 Conclusion and Future Scope

This paper proposes a novel approach to perform early analyses [14] with respect to performance and perform an effective resource management. The analyses are performed for an embedded real-time control system, Speed Control Unit of PowerBoat Autopilot System. The graphical representations of the architecture and the metrics generated, helps in better understanding of the system. It also uncovers the flaws, validating architecture against requirements, efficient resource utilization and management, by making engineering processes more effective. The formal foundations of AADL and its relying methodologies can thus be successfully adapted in validating the architectures of similar Embedded Real-Time Safety-Critical Systems in various application domains such as aerospace, spacecrafts, automobiles, telemedicine etc. In purview of futuristic developments, as a scope we can include behavioral and error annexes [15] constructs in validating the architectural models and also have the various standards protocols such as ARINC653 [16], ARINC429 included in the system architecture. Fault models can be constructed and evaluated.

Acknowledgments The authors would like to thank the Director, CSIR – NAL, Bengaluru for supporting this work.

References

1. Feiler, P.H., Gluch, D.P.: Model-based engineering with AADL—an introduction to the SAE architecture analysis & design language. Addison—Wesley, Pearson Education, Inc., September 2012. ISBN:978-0-321-88894-5
2. Joshi, A., Vestal, S., Binns, P.: Automatic generation of fault trees from AADL models. In: Workshop on Architecting Dependable Systems (DSN'07), Critical Systems Research Group, June 2007
3. Anderson, S.O., Bloomfield, R.E., Cleland, G.L.: Guidance on the use of formal methods in the development and assurance of high integrity industrial computer systems. In: Working Paper 4001, European Workshop on Industrial Computer Systems Technical Committee (Safety, Reliability and Security), pp. 13–29, June 1998
4. Feiler, P.H., Lewis, B., Vestal, S., Colbert, E.: An overview of the SAE architecture analysis & design language (AADL) standard: a basis for model-based architecture-driven embedded systems engineering. In: Architecture Description Languages, The International Federation for Information Processing, vol. 176, pp. 3–15. (2005). ISBN:978-0-387-24586-8
5. Milner, R.: A calculus of communicating systems. Springer - Verlag NewYork, Inc., Secaucus (1982). ISBN 0387102353
6. Knight, J.C.: Safety critical systems: challenges and directions. In: IEEE Proceedings of 24th International Conference on Software Engineering, pp. 547–550, May 2002. ISBN:1-58113-472-X
7. Hoare, C.A.R.: Communicating sequential processes. Prentice—Hall Inc., pp. 181–204. (1985). ISBN:978-0131532717
8. Robati, T., El Kouhen, A., Gherbi, A., Hamadou, S., Mullins, J.: An extension for AADL to model mixed-criticality avionic systems deployed on IMA architectures with TTEthernet. In: 1st Architecture Centric Virtual Integration Workshop @ MODELS 2014, SEI-CMU, pp. 14–27, September 2014
9. Muhammad, N., Vandewoude, Y., Berbers, Y., van Loo, V.: Modelling embedded systems with AADL: a practical study. New Advanced Technologies, Aleksandar Lazinica (Ed.), pp. 250–265 (2010). ISBN:978-953-307-067-4
10. Raghav, G., Gopalswamy, S., Radhakrishnan, K., Hugues, J., Delange, J.: Model based code generation for distributed embedded systems. In: Proceedings of European Congress on Embedded Real-Time Software (ERTS 2010). (2010)
11. Weiss, K.A., Woodham, K., Feiler, P.H., Gluch, D.P.: Model-based software quality assurance with the architecture analysis and design language. AIAA Infotech@Aerospace Conference, AIAA 2009–2034, April 2009
12. Hudak, J.J., Feiler, P.H, Gluch, D.P.: Bruce a lewis. In: Embedded System Architecture Analysis Using SAE AADL. Technical Report, CMU/SEI-2004-TN-005, Performance-Critical Systems Initiative, June 2004
13. http://www.eclipse.org
14. Fitzgerald, J.S., Larsen, P.G.: Modelling Systems: Practical Tools and Techniques in Software Development, 2nd edn., pp. 71–96. Cambridge University Press (2009). ISBN:978-0-521-89911-6
15. Architecture Analysis & Design Language (AADL) Annex, vol. 1: Annex E: Error Model Annex, SAE International Standards: AS5506/1. http://www.sae.org/technical/standards/ AS5506/1 (2006)
16. Delange, J., Plantec, A., Kordon, F.: Validate, simulate, and implement ARINC653 systems using the AADL. In: Proceedings of the ACM SIGAda Annual International Conference on Ada and Related Technologies (SIGAda'09), vol. 29, Issue. 3, pp. 31–44, December 2009. ISBN:978-1-60558-475-1
17. Sagent, R.G.: Validation and verification of simulation models. In: Proceedings of the 2010 Winter Simulation Conference(WSC), pp. 166–183, December 2010. ISBN:978-1-4244-9866-6

18. Feiler, P.H.: Modeling of system families. In: Technical Report, CMU/SEI-2007-TN-047, Performance-Critical Systems Initiative, Software Engineering Institute, July 2007. http://repository.cmu.edu/sei/339/
19. Architecture Analysis & Design Language (AADL). http://www.aadl.info Wiki: https://wiki.sei.cmu.edu/aadl
20. Feiler, P.H., Lewis, B.: SEI technology highlight: AADL and model-based engineering. In: Industry Standard Notation for Architecture-Centric Model-Based Engineering, SEI Research and Technology Highlight: AADL and MBE, January 2010

Performance Evaluation of Rule Learning Classifiers in Anomaly Based Intrusion Detection

Ashalata Panigrahi and Manas Ranjan Patra

Abstract Intrusion Detection Systems (IDS) are intended to protect computer networks from malicious users. Several data mining techniques have been used to build intrusion detection models for analyzing anomalous behavior of network users. However, the performance of such techniques largely depends on their ability to analyze intrusion data and raise alarm whenever suspicious activities are observed. In this paper, some rule based classification techniques, viz., Decision Table, DTNB, NNGE, JRip, and RIDOR have been applied to build intrusion detection models. Further, in order to improve the performance of the classifiers, six rank based feature selection methods, viz., Chi squared attribute evaluator, One-R, Relief-F, information Gain, Gain Ratio, and Symmetrical Uncertainty have been employed to select the most relevant features. Performance of different combinations of classifiers and feature selection techniques have been studied using a set of performance criteria, viz., accuracy, precision, detection rate, false alarm rate, and efficiency.

Keywords IDS · Decision table · DTNB · Data mining · NNGE

1 Introduction

The unprecedented expansion in Internet connectivity has led to a plethora of web based services catering to a wide range of user groups. This has evoked security concerns for protecting personal and sensitive data from misuse. As more and more users are getting connected to the Internet, incidences of malicious users fiddling with network resources are becoming more frequent. Thus, there is a need for

A. Panigrahi (✉) · M.R. Patra
Department of Computer Science, Berhampur University,
Berhampur 760007, India
e-mail: ashalata.panigrahi@yahoo.com

M.R. Patra
e-mail: mrpatra12@gmail.com

© Springer India 2016   97
H.S. Behera and D.P. Mohapatra (eds.), *Computational Intelligence in Data Mining—Volume 2*, Advances in Intelligent Systems and Computing 411, DOI 10.1007/978-81-322-2731-1_9

building security infrastructure to protect against malicious users. An Intrusion detection system is a widely accepted solution for detecting intrusive behavior of network users. Intrusion detection can be broadly classified as: misuse and anomaly detection. Misuse or signature based IDS detect intrusions based on known intrusions or attacks. It performs pattern matching of incoming packets and/or command sequences to the signatures of known attacks. But, anomaly based IDS has the ability to detect new types of intrusions. However, its major limitation lies in discovering boundaries between normal and abnormal behaviour due to deficiencies in abnormal samples in the training phase. Different techniques have been proposed to build intrusion detection systems but the challenge lies in dealing with issues like huge volume of network traffic, identifying boundaries between normal behavior and attacks, imbalanced data distribution, and need for continuous adaptation to a constantly changing network environment [1]. Data mining based IDSs have to deal with these limitations.

2 Literature Survey

Affendey et al. [2] proposed machine learning algorithms for intrusion detection system which compared the performance of C4.5 algorithm with SVM in detecting intrusions and the results revealed that C4.5 performed better than SVM in terms of intrusion detection and false alarm rate. Jain et al. [3] proposed classification of intrusion detection based on various machine learning algorithms viz. J48, Naïve-Bayes, One-R, and Bayes Net. They found that the Decision Tree algorithm J48 is most suitable as it has low false positive rate. Rames et al. [4] have analyzed the performance of ID3 and J48 on a reduced data set obtained by filter based feature selection. ID3 produced more accurate results compared to J48. Hota et al. [5] proposed an IDS based on decision tree technique with feature selection. Gain ratio rank based feature selection method was applied on NSL-KDD dataset, and Random forest model was used for classification. Empirical results show that random forest model produced an accuracy of 99.84 %. Jain et al. [6] have proposed an intrusion detection method using information gain, NB and Bayes Net. They reduced the features of the dataset using information gain of the attributes. After feature reduction the data was analyzed using NB and Bayes Net algorithms.. Bayes Net with an accuracy rate of approximately 99 % was found to perform much better than NB in detecting intrusions. Atefi et al. [7] have proposed a hybrid model using SVM and GA (Genetic Algorithm). They compared true negative and false positive rates between SVM and hybrid model SVM+GA. Hybrid model recorded low false negative rate of 0.5013 % and high true negative value of 98.2194 %. The result shows high accuracy of the hybrid model. A hybrid classification model using evolutionary computation based techniques was proposed in our earlier work [8]. The results show that AIRS1 classifier with best first search feature selection gives highest accuracy and AIRS2 classifier with Gain Ratio feature selection gives lowest false alarm rate. Similarly, using a neural network based classification model that employed entropy based feature selection [9], it was shown that PART classification with symmetrical uncertainty feature selection gives highest accuracy, highest detection rate and low false alarm rate.

3 Rule Learning Classification Algorithms

3.1 Decision Table (DT)

Two variants of decision table classifiers [10] are DTM (Decision Table Majority) and DTL (Decision Table local). A DTM decision table has two components: a schema which is a list of features and a body consisting of a multi set of labeled instances. To build a DTM, the induction algorithm must decide which features to include in the schema and which instances to store in the body.

Given an unlabeled instance U, let I be the set of labeled instances in the DTM exactly matching the given instance U, where only the features in the schema are required to match and all other features are ignored. If I = Φ, it returns the majority class in the DTM; otherwise it returns the majority class in I. Unknown values are treated as distinct values in the matching process.

3.2 DTNB Classifier

In decision table/Naïve Bayes classifier [11], at every point in the search, the algorithm estimates the value of dividing the features into two disjoint subsets: one for the decision table and the other for Naïve Bayes. A forward selection search is used at each step. The selected features are exhibited by Naïve Bayes and the remainder by the decision table. At each step, the algorithm drops a feature entirely from the model.

Assuming X^A as the set of features in the DT and X^B in NB, the overall class probability is computed as

$$Q(y|X) = \alpha \times Q_{DT}(y|X^A) \times Q_{NB}(y|X^B) / Q(y) \qquad (1)$$

where $Q_{DT}(y|X^A)$ and $Q_{NB}(y|X^B)$ are the class probability estimates obtained from the DT and NB respectively; α is a normalization constant, and Q(y) is the prior probability of the class.

3.3 Non-nested Generalized Exemplars (NNGE)

NNGE [12] performs generalization by merging exemplars, forming hyper-rectangles in feature space that represent conjunctive rules with internal disjunction. NNGE forms a generalization each time a new example is added to the database, by joining it to its nearest neighbor of the same class and does not allow hyper-rectangles to nest or overlap. It also prevents the overlap by testing each prospective new generalization to ensure that it does not cover any negative examples. NNGE uses a modified Euclidean distance function that handles hyper-rectangles. It uses dynamic feedback to adjust exemplar and feature weights after each new example is classified.

Let us consider a learning process starting from a set of m examples or training instances $\{T^1, T^2,..., T^m\}$, each one being characterized by the values of n features which can be nominal or numeric. The aim of the learning process is to construct a set of generalized hyper-rectangles (exemplars), $\{H^1, H^2,...,H^p\}$.

NNGE Algorithm

For each example T^i in the training set do:
Find the hyperrectangle H^p which is closest to T^i
If $D(H^p, T^i) = 0$ THEN
    IF class$(T^i) \neq$ **class**(H^p) THEN Split(H^p, T^i)
    ELSE H':= Extend(H^p, T^i)
    IF H' overlaps with conflicting hyperrectangles
    THEN add T^i as a non generalized exemplar
ELSE $H^p := H'$

NNGE classifies new examples by determining the nearest neighbor in the exemplar/hyper-rectangle database using Euclidean Distance function. The distance $D(T, H)$ between Training instances $T = (T_1, T_2,...,T_m)$ and an exemplar or hyper-rectangle H is given as

$$D(T, H) = W_H \sqrt{\sum_{i=1}^{n} \left[W_i \frac{d(Ti - Hi)}{T_{max}^i - T_{min}^i} \right]^2}. \qquad (2)$$

where T_i is the ith feature value in the example, H_i is the ith feature value in the hyper-rectangle, W_H is the exemplar weight and W_i is the feature weight, T_i^{min} and T_i^{max} are the range of values over the training set which correspond to attribute i of numeric features.

3.4   JRip

JRip algorithm [13] consists of four phases: Growth, Pruning, Optimization, Selection. In the growth phase, it produces a sequence of individual rules by adding predicates until the rule satisfies the stopping criteria. The rules that reduce the performance of the algorithm are pruned in the second phase. In the optimization step, each rule is optimized by adding attributes to the original rule or generates a new rule using phase 1 and phase 2. In the last stage, the best rules are retained and others are ignored from the model.

3.5 Ripple Down Rule Learner (RIDOR)

RIDOR learns rules with exceptions by generating the default rule, and exceptions are generated using incremental reduced error pruning with smallest error rate [14, 15]. The exceptions with minimum support are selected and the best exceptions are found for each exception. A ripple down rule is a list of rules, each of which may be connected to another ripple down rule, specifying exceptions. An expert can create a new rule and add to the list, which have effects in the given context of the parent rule. Thus, ripple down rules form a binary decision tree that differs from standard decision trees in which all decisions are made at root nodes. The rules are never modified or deleted but they are locally patched.

4 The Proposed Model

The objective of the proposed model is to apply different rule learning algorithms to build intrusion detection system that exhibit high detection rate and low false alarm rate. The model is depicted in Fig. 1. In the first step, NSL-KDD dataset is

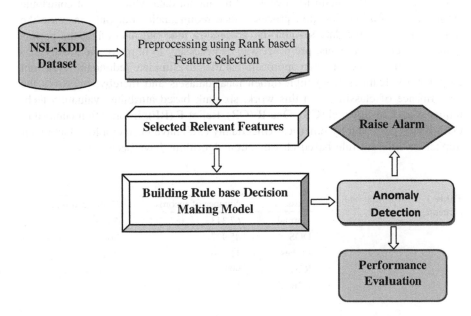

Fig. 1 Rule learning based classification model

preprocessed using six rank based feature selection methods viz. Chi Squared attribute evaluator, One R, Relief F, Information Gain, Gain Ratio, and Symmetrical Uncertainty. Next, the reduced dataset is classified using five rule learning algorithms viz. JRip, Decision Table, DTNB classifier, NNGE, and RIDOR.

5 Experimental Setup and Result Analysis

5.1 NSL-KDD Dataset

The NSL-KDD intrusion dataset (a new version of KDD CUP 99 dataset) [16] has been used for our experimentation. The data set consists of 41 feature attributes out of which 38 are numeric and 3 are symbolic. The total number of records in the data set is 125,973 out of which 67,343 are normal and 58,630 are attacks which fall into 24 different attack types and can be classified into four attack categories namely, Denial of Service (DOS), Remote to Local (R2L),User to Root (U2R), and Probing (Table 1).

5.2 Feature Selection Process

Normally, data sets contain irrelevant and redundant data which do not contribute significantly to a classification process. Such redundancies not only increase the dimensionality of the data set and the processing time but also affect the performance of a classifier. Thus, pre-processing is necessary to improve the quality of a dataset. Feature selection is an approach that tries to eliminate redundant, irrelevant and/or noisy features from high dimensional datasets and thereby, improves the performance of classifiers. In this work, six rank based attribute evaluation techniques, viz., Chi-Squared (CS), One-R (OR), Relief-F (RF), Information Gain (IG), Gain Ratio (GR), and Symmetrical Uncertainty (SU) have been employed and their impact on different rule based classification algorithms have been studied.

Table 1 Distribution and percentage of records

Class	Number of records	% of occurrence
Normal	67,343	53.48
DOS	45,927	36.45
Probes	11,656	9.25
R2L	995	0.78
U2R	52	0.04

5.3 Confusion Matrix

An intrusion detection model can be evaluated by its ability to make accurate prediction of attacks. Intrusion detection systems mainly discriminate between two classes, attack class and normal class. The confusion matrix reports the number of False Positives (FP), False Negatives (FN), True Positives (TP), and True Negatives (TN). Based on these values the following performance measurements can be made:

$$\text{Accuracy} = \frac{TP + TN}{TN + TP + FN + FP}. \tag{3}$$

$$\text{Precision} = \frac{TP}{TP + FP}. \tag{4}$$

$$\text{Recall/Detection Rate} = \frac{TP}{TP + FN}. \tag{5}$$

$$\text{False Positive Rate (FPR) or False Alarm Rate} = \frac{FP}{TN + FP}. \tag{6}$$

$$\text{Efficiency} = \frac{Total\,Detected\,Attack}{Total\,Attack} \times 100 \tag{7}$$

$$\text{Rate of Attack} = \frac{Number\,of\,attack\,detected\,correctly}{Total\,number\,of\,attacks} \tag{8}$$

6 Result Analysis

Different combinations of 5 classifiers (as mentioned in Sect. 3) with 6 rank based feature selection (as mentioned in Sect. 5.2) were applied on the NSL-KDD intrusion dataset (detailed in Sect. 5.1). Performance measurements were made using the criteria mentioned in Sect. 5.3 and the results are tabulated. 10-fold cross validation has been used for training and testing data in each case. Table 2 depict the attack rate of four attacks.

It is observed that the rate of attack depends on the number of instances, irrespective of the classification as well as feature selection technique used.

Table 3 depict the performance of five rule learning techniques based on the parameters such as accuracy, precision, detection rate, false alarm rate, and efficiency. JRip classification gives lowest false alarm rate irrespective of the feature selection method. JRip technique with symmetrical uncertainty feature selection gives highest accuracy of 99.8285 % and highest detection rate of 99.797 % (Table 4).

Table 2 Attack rate of four attacks

Feature selection methods	Classifiers	Rate of attack			
		DOS	Probes	R2L	U2R
ChiSquared evaluator	JRip	99.9521	99.4767	95.8794	46.1538
	DT	99.7082	96.4911	85.1256	44.2308
	DTNB	99.2401	97.3404	94.3718	55.7692
	NNge	99.9281	98.7388	89.4472	46.1538
	Ridor	99.9303	99.1764	93.3668	48.0769
OneR	JRip	99.9717	99.3994	94.8744	46.1538
	DT	99.7266	96.534	84.9246	34.6153
	DTNB	99.46	97.4691	93.3668	34.6153
	NNge	99.939	98.8075	89.4472	42.3077
	Ridor	99.9303	99.1592	94.2713	44.2308
Relief-F	JRip	99.6669	98.8589	91.3568	25
	DT	96.6468	87.9547	87.5377	5.7692
	DTNB	95.5778	96.4996	91.6583	42.3078
	NNge	99.6821	98.4128	87.5377	30.7692
	Ridor	99.4578	98.653	89.7487	15.3846
Information gain	JRip	99.9673	99.5367	94.4724	48.0769
	DT	99.7278	96.4911	85.3266	44.2308
	DTNB	99.3903	97.3318	93.1658	46.1538
	NNge	99.9368	98.8589	90.0502	44.2308
	Ridor	99.9368	99.2364	94.7739	46.1538
Gain ratio	JRip	99.9477	91.4894	95.5779	46.1538
	DT	99.7714	94.1232	92.1608	7.6923
	DTNB	99.7278	94.1661	92.4623	9.6154
	NNGE	99.926	98.8504	95.3769	53.8461
	Ridor	99.9368	95.2814	95.8794	40.3846
SU	JRip	99.95	99.5796	99,7883	55.7692
	DT	98.0251	96.4911	85.3266	44.2308
	DTNB	99.4056	97.1774	94.1708	42.308
	NNge	99.939	98.7817	89.4472	42.308
	Ridor	99.9325	99.2107	94.1708	46.1538

Table 3 Comparison of rule learning classifiers with rank based feature selection method

Feature selection method	Test mode	Classifier techniques	Evaluation criteria					
			Accuracy in %	Precision in %	Recall or detection rate in %	False alarm rate in %	Efficiency in %
Chi squared attribute evaluator	10-fold cross-validation	**JRip**	**99.823**	**99.8277**	**99.7919**	**0.15**	**99.7407**
		DT	99.4038	99.7576	98.9598	0.2094	98.7719
		DTNB	98.1821	96.5343	99.6725	3.1154	98.7515
		NNge	99.6396	99.5959	99.6299	0.3519	99.4661
		Ridor	99.7118	99.698	99.6827	0.2628	99.623
OneR	10-fold cross-validation	**JRip**	**99.8238**	**99.8481**	**99.7731**	**0.1321**	**99.7288**
		DT	99.3983	99.7456	98.9596	0.2198	98.7822
		DTNB	98.3425	96.9197	99.6043	2.756	98.9033
		NNge	99.6301	99.5536	99.652	0.389	99.4849
		Ridor	99.7436	99.7627	99.6862	0.2064	99.6316
ReliefF	10-fold cross-validation	**JRip**	**99.4896**	**99.4981**	**99.4047**	**0.4366**	**99.299**
		DT	97.4129	98.7498	95.6524	1.0543	94.6836
		DTNB	96.4405	94.9628	97.5251	4.5038	95.6473
		NNge	99.4356	99.4113	99.3757	0.5123	99.1625
		Ridor	99.3308	99.2265	99.3365	0.6742	99.212
Information gain	10-fold cross-validation	**JRip**	**99.8284**	**99.8447**	**99.7868**	**0.1351**	**99.7424**
		DT	99.4102	99.7542	98.9766	0.2123	98.7907
		DTNB	98.0861	96.4029	99.6043	3.2357	98.8282
		NNge	99.6563	99.6113	99.6503	0.3386	99.5054
		Ridor	99.7491	99.7458	99.7152	0.2212	99.6623

(continued)

Table 3 (continued)

Feature selection method	Test mode	Classifier techniques	Evaluation criteria				
			Accuracy in %	Precision in %	Recall or detection rate in %	False alarm rate in %	Efficiency in %
Gain ratio	10-fold cross-validation	JRip	99.0633	**99.8162**	98.1682	**0.1574**	98.1443
		DT	99.1974	99.7599	98.5127	0.2064	98.4376
		DTNB	99.0665	99.76	98.5383	0.207	98.4189
		NNge	**99.742**	99.7423	**99.7032**	0.2242	**99.5941**
		Ridor	99.3737	99.7232	98.9289	0.2391	98.8896
SU	10-fold cross-validation	**JRip**	**99.8285**	**99.8345**	**99.797**	**0.144**	**99.7493**
		DT	99.4102	99.7542	98.9766	0.2123	98.7907
		DTNB	98.0353	96.3164	99.5872	3.3159	98.8231
		NNge	99.6412	99.6061	99.623	0.343	99.4798
		Ridor	99.765	99.8002	99.6947	0.1737	99.6435

Table 4 A comparative analysis of our results with that of results obtained by other authors/works

Author	Feature selection method	Classifier techniques	Accuracy	Detection rate/recall	False alarm rate
Chen et al. [17]	Rough set	SVM	89.13 %	86.72 %	13.27 %
Jain et al. [6]	Information gain	Bayes Net	99.1073 %	92.7 %	0.052 %
Chandrasekhar et al. [18]	PSO	SVM	95 %	Not provided	Not provided
Kavitha et al. [19]	Best first search	ENLCID	Not provided	99.02 %	3.19 %
Elngar et al. [20]	PSO discretization method	Hidden Naïve Bayes	98.2 %	Not provided	Not provided
Panigrahi et al. [8]	Best first search	AIRS1	94.2757 %	90.6549 %	2.5719 %
Panda et al. [21]	PCA	Discriminative multinomial Naïve Bayes	94.84 %	Not provided	3 %
Panigrahi et al. [9]	Gain ratio	PART	99.4165	98.9442	0.1722
Present work	SU	JRip	99.8285	99.797	0.144

7 Conclusion

In this paper the analysis is based on different rule learning classification techniques with six rank based feature selection methods. JRip classification gives lowest false alarm rate irrespective of the feature selection method. JRip technique with symmetrical uncertainty feature selection gives highest accuracy of 99.8285 % and highest detection rate of 99.797 %.

References

1. Banzhaf, W., Wu, S.X.: The use of computational intelligence in intrusion detection systems: a review. Appl. Soft Comput. **10**, 1–35 (2010)
2. Affendey, L.S., Ektefa, M., Memar, S., Sidi, F.: Intrusion detection using data mining techniques. In: Proceedings of IEEE International Conference on Information Retrieval & Knowledge Management, Exploring Invisible World, CAMP' 10, pp. 200–203. Shah Alam, Selangor (2010)
3. Jain, Y.K., Upendra: An efficient intrusion detection based on decision tree classifier using feature reduction. Int. J. Sci. Res. Publ. **2**, 1–6 (2012)
4. Rames, T., Revathi, M.: Network intrusion detection system using reduced dimensionality. Indian J. Comput. Sci. Eng. (IJCSE) **2**, 61–67 (2010)

5. Hota, H.S., Shrivas, A.K., Singhai, S.K.: Article: an efficient decision tree model for classification of attacks with feature selection. Int. J. Comput. Appl. 42–48 (2013)
6. Jain, Y.K., Upendra: Intrusion detection using supervised learning with feature set reduction. Int. J. Comput. Appl. **33**, 22–31 (2011)
7. Atefi, A., Atefi, K., Dak, A.Y., Yahya, S.: A hybrid intrusion detection system based on different machine learning algorithms. In: Proceedings of the 4th International Conference on Computing and Informatics, ICOCI 2013, pp. 312–320. Sarawak, Malaysia (2013)
8. Panigrahi, A., Patra, M.R.: An evolutionary computation based classification model for network intrusion detection. In: International Conference on Distributed Computing and Internet Technology (ICDCIT-2015), pp. 318–324. Lecture Notes in Computer Science, Springer Verlag, Bhubaneswar, India (2015)
9. Panigrahi, A., Patra, M.R.: An ANN based approach for network intrusion detection using entropy based attribute reduction. Int. J. Network Secur. Appl. (IJNSA). **7**, 15–29 (2015)
10. Kohavi, R.: The power of decision tables. In: Proceedings of the European Conference on Machine Learning (ECML), pp. 174–189. Lecture Notes in Artificial Intelligence. Springer Verlag, Heraclion, Crete, Greece (1995)
11. Frank, E., Hall, M.: Combining Naïve Bayes and decision tables. In: Proceedings of the 21st Florida Artificial Intelligence Society Conference (FLAIRS), pp. 318–319. Florida, USA (2008)
12. Salzberg, S.A.: Nearest hyperrectangle learning method. Mach. Learn. 277–309 (1991)
13. Cohen, W.W.: Fast effective rule induction. In: Proceedings of the Twelfth International Conference on Machine Learning, pp. 115–123 (1995)
14. Sharma, P.: Ripple-down rules for knowledge acquisition in intelligent system. J. Technol. Eng. Sci. **1**, 52–56 (2009)
15. Compton, P., Kang, B., Preston, P.: Local patching produces compact knowledge bases, pp. 104–117. A Future in Knowledge Acquisition. Springer Verlag, Berlin (1994)
16. Bagheri, E., Ghorbani, A., Lu, W., Tavallaee, M.: A detailed analysis of the KDD CUP 99 data set. In: Proceedings of the 2009 IEEE Symposium on Computational Intelligence in Security and Defense Applications (CISDA 2009), pp. 1–6. Ottawa (2009)
17. Chen, R.C., Cheng, K.F., Hsieh, C.F.: Using rough set and support vector machine for network intrusion detection. Int. J. Netw. Secur. Appl. (IJNSA) 1–13 (2009)
18. Chandrasekar, A., Vasudevan, V., Yogesh, P.: Evolutionary approach for network anomaly detection using effective classification. Int. J. Comput. Sci. Netw. Secur. **9**, 296–302 (2009)
19. Karthikeyan, S., Kavitha, B., Sheeba Maybell, P.: Emerging intuitionistic fuzzy classifiers for intrusion detection system. J. Adv. Inf. Technol. **2**, 99–108 (2011)
20. Dowlat, A., Elngar, A.A., Ghaleb, F.M., Mohamed, A.: A real-time anomaly network intrusion detection system with high accuracy. Inf. Sci. Lett. An Int. J. 49–56 (2013)
21. Abraham, A., Panda, M., Patra, M.R.: Discriminative multinomial Naïve Bayes for network intrusion detection. In: Sixth International Conference Information Assurance and Security, pp. 5–10 (2010)

Evaluation of Faculty Performance in Education System Using Classification Technique in Opinion Mining Based on GPU

Brojo Kishore Mishra and Abhaya Kumar Sahoo

Abstract Large amount of data is available in the form of reviews, opinions, feedbacks, remarks, observations, comments, explanations and clarifications. In Education system, main focus is given to quality of teaching. That quality depends on coordination among teacher and student. Feedback analysis is more important to measure the faculty performance. Performance of faculty should be evaluated so that we can enhance our education quality. To measure the performance of faculty, we use classification technique by using opinion mining. We also use this technique on GPU architecture using CUDA-C programming to evaluate performance of a faculty in very less time. This paper uses opinion mining concept with GPU to extract performance of a faculty.

Keywords Classification · CUDA-C · Education system · GPU · Opinion mining

1 Introduction

In today's competitive market, education institute wants to give best education facility to students and tries to prove that they are the best among all Institutes. Every parent wants to send their children to best institute for learning. So education market is so competitive that it is very difficult to choose the best institute. Value of Institute is measured by student's feedback. Students always compare with their institute with other institute. Students give their feedback to teachers at the end of semester through online or offline form. According to student's feedback, the performance of a teacher is evaluated and management decides how much improvement is necessary for that teacher. By doing this, a teacher knows about his

B.K. Mishra (✉) · A.K. Sahoo
C.V. Raman College of Engineering, Bhubaneswar, India
e-mail: brojokishoremishra@gmail.com

A.K. Sahoo
e-mail: abhayakumarsahoo2012@gmail.com

© Springer India 2016
H.S. Behera and D.P. Mohapatra (eds.), *Computational Intelligence in Data Mining—Volume 2*, Advances in Intelligent Systems and Computing 411, DOI 10.1007/978-81-322-2731-1_10

109

or her performance and changes own teaching style as per performance report for improvement of teaching quality [1]. After passing out their registered course or progressing themselves to next year, the feedback provided by those students are unstructured in their own language. This information helps the institute to improve course curriculum, teaching methods and student's activity [2].

This student feedback can be considered in the field of opinion mining. Opinion mining is the process of extract subjective information by using text analysis, natural language processing and computational intelligence. It deals with computational treatment of opinion sentiment and subjectivity in text. From multiple opinions it is very difficult to find conclusion, either may be positive or negative. So mining or analysis of opinion is required. Person's feeling or attitude towards a particular subject is called as opinion. We use this opinion mining concept on education system for finding performance of a faculty. To get faster result about performance of a faculty, we add classification technique which is applied on Graphics Processing Unit (GPU) architecture.

This paper is summarized as follows: Sect. 2 represents Opinion Mining and its concepts. Section 3 explains about classification technique. Section 4 tells related to feedback system of education organization. Section 5 describes about GPU architecture. Section 6 gives proposed algorithm to find the performance of a faculty. Section 7 presents result analysis. Section 8 concludes the work and presents future work.

2 Opinion Mining and Its Concepts

Opinion Mining aims to determine the attitude of a speaker or a writer with respect to some topic or the overall contextual polarity of a document. By using different types of data mining techniques, we can analyze person's emotions, remarks, opinions and appraisals through opinion mining [3].

Here we are proposing content based clustering technique to analysis of feedback and remarks given by students. The steps are:

Step 1. Data Preprocessing: In data preprocessing step we clean the data provided by university student feedback system. Feedback forms fully filled by students are taken for consideration.

Step 2. Ontology Learning: This method selects features from unique words that effects in the review or feedback. Ontology learning framework includes term extraction, ontology building and ontology pruning.

Step 3. Grouping similar or synonym words Lexical similarity based on WordNet is commonly used in the NLP (Natural Language Processing) to determine the similarity of two words [4, 5].

Step 4. Tokenization: This method is considerable after collecting list of words which are effective in review of feedback analysis.

TF and IDF calculation is used to select a certain numbers of top ranked terms based on term frequency–inverse document frequency (tf–idf) computation to form a document vector for each review.

Step 5. Assign weight for each feedback or remark: We assign all weights to each review based on wordlist and their weights. Then we find to calculate cumulative weights to each review.

3 Classification Techniques

Classification is a form of data analysis that can be used for extracting models describing important classes or to predict future data trends. Decision tree acts as predictive model for classifying dataset into particular class labels.

Data Classification involves two steps: (a) Building the Classifier or model (b) Using Classifier for Classification.

a. **Building the Classifier or Model**
Learning step is the first step of this model. The classifier is built by classification algorithm. The training set consists of database tuples along with associated class labels. The classifier is built from training set. Each tuple containing training set is treated as a class. We can say that these tuples can be referred as sample or object.

b. **Using Classifier for Classification**
This is the second step of data classification. In this step classifier is used for classify the test data which is based on classification rules. If the accuracy is acceptable, then classification rules can be applied to new data tuples.

4 Feedback System of Education Organization

Learning capability of students is mainly dependant on teacher's performance in term of student's feedback. By introducing student's feedback system, teacher can understand the difficulty of student. This feedback system also enhances teaching quality. The irregular student should not involved in feedback system. In education system, participation among teacher and student is main key factor. Student's feedback is collected through different format like use of clickers, use of feedback form and online feedback system.

In [6] the author attempted to find out how student response systems impact student learning in large lectures [6], where students have less chance to ask questions because of the class size [7]. Feedback can be collected through a variety of SRSs like questions, clickers, short message and mobile phone. One common

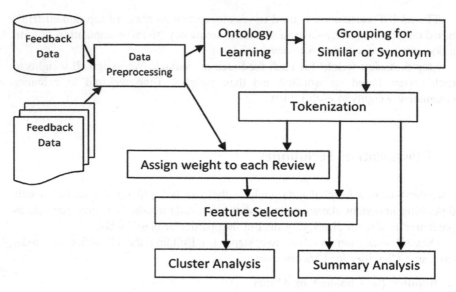

Fig. 1 Student's feedback analysis model

disadvantage of SRS is the cost of the clickers and mobile phones, but nowadays it is very rare to find a student without a mobile so there is not an extra cost for the student or the university (Fig. 1).

5 Graphics Processing Unit (GPU)

5.1 GPU Architecture

Now-a-days GPU has taken a major role in parallel computing field. This is high computational functional unit which is based on multithreaded concept. GPUs are also important to compute a heavy task in parallel. NVIDIA's GPU is compatible on CUDA programming model which provides parallel platform for computing high-intensive task. This GPU card consists of several no. of streaming multi processor units (SMs). No. of SMs depends on different version of GPU cards. NVIDIA Tesla C2070 consists of 448 parallel cores which are based on 14 no. of SMs as shown in Fig. 2. This card has read only memory, L2 cache memory and shared memory which are shared by all the SMs on the GPU. A set of local 32-bit registers is available for each SMs [8]. All the SMs are connected and communicated through global or device memory. The host can read or write on to this device memory. According to programmer's requirement, shared memory is allocated. The transistors used in GPU are used for floating-point operations and optimizing memory bandwidth in case of NVIDIA'S graphics card.

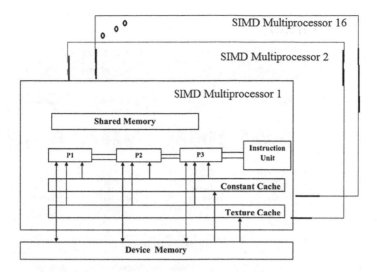

Fig. 2 A set of SIMD stream multiprocessors with memory hierarchy

5.2 CUDA Programming Model

In the programming level, CUDA model is a parallel model used for computing large-intensive tasks. This programming model is based on threads running in parallel. Workspace area of GPU is called as kernel which is launched by CPU. Computation is organized as grid of thread blocks which consists of threads as shown in Fig. 2. Collection of threads is treated as block where as collection of blocks are treated as grid. 32 consecutive threads in a particular block is known as wrap which run in parallel. Each SM executes one or more threads blocks concurrently. A block consisting of threads that runs on the same SM. Different threads of a particular block can use the same shared memory, but different thread of different block cannot share same shared memory. CUDA Programming model consists of set of C language extensions and runtime library that provides APIs to control the GPU. In this way, Programmers can use CUDA programming model to exploit the parallel power of GPU for parallel computing (Figs. 3 and 4).

5.3 Decision Tree Model on Opinion Mining Using GPU

Opinion mining takes a major role to find the valuable information from the student's feedback about faculty's performance. We classify the opinion from the collected feedback data sets into class labels using decision tree model. Here GPU plays a vital role for measuring the computation time on Student feedback system

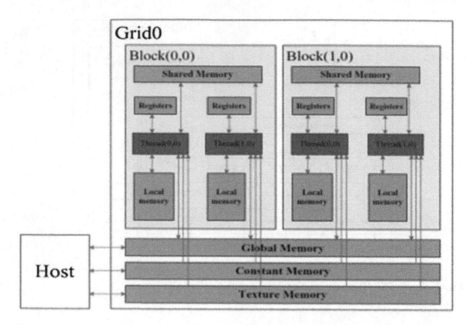

Fig. 3 Architecture of device memory in GPU

Fig. 4 Memory spaces on
CUDA device

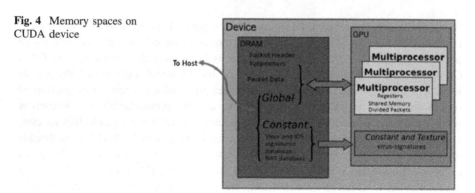

along with the result. Decision tree can be designed using the result obtained from
GPU as training data sets, which may be helpful for the higher authorities for taking
an appropriate decision.

6 Proposed Algorithm

In this algorithm, we present how to measure the performance of a teacher. Some
parameters have to be concerned to evaluate a teacher. Here we consider ten
parameters. The steps are:

1. One student feedback form is designed. That form describes ten parameters. Students will give mark out of five based on parameters.
2. Calculate each teacher's average mark. That average mark depends on no of students.
3. According to average mark, we can do category among teachers. That category is of four types: increment only, award and increment, termination notice and warning.
4. After knowing teacher's category, Education system can improve performance of teacher.

   Our algorithms are been divided into two major parts based on their working. The first algorithm is based on static part, where all average marks of all teachers are calculated. The second algorithm is based on dynamic part, where teacher's performance is compared and category is also identified.

6.1 CPU Implementation

1. Collect your CUDA device properties by using the 'cudaDeviceProp' structure, determine the available amount of shared-memory per block, DRAM, Constant cache memory available with the device.
2. After collecting student's feedback, these different marks in different subjects of a particular teacher are stored.
3. Average mark obtained by individual teacher is calculated.
4. Then each average mark is stored in an array. That array is passed to GPU for finding different category among teachers.
5. Call the GPU kernel from CPU using <<<number of blocks, number of threads, dynamic memory per block, stream associated>>> (parameters) syntax.
6. Send array containing average marks and teacher's category file to GPU.
7. Collect back the result from the GPU using the same cudaMemcpyAsync () with teacher's category.

6.2 GPU Implementation

8. Inside GPU kernel function create two shared variable to store the contents of average mark and teacher's category file parameters (from the global memory) using __shared__.
9. Divide and copy the contents from DRAM to shared memory for each block.
10. Now each thread calculates individual teacher's average mark by using teacher's category file with the help of shared memory.

11. Write the result into device memory and send back to CPU.
    NOTE: Indexes of the 1-d array is associated with each teacher's average mark sent to GPU.

12. END

7 Experimental Analysis

This proposed algorithm shows following results by which we can measure the performance of a teacher. The following Fig. 5 shows feedback form which is filled by students (Fig. 6).

The following figure describes ten parameters on which we can evaluate teacher's performance. The result generated from above feedback entry form i.e. Used in below Fig. 7.

In this way we have collected opinions from twenty students about six faculty members of respective subjects. At the last, we calculate total marks secured by individual faculty member through twenty students. Then we find average mark obtained by individual faculty member, on which we determine different category among faculty members. This following Fig. 8 can help the education system for making decision about improvement of teaching methods and other activities.

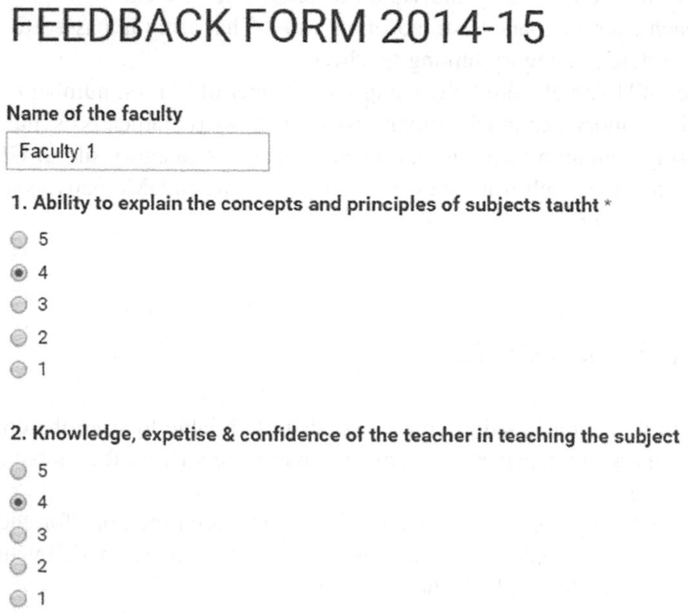

FEEDBACK FORM 2014-15

Name of the faculty

Faculty 1

1. Ability to explain the concepts and principles of subjects tautht *

○ 5
◉ 4
○ 3
○ 2
○ 1

2. Knowledge, expetise & confidence of the teacher in teaching the subject

○ 5
◉ 4
○ 3
○ 2
○ 1

Fig. 5 Feedback entry form

Marks	Remarks
5	Excellent
4	Very good
3	Good
2	Average
1	Poor

Fig. 6 Opinion remarks against faculty's performance

			FEED BACK FORM 2014–15																			
DEPARTMENT	**INFORMATION TECHNOLOGY**									**SEMESTER**	**SIXTH**											
Subject	**Facult Name**	**Parameters**	s1	s2	s3	s4	s5	s6	s7	s8	s9	s10	s11	s12	s13	s14	s15	s16	s17	s18	s19	s20
		Ability to Explain the Concepts and Principles of subjects taught	3	4	1	5	4	4	3	5	5	4	3	5	4	1	4	5	5	4	5	4
		Knowledge, expertise & confidence of the teacher in teaching the subject	4	3	5	5	4	4	5	4	4	3	5	5	4	5	4	3	1	4	3	3
		Ability to clear doubt in the class room and outside class	5	4	4	4	3	5	4	4	5	4	4	3	5	4	1	4	3	5	4	4
		Ability to corelate concepts with examples	3	5	4	5	4	4	3	5	5	4	5	5	4	4	3	5	4	3	5	4
		Communication skill and clarity	5	5	4	5	4	3	5	4	5	4	4	5	4	4	3	5	4	4	3	3
Sub 1	Faculty1	Punctuality & Regularity in taking class & tim management with respect to syllabus coverage	4	5	4	4	3	4	4	3	5	5	4	5	5	4	5	4	3	5	4	4
		Interaction / Discussion with the students in class room	5	4	4	3	3	4	4	4	3	5	4	4	1	4	4	3	5	5	4	5
		Attitude towards students & monitoring activities.	4	3	3	4	4	4	3	5	4	5	4	3	5	5	5	4	1	4	3	5
		Motivating students & creating interest on subjects taught.	3	1	4	3	3	4	4	3	5	4	4	5	4	4	3	5	4	4	3	3
		Timely evaluation of internal assessment papers, showing the same to students & discussions there on.	4	5	5	4	4	4	4	4	3	5	4	4	5	5	5	4	5	4	3	5
			40	39	38	42	36	40	39	41	44	43	41	44	41	40	37	42	35	42	37	40

Fig. 7 Feedback form filled by twenty students of one faculty member

Fig. 8 Different category of different faculty members according to average marks

	A	B	C	D
1				
2	Subject	Facult Name	avg	Remark
3	Sub 1	Faculty 1	40.05	Increment only
4	Sub 2	Faculty 2	33.65	Termination notice
5	Sub 3	Faculty 3	39.45	Warning
6	Sub 4	Faculty 4	35.75	Warning
7	Sub 5	Faculty 5	45.25	Award & Increment
8	Sub 6	Faculty 6	38.35	Warning

After calculating average marks obtained by individual faculty member, then we can categorize them by using decision tree. Figure 9 shows the performance report of faculty according to specified criteria.

The above Fig. 10 shows a comparative study on computational time of CPU and GPU by a graph where x-axis represents no of faculty's result analysis and y-axis represents about time execution difference between CPU and GPU.

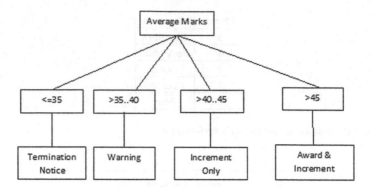

Fig. 9 Decision tree of Performance evaluation for faculty members

Fig. 10 Time execution
difference between CPU and
GPU for performance
evaluation of different no of
faculty members

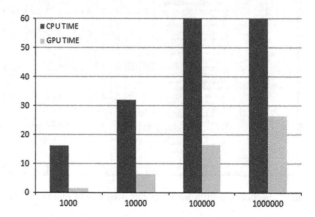

8 Conclusion

In this paper, a brief background of educational opinion mining and parallel
computing was presented for collecting feedback from students. We stated that
opinion mining has taken major role in educational system. From experimental
analysis, we conclude that education system can be improved by opinion mining on
student feedback so that teacher can improve his or her teaching quality. By using
GPU architecture, parallel algorithm implemented by CUDA language could give
the performance result of faculty in less time as compared to CPU by using GPU
Tesla C1070 card. We can also optimize this parallel algorithm by using different
version of NVIDIA graphics card.

References

1. Trisha, P., Jaimin, U., Atul, P.: Sentiment analysis of parents feedback for educational institutes. Int. J. Innov. Emerg. Res. Eng. 2(3), e-ISSN-2394–5494
2. Balakrishnan, R., Rajkumar, K.: Extracting features and sentiment words from feedbacks of learners in academic environments. In: 2012 International Conference on Industrial and Intelligent Information (ICIII 2012), IPCSIT vol. 31 © IACSIT Press, Singapore (2012)
3. Jai Prakash, V., Bankim, P., Atul, P.: Web mining: opinion and feedback analysis for educational institutions. Int. J. Comput. Appl. (0975–8887) 84(6) (2013)
4. Zhongwu, Z., Bing, L., Hua, X., Peifa, J.: Clustering product features for opinion mining WSDM'11, Feb 9–12, 2011. Hong Kong, China. Copyright 2011 ACM
5. Pedersen, T.: Information content measures of semantic similarity perform better without sense-tagged text. In: Proceedings of NAACL HLT (2010)
6. Denker, K.J.: Student response systems and facilitating the large lecture basic communication course: assessing engagement and learning. Commun. Teach. 27.1 (2013)
7. Gehringer, E.F.: Ac 2012-4769: applications for supporting collaboration in the classroom (2012)
8. Mayank, T., Abhaya, K., Sahoo, R.M.: Efficient implementation of Apriori algorithm on HDFS using GPU. In: 2014 International Conference on High Performance Computing and Applications, pp. 1–7. ISBN: 978-1-4799-5957-0
9. Akshata Shirish, N., Roshan, F.: An study and analysis of opinion mining and sentiment analysis. Int. J. Adv. Res. Comput. Eng. Technol. (IJARCET) 3(12) (2014)
10. Tushar, G., Lata, R.: Featured based sentiment classification for hotel reviews using NLP and Bayesian classification. In: 2012 International Conference on Communication, Information & Computing Technology (ICCICT), Oct 19–20, Mumbai, India (2012)
11. Vladimir, O., Asle, P.: Ontology based semantic similarity comparison of documents. In: Proceedings of the 14th International Workshop on Database and Expert Systems Applications (DEXA'03) © 2003, pp. 735–738. IEEE
12. Wu, D., Li, X., Zhang, C.: The design of ontologybase semantic label and classification system of knowledge elements. In: 2011 International Conference on Uncertainty Reasoning and Knowledge Engineering, 978-1-4244-9983-0/11 ©2011, pp. 95–98. IEEE
13. Bishas, K., Aarpit, S., Sanjay, S.: Web opinion mining for social networking sites. In: CCSEIT-12 October 26–28, 2012, Coimbatore (Tamil Nadu, India) (2012)
14. Nabeela, A., Mohamed, M.G., Mihaela, C.: SA-E: Sentiment analysis for education
15. Poulos, A., Mary, J.M.: Effectiveness of feedback: the students perspective. Assess. Eval. Higher Educ. 33.2, 143–154 (2008)

A Constraint Based Question Answering over Semantic Knowledge Base

Magesh Vasudevan and B.K. Tripathy

Abstract The proposed system aims at extracting meaning from the natural language query for querying the semantic knowledge sources. Semantic knowledge sources are systems conceptualized with Ontology. Characterization of a concept is through other concepts as a constraint over other. This very method to extract meaning from the natural language query has been experimented in this system. Constraints and entities from the query and the relationship between the entities is capable of transforming natural language query to a SPARQL (a query language for Semantic Knowledge sources). Further the SPARQL query is generated through recursive procedure from the intermediate query which is more efficient that mapping with patterns of the question. The system is compared with other systems of QALD (Question Answering over Linked Data) standard.

Keywords Answer extraction · Information retrieval · Natural language query · Ontology · Linked data

1 Introduction

The information retrieval systems identify the set of documents that match with the user query. Question answering systems must return the exact answer rather than retrieving set of documents based upon the ranking. The proposed system is queried over the Linked data. The linked data is built upon Ontology, which conceptualizes the knowledge available over any domain. Ontologies are represented using OWL (Web Ontology Language) and RDF (Resource Description Framework) is used to achieve the linked data with the semantics added by OWL. The ontologies are

M. Vasudevan (✉) · B.K. Tripathy
School of Computing Science and Engineering, VIT University, Vellore, India
e-mail: mageshv88@gmail.com

B.K. Tripathy
e-mail: tripathybk@vit.ac.in

© Springer India 2016
H.S. Behera and D.P. Mohapatra (eds.), *Computational Intelligence in Data Mining—Volume 2*, Advances in Intelligent Systems and Computing 411, DOI 10.1007/978-81-322-2731-1_11

121

designed to meet the requirements of particular domain, or auto generate from the large open domain knowledge. The latter method is used to generate DBPedia [1] a large semantic knowledge base generated from Wikipedia. The DBPedia classes, Individuals and the relationship between them has to be mapped to the natural language question.

Ontology is conceptualization of a domain; domain can be either closed or open. In this system we are using an existing open domain knowledge source Dbpedia. Ontology enables meaning of the information through concepts, roles and individuals; concepts are analogous to class, individuals to instances and roles to relationship. Information is human interpretable but there is necessity for machines to understand the information the way humans perform. The larger number of present information retrieval systems concentrate on the text content rather the meaning of the query user has given. To enable efficient retrieval of information we need to add semantics to the content which ontology with Description Logics as knowledge representation language can enable semantics, through which systems can be designed to understand the ontology, further enabling the systems to exactly retrieve the information content it requires.

The concept has to be defined in terms of Description logics, where the concepts are defined by conjunction, disjunction and negation of other concepts. Roles relate the concepts with Domain and Range of the roles are specified. This is usually termed as Object property or Data property. Object property relates with individuals while the Data property relates with XML data types.

The retrieval of the data from an RDF based knowledge sources can be performed through SPARQL (SPARQL Protocol And RDF Query Language). It can recursively retrieve data by representing the triple pattern that must be retrieved. The pattern can be of union, intersection and other set operations. SPARQL allows, filter and federated query mechanisms. DBpedia is a large automated extraction of semantic knowledge base from Wikipedia.

Section 2 explores the related work associated with the proposed system. Section 3 discusses in detail about the components of the system. Section 4 evaluates the system with QALD standard and we conclude the work in Sect. 5.

2 Related Work

Systems of Question answering in linked data, have either approached in matching maximum match or template based, where the scope of the questions is limited. In the proposed system a model has been defined to extract the abstract meaning of the query and directly incorporate it with the Ontology based knowledge sources.

The In [2] the question is directly transferred to templates, then the entities are instantiated according to the template. The predicates are mapped based on the BOA [3], which is a large corpus with predicates representing the same meaning can be mapped. Finally based upon the slots filled the query is ranked based upon the average of scores that instantiated the template. In [4] a pipelined architecture is used, where the LTAG (Lexicalized Tree Adjoining Grammar) is used to generate templates and further the string and semantic similarity functions are performed to choose the template, which is most possible to generate the answer. [5] uses wikiFramework, to match the pattern found in question with the structure of Wikipedia infobox and sentences, which leads to identification of multiple representation of same predicates to be mapped; it uses answer type and named entity recognition for mapping the KB literals with the natural language question. In [6] the question is transferred to triple pattern, then the relational patterns are matched to find the final triples to generate the SPARQL [7]. Uses the triple extraction from the question, finally few rule of thumb and string similarity measures are performed to achieve the results. In [8] Resource disambiguation is done using Hidden Markov Models and the patterns are matched with bootstrapped relations extracted, finally query graph is constructed to generate SPARQL. In [9] the synfragments of the give natural language query is used to transform into SPARQL, synfragments are the minimal concepts related to the KB, further the HSO similarity measures are used to identify the predicates.

There are many different approaches in the existing systems, we have differentiated based upon the Bootstrapping method and direct semantic similarity search methods. The bootstrapping will work best in case of symmetric relations but semantically related relations of concepts; it too needs similarity measures be applied to identify the relation from a pool of strings. The semantic similarity methods for direct phrase matching have been performed; we have additionally considered the derivationally related forms of the lemmas, which leads to better similarity due to the change in the POS (Part of Speech) in predicate and relation in ontology. The clue vector to identify the relations, with the help of question's intent has been included. Further the system mainly looks at the concepts in the question which needed to be mapped with the concepts in knowledge base, where it can be complex concepts with relations.

3 Proposed System Description

The aim of the proposed system is to generate the SPARQL for a given natural language query. The system must associate the nouns from the query to a resource and further narrow down by applying the constraints and roles played by the nouns. The search for resource are two ways; in case of Noun then the search must be of types the resources are associated and in case of Proper Noun the search will be of resources with maximum match with the rdfs:label predicates value.

> **Example (1)** Who wrote the book The pillars of the Earth?;
> Book − class; The pillars of the Earth - Individual; wrote − relationship
> <The Pillars of the Earth> rdf:type <Book>; <The Pillars of the Earth> author? <answer>

The items that must be identified in the natural language are the Entities, constraint over the entities and relationship between them. In the Example (1), the entities are book and the pillars of the Earth, the relationship is wrote. In the Example (2), the class is movies, the constraint over the class is Indian, which modifies the movie or can be expressed as restricting the movies to Indian. The constraints can be like adjective, superlative adjective (most, highest, largest etc.) and comparative adjective (higher, larger etc.). Each step is discussed in detail in the coming sections. The procedure to generate SPARQL is given in Procedure 1.

Example (2) *Give me all Indian movies produced by Warner Bros?*
?<answer_movie> location <india>; ?< answer_movie> producer <Warner Bros>

```
PROCEDURE 1. To Generate SPARQL from the NL Query.
Generate_sparql(Q):
Input: Q := Natural Language Query.
Output: SPARQL
Step1:   Initialization
         CE={},R={},KB_R={},KB_P={},
         AETYPE=None,IQ=None,SPARQL=None.
Step2:   CE := FSM(Q). // FSM in Section 3.1
Step3:   R := REL_EXTRACT(CE,Q). // Procedure 2.
Step4:   AETYPE := nbmodel.predict(Q).
Step5:   for each ce in CE:
              KB_R = KB_R ⋃ (ce,get_resource(ce))

              //get_resource:Procedure 3.
Step6:   for each r in R:
              KB_P = KB_P ⋃ (r,get_property(r,KB_R(ce)))

         //get_property: Procedure 4.
Step7:   IQ = generate_IQ(KB_R,KB_P,AETYPE)
Step8:   SPARQL = gen_SPARQL(IQ) // Section 3.5.
Step9:   Return SPARQL.
```

3.1 Entities and Constraints Extraction

The entities and constraints are the Noun, Adjectives and combination of these with the cardinals and prepositions form constraints of a noun. The identification of the tags for each word is achieved through the Part of Speech tags, the POS tagging is done with the state of the art tagger, Stanford NLP. The tagged question is passed through a Finite State Machine to extract the constraints like <highest> mountain, <more than 500,000> employees Example (3), <larger than> Agra etc. (constraint is enclosed in < >). The Finite State Machine designed for this purpose is in Fig. 1. Tokens are from Penn Treebank Tag set. **Tokens.** JJ/RB -> Adjective or Adverb; NOUN -> NN/NNS/NNP/NNPS;CD -> Cardinal Number;OT -> Any Other POS; PREP -> Preposition (Of,In,than); <Q> -> Start Token; </Q> -> End Token; **States.** START -> Start FSM with i = 0;1 -> Start Registering Constraint Ci.;2 -> Append

Fig. 1 FSM to extract
constraints and entities

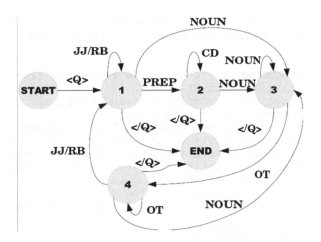

Constraint.;3 -> Start Registering Entity.;4 -> End Entity and Increment i for next
constraint & entity.;END -> Finish.

Example (3) *Give me websites of companies with more than 500000 employees.E1
– website,E2 – company,E3 – <more than 500,000> employee.*
Example (4) *Give me all female Indian astronauts. E1 – <female, Indian>
astronaut.*

In Example (4), the female/JJ and Indian/JJ are considered as constraints because
it modifies the noun astronaut (JJ POS represents Adjective). We have included a
location and Gender constraint component which will handle the location and
gender with the concepts in the knowledge base.

3.2 Answer Entity Type Identification and Relations Extraction

We identify the answer entity type like count, resource, list, date and Boolean. The
prediction is based on the multinomial naive Bayes classifier. The syntactic
structure and few other features like Wh-phrase, Wh-modifiers etc. Where extracted
from questions given in DBPedia train set were used to train the classifier. The
classifier trained with accuracy *0.98*. The predicted type are useful in constructing
the sparql and as clues in searching relationship of the entities. Which will be
discussed further in later sections.

Relations extraction is nothing but extracting the words between the entities,
these may be preposition, verb or combination of these. These relations are
extracted based on the Stanford dependency parser, which generates dependency
between nouns and verbs, by model trained from a large manually annotated corpus. We first present the Entities as Nodes and make edges between nodes based
upon the path given by the dependency parser. The edges are undirected for now,

when we match the properties in DBPedia we make directed edges based upon the triple. Procedure 2 is the outline of generating relations.

PROCEDURE 2. To generate relations from the constraints and entities with NL query.
Input: CE := Constraint and Entities, Q:= NL Query.
Output: Set of Relations R.
Step1: Generate Stanford Typed Dependencies.
Step2: G(V,E):= Graph from the nodes and connected dependency from the Stanford Dependencies.
Step3: Initialize R := |CE| X |CE| matrix.
Step4: for each ce as i in CE:
        for each ce as j in CE:
            if there Exists edge E for V_i and V_j:
                R(i,j) := E (relation)
Step5: Return R.

3.3 Match Entities with KB Literals

In our case the entities identified in the natural language questions have to be mapped with the resource or type in the knowledge base. We have considered the wiki redirects and wiki disambiguation links in identifying the Resources in DBPedia. In Example (6) the type monument as <http://dbpedia.org/ontology/Monument> and the Resource India as <http://dbpedia.org/resource/India> have to be mapped. For the mapping of entities we have to index the rdfs:labels of the classes and resources. The extracted labels with their subject are indexed using Solr. Edge Ngram and Double metaphone indexing is done over the labels, in case of edge ngram the string is split with ngrams with beginning of the words; here the minimum is 3 and maximum is unbounded. The double metaphone goes with series of changes to the given word to produce the phonetic form. The documents are retrieved individually based upon edge ngram and double metaphone, then the choice of document is made upon the given formula where edge ngram is boosted twice; given set of D documents,

PROCEDURE 3. Match KB Literals.
Input: CE := Constraints and Entities.
Output: Set Resource Identifier Nodes for the items in CE.
Step1: Resource={}
Step2: for each ce in CE:

$$K = \bigcup_D 2* \text{ Edge-Gram} + \text{Phonetic.}$$

        Resource(ce) = Max(K)
Step3: for each r in Resource:
        If levenshtein_distance(r,ce) > Threshold:
            Raise "Resource not found Exception".
Step4: return Resource.

Further the best match is retrieved based on Levenshtein distance. When the entity is noun we search in the type documents and when the entity is proper noun we search in the resource documents.

Example (5) *who is the author of Samkhya Philosophy ? <author,of,Samkhya Philosophy>*
Example (6) *give me all monuments in India ? <monuments,in,India>*

3.4 Matching the Properties with KB Properties

The property matching is the important phase where the directed links and whole structure of the intermediate query depends. We reduce the search with domain and range of the properties, we retrieve all properties of the resource or type associated and restrict to the path between these two.

Example (7) *give me all books written by actors ? <book,written by, actor>*

In Example (7), the properties connecting book and actor are retrieved, the triple representation is given as *?book rdf:type dbtype:Book.; ?actor rdf:type dbtype: Actor.;?book ?prop1 ?actor.;?actor ?prop2 ?book.;?prop1 rdfs:label ?property1.;?prop2 rdfs:label ?property2.* The restricted properties are then semantically matched with the multiple semantic similarity measures, where the search is made more detail with the clue vector generated. The important thing to notice here is the match of phrases, two similar phrases have to be semantically matched with the place of birth, date of birth, source country, but the property we are matching may not be in phrase as in Example (8), So we need a clue vector which will reflect the underlying meaning represented in the question, which plays an important role in identifying the relationship between the entities that are represented in triple form in the knowledge base. The search for predicate has predefined constraints, which need to be applied while searching predicates, the addition of such constraints which is extracted from question is represented as clue vector.

The clue vector consists of [type related, isCardinal, isDate, isLocation]; though the clue vector has been limited now, we are working on a broader version of clue vector where it can reflect the entire possibilities of the question's semantics. The clue vector for Example (8) will be [None,False,True,False]. So, the phrase searched for will be *'date born'*. Interestingly in Example (6) the word to be searched is *'in'* which has no detail to match, but the NER of resource associated is location, so the clue vector will be [monument,False,False,True]. So the search phrase can contain matches with "*India has location of monument X*" or "*X located in India*". The purpose of cardinal is to search the total,count,number etc. in case where we search for "organization with more than 50,000 employees" here there may be nodes which will have link as "Org_X has employee Person_X" or "Org_X total employees Cardinal_N", there has to be resolution with searching properties where the object is resource or literal. If literal the number, count or total words are matched. The clue vector for Example (9) is [country,False,False,False], the search predicate will be '*start country*' and the predicated matched is '*source country*'. In Example (10) the relation searched is *written* where the case to be matched is programming language the clue vector gives the possibility to match the predicate due to the type being searched is programming language. The important thing to be noticed is that the clue vector is an optional match, where the details of the relations between the DBPedia concepts or individuals is not missed.

Example (8) *Where was Mahatma Gandhi born? <born,Mahatma Gandhi>*
Example (9) *In which country does river Nile start? <country,start,riverNile>*
Example (10) *In which programming language is GIMP written?*

The similarity measures used here is path, wup and lch [10] combination of these three is used and further the common rank is retrieved. Path similarity works based on the shortest distance in the wordnet taxonomy. The lch similarity works with path distance and the maximum depth of the wordnet synsets. The wup similarity works on the least common subsumer and the depth measures. Due to the failure of these three combinations we used additional approach where the derivationally related lemmas are taken into consideration, in cases like '*inhabitants*' and '*population*' these were more prone to noise in search for predicates, when derivationally related lemmas were considered of all synsets of a word, then the match was more accurate.

The match for phrases is based on exclusive match between the properties, when '*start location*' and '*source country*' are matched the *start-source* and *location-country* are exclusively matched. Average of the ranks of all similarity measure is performed to get the best matched predicate. Given the set of properties P, with property to find as FP, where each P is associated with M words and FP is associated with N words. The Procedure 4 explains the Property matching.

```
PROCEDURE 4. Match the property in KB for the relations.
Input: R:= Resources, REL:= Relation.
Output: Set of Properties for Relations.
Step1: FP := Property to Find,
       property := None.// matched property.
       P:=None. //Set of properties for Resource R.
Step1: Construct Clue Vector from the given Resources and
Relation.
Step2: P:= Restricted properties for Resource r from through
clue vector.
Step3: for each p_i in P:
```

$$Sim(P_i, FP) = \frac{Max\{\sum_{j=0}^{M} \sum_{k=0}^{N} similarity(P_{ij}, FP_k)\}}{|M-N|+1}$$

$$where, j = 0,...,M \ and \ k = 0,...,N.$$

```
Step4: property := Average(Rank (Sim_LCH(P_i+FP))+ Rank
(Sim_PATH(P_i+FP))+Rank(Sim_WUP(P_i+FP))+Rank(Sim_LDF(P_i+FP)))
Step5: Return property.
```

3.5 Generating Intermediate Query and SPARQL

Finally the intermediate query is generated based upon the domain or range direction of the properties retrieved, the sample query is in Example (11). In case of (11) Software, developer and Organization, the direction of edges is software to developer and developer to Organization. The relationships are individually conceptualized to achieve triple structure and finally parsed based on the IQ represented.

Example (11) which softwares has been developed by organizations founded in California?
   IQ.(LIST(<http://dbpedia.org/ontology/Software><http://dbpedia.org/ontology/developer>(<http://dbepdia.org/ontology/Organization><http://dbpedia.org/ontology/foundingPlace><http://dbpedia.org/resource/California>*)))*

From the Intermediate Query, the SPARQL is generated based upon the links between the individual concepts recursively. We generate rdf:type statements for direct classes in ontology, further this link of ontology concepts is recursively connected to the next concept defined in the intermediate query. When cardinal constraints occur for particular property, the Filter statements and for Superlative the order by statements are appended. If the answer is of type single, and retrieves more than one answer the SPARQL is dropped and responds with fail statement. In case of Count and Boolean type answers the appropriate SPARQL statements are chosen.
   SPARQL for Example (11),

```
   SELECT ?answer WHERE{
          ?answer rdf:type <http://dbpedia.org/ontology/Software>.
          ?answer <http://dbpedia.org/ontology/developer> ?link1.
          ?link1 rdf:type <http://dbpedia.org/ontology/Organization>.
          ?link1                     <http://dbpedia.org/ontology/foundingPlace>
<http://dbpedia.org/resource/California>.}
```

4 Evaluation

The system is evaluated based on the QALD-2 Benchmark [11]. The evaluation parameters are precision, recall and F-Measure [11]. We chose 72 DBPedia namespace queries and refined based upon the system design to capture the structure where 14 questions cannot be parsed, finally 58 questions were chosen, 35 questions were answered and 4 questions were partially answered. The mean precision of the system is **0.67** and the recall value is **0.71**. The F-Measure of the system in **0.69**. The comparison of results with other systems in give in Table 1, the

Table 1 Comparison of result of proposed system with other systems

System	Answered	Coverage	Correct	Partial	R	P	F	F'
Proposed system	58	0.58	**35**	4	0.71	**0.67**	**0.69**	**0.63**
SemSek	80	0.80	32	7	0.44	0.48	0.46	0.58
MHE	**97**	**0.97**	30	12	0.36	0.40	0.38	0.54
QaKis	35	0.35	11	4	0.39	0.37	0.38	0.36
BELA	31	0.31	17	5	**0.73**	0.62	0.67	0.42

comparison of systems from QALD-2 challenge provided by [4] (Further reference for evaluation and system results) with BELA is compared including the F' measure.

The system is capable of identifying structure for recursive queries, where the entities are available explicitly like "give me software developed by organizations founded in California" but in case of expansion dynamically "give me parents of wife of Juan Carlos I" it fails. Need to address change in the entity on dynamic expansion when the search for predicate is carried out. There are cases where the similar question has answers in different relations, case where "which is the largest city in Australia", our system approaches, with identifying cities in Australia and applying the "largest" superlative adjective; while the expected method is that the largest city is direct property of resource Australia, but in case of "which is the largest city in Germany" and few other cases works correctly; In such cases system should have insight about the returned results and change according to it, where intelligence component is required to handle such cases.

The functional box, which is highly needed in cases of performing operations on numeric or date. Where the age from date of birth, numeric units conversions have to be performed. The complexity in identifying whether entity represents a concept in Ontology or a Relationship between entities to represent the concept system searches is very high, which dynamically changes with the change of policies at different layers in the system.

Agent based global state is the most suitable solution for question answering systems. Agent based QA with generalization of clue vectors as global state and each entity is associated with frame containing the possibilities of KB representation.

5 Conclusion

Question answering can be seen as a tool to evaluate the efficiency in retrieving information from a knowledge base. Semantic knowledge bases, have been in consideration for artificial intelligence systems, due to its structure which makes information meaningful. Future of web will be one such global access; it could be achieved by evolution of semantic web. Human like computing system, has to perceive the world through their sensors, due to the strong dependency of natural language and semantic knowledge bases, the information perceived by systems have to be on natural language, and the communication to the user have to be in natural language. This gives a way, where question answering systems can be used to evaluate the efficiency of knowledge base, in conceptualizing or perceiving the given natural language input. The system has achieved better F' measure than the existing systems; the response time must be increased and dynamic change of approach based upon global state must be adopted.

References

1. Auer, S., Bizer, C., Kobilarov, G., Lehmann, J., Cyganiak, R., Ives, Z.: Dbpedia: a nucleus for a web of open data, pp. 722–735. Springer, Berlin Heidelberg (2007)
2. Unger, C., Bühmann, L., Lehmann, J., Ngonga Ngomo, A.C., Gerber, D., Cimiano, P.: Template-based question answering over RDF data. In: Proceedings of the 21st International Conference on World Wide Web, pp. 639–648. ACM, April 2012
3. Gerber, D., Ngomo, A.C.N.: Bootstrapping the linked data web. In: 1st Workshop on Web Scale Knowledge Extraction@ ISWC, vol. 2011
4. Walter, S., Unger, C., Cimiano, P., Bär, D.: Evaluation of a layered approach to question answering over linked data. In: The Semantic Web–ISWC 2012, pp. 362–374. Springer, Berlin, Heidelberg
5. Cabrio, E., Aprosio, A.P., Cojan, J., Magnini, B., Gandon, F., Lavelli, A.: Qakis@ qald-2. In: Proceedings of Interacting with Linked Data (ILD 2012) [37], pp. 87–95
6. Hakimov, S., Tunc, H., Akimaliev, M., Dogdu, E.: Semantic question answering system over linked data using relational patterns. In: Proceedings of the Joint EDBT/ICDT 2013 Workshops, pp. 83–88. ACM
7. He, S., Liu, S., Chen, Y., Zhou, G., Liu, K., Zhao, J.: Casia@ qald-3: a question answering system over linked data. In: Proceedings of the Question Answering over Linked Data lab (QALD-3) at CLEF 2013
8. Shekarpour, S., Ngonga Ngomo, A.C., Auer, S.: Question answering on interlinked data. In: Proceedings of the 22nd International Conference on World Wide Web, pp. 1145–1156. International World Wide Web Conferences Steering Committee, 2013 May
9. Dima, C.: Intui2: a prototype system for question answering over linked data. In: Proceedings of the Question Answering over Linked Data lab (QALD-3) at CLEF 2013
10. Varelas, G., Voutsakis, E., Raftopoulou, P., Petrakis, E.G., Milios, E.E.: Semantic similarity methods in wordNet and their application to information retrieval on the web. In: Proceedings of the 7th annual ACM international workshop on Web information and data management, pp. 10–16. ACM, November 2005
11. Lopez, V., Unger, C., Cimiano, P., Motta, E.: Evaluating question answering over linked data. In: Web Semantics: Science, Services and Agents on the World Wide Web, vol. 21, pp. 3–13 (2013)

ACONN—A Multicast Routing Implementation

Sweta Srivastava and Sudip Kumar Sahana

Abstract In a communication network the biggest challenge with multicasting is minimizing the amount of network resources employed. This paper proposes an ant colony optimization (ACO) and neural network (NN) based novel ACONN implementation for an efficient use of multicast routing in a communication network. ACO globally optimize the search space where as NN dynamically determine the effective path for multicast problem. The number of iteration and complexity study shows that the proposed hybrid technique is more cost effective and converges faster to give optimal solution for multicast routing in comparison to ACO and Dijkstra's algorithm.

Keywords Ant colony optimization · Neural network · Multicast routing · Dijkstra's algorithm

1 Introduction

Multicast routing involves the transportation of information between a single sender and multiple receivers. In a Packet Switched network there are several paths from the source to the destination node through some intermediate nodes. In such cases, network routing has significant impact on the network performance. The data flow in a real time network should be congestion free, reliable, adaptive with balanced load and QoS [1, 2] should be maintained.

There are several techniques available [3–9] for multicasting in networks. Frank et al. [10] described six techniques: flooding where packets are broadcast over all links, separate addressing where a separately addressed packet is sent to each

S. Srivastava (✉) · S.K. Sahana
Department of CSE, Birla Institute of Technology, Mesra, India
e-mail: ssrivastava.rnc09@gmail.com

S.K. Sahana
e-mail: sudipsahana@bitmesra.ac.in

© Springer India 2016   133
H.S. Behera and D.P. Mohapatra (eds.), *Computational Intelligence in Data Mining—Volume 2*, Advances in Intelligent Systems and Computing 411, DOI 10.1007/978-81-322-2731-1_12

destination, multidestination addressing where variable sized packet headers are used to send a few multiply addressed packets, partite addressing where destinations are partitioned by some common addressing locality and packets are sent to each partition for final delivery to local hosts, singletree forwarding in which a single tree spans all nodes of the group, and multiple-tree forwarding where each member may have a different spanning tree.

Over the past decades, several researches have been performed to solve multicast routing problems using conventional algorithm, such as Dijkstra's algorithm, exhaustive search routing and greedy routing. A setting of routing table was realized by Yen [11] using k-shortest path algorithm. Due to the high degree of complexity, it is not practical to use these algorithms in real-time multicast routing. Recently some heuristic algorithms have been proposed. Wang and Xie [12] proposed an application of the basic ACO to multi-constraint QoS multicast routing, where a routing table is set for every pair of source–destination nodes. Route to multiobjective cannot be found at the same using ACO. Yuan and Hai [13] proposed a variation of ACO, CACO to overcome this limitation by copying some ants to find the routing from the contrary direction when a ant reaches a receiver node. Wang et al. [14] proposed an ant colony algorithm with orientation factor and applied it to multicast routing problem with the constraints of delay variation bound. Orientation factor enabled the ant to get rid of the initial blindness. Pan et al. [15] designed an ACO-based multicasting routing algorithm where a source-based approach was adapted to build a multicast tree among all the multicast members. Singh et al. [16] provided taxonomy of various ACO algorithms for efficient routing in MANET. An evolutional scheme for the original multicast tree was designed in [17, 18] to find a more optimal multicast tree. Load balancing was neglected in the said literature which resulted in system failure.

Neural Network is proven to perform well to solve real time dynamic multicast routing problem with adaptive powerful parallel computing ability. The convergence speed of NN is also fast. Liu and Wang [19] solved delay constant multicast routing problem the transiently chaotic neural network which was found to be more capable than Hopfield neural network to reach global optima. Pour et al. [20] used the concept of hybrid approach of ACO and NN for direction of arrival estimation which encounters an interpolation of a complex nonlinear function. They used Continuous ACO to train a neural network. Direction of arrival estimation will not serve the purpose of solving multicast routing fully.

The main objective is to find out the optimum path for data transmission in destinations with diversities using multidestination addressing so that the propagation delays during transmission is minimized using a hybrid of ACO and NN.

The paper is organized as follows: Sects. 2 and 3 discusses ACO and neural network for multicast routing respectively. Section 4 explains the proposed model and implementation of the model is described in Sect. 5, result is discussed in Sect. 6 followed by conclusion drawn from the work.

2 Ant Colony Optimization

The ant colony algorithm [21] is a random search algorithm, which gets optimal solution through simulating the process of ants' food seeking behavior. The most important aspect of collaborative behavior of several ant species is their ability to find shortest path between nest and the food sources. They deposit a trail of pheromone on the path which attracts other ants to follow the trail or they can take a new path. Most ants tend to follow the path with maximum pheromone deposit. There are three major steps in ACO.

State Transition Rule: Ants prefer to move to nodes which are connected by short edges with a high amount of pheromone.

Local pheromone updating rule: While building a solution, ants visit edges and deposit some amount of pheromone.

Global pheromone updating rule: Once all ants have build their tours, pheromone is updated on all edges

ACO has been successfully implemented for different multiobjective optimization problems [12, 22–25]. The pseudo code in Fig. 1 illustrates working of ACO.

3 Neural Network

The network is modeled as a group of fully intraconnected neural networks. These intraconnected neural networks are interconnected by local connections in the routing system. The system is organized by interconnecting these groups of fully connected small neural networks, each of which resides in a node and can decide the optimum path independent of other nodes.

Using neural network [26], for a communication network with N number of nodes the routing system consists of N intraconnected small size neural networks placed at each node of the communication system. Each communication link is modeled as a neuron and all the neurons i.e. communication links approaching each node are fully connected.

Within each node the neurons form a small but fully intraconnected neural network. The intraconnected neural network requires only local information related to the output of each neuron of the same node, the load of the neighboring nodes and the remaining distances of packets from destinations. All this information can be obtained from the neurons interconnecting the corresponding nodes in the communication system. The decision of the direction of data packet depends on the results of the trade off of this information which is achieved by the neural network used. As the dynamics of each intra-neural network is independent of other intra-neural networks situated at other nodes in the communication network, the system realizes a parallel adaptive decentralized network routing. Also since each neuron updates with a unique state equation independent of others, the system is highly simplified.

Begin
Initialize
While stopping criterion not satisfied **do**
Repeat
For each ant **do**
Choose next node by applying state transition rule

$$P_k(r,s) = \begin{cases} \dfrac{\tau(r,s).[\eta(r,s)]^\beta}{\sum_{u \in Jk\,(r)} \tau(r,u)[\eta(r,u)]\,\beta} & , \text{if } s \in J_k(r) \quad\quad (1) \\ \\ 0 & , \text{otherwise} \end{cases}$$

Apply local pheromone update
$$\tau(r,s) \leftarrow (1-\rho).\,\tau(r,s)+\rho.\Delta\,\tau(r,s) \qquad\qquad (2)$$

(Where $0<\rho<1$ is a pheromone evaporation parameter,
$\Delta\tau(r,s)= 0\tau$ where τ is pheromone. We here assume ρ as 0.5.)

End for
Until every ant had built a solution
Update best solution
Apply global pheromone update
$$\tau(r,s) \leftarrow (1-\alpha)^*\,\tau(r,s)+\alpha^*\Delta\tau(r,s) \qquad\qquad (3)$$

Where $0<\alpha<1$ is pheromone decay parameter and we
assume $\alpha=0.2$

$$\Delta\,\tau(r,s)= \begin{cases} 1/L_{gb} \text{ if } (r,s) \in \text{ global best tour} \\ 0 \text{ otherwise} \end{cases}$$

Where L_{gb} is length of globally best tour

End while
End

Fig. 1 Pseudo code for ant colony optimization

The network routing system is implemented with the constraints that can be stated as follows:

1. Data packets are transmitted as one packet at a time from each node to another node.
2. Data packets are transmitted to other nodes in order to minimize the variation of network load.
3. Data packets are transmitted in order to minimize the remaining distance from their destination node.

There are three types of neural network: Hopfield Neural Network, Feed-forward Neural Network, and Recurrent Neural Network. The network used here is recurrent neural network and the learning approach is supervised [26]. Recurrent network contains at least one feedback connection loop so the activation can flow in a loop. That enables the network to do temporal processing and learn sequences. Neural network for multicast routing is explained bellow.

For each neuron i, the firing rate of neuron is given by:

$$R_i = \sum \left(W_{ij}^+ + W_{ij}^- \right) \tag{4}$$

The probability that neuron i is excited is given by

$$q_i = \frac{\lambda_i^+}{R_i + \lambda_i^-} \tag{5}$$

4 Proposed Model

A simple network for solving multicast routing problem is as shown in Fig. 2.

Vertex represents the nodes of network and the edges represent the links between the nodes. Here in this figure 'r', 'v' and 's' are the source node respectively and 'u' and 't' are destination nodes. The cost and pheromone intensity of the paths are given in Table 1.

Fig. 2 Sample network

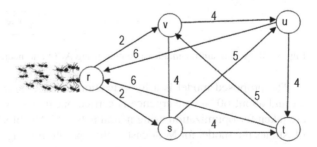

Table 1 Cost and pheromone intensity of the paths

Path	Cost	Pheromone intensity
r,v	2	4
r,u	2	4
r,u	6	1.4
r,t	6	1.4
v,u	4	2
v,s	4	2
u,t	4	2
s,t	4	2
v,t	5	1.6
s,u	5	1.6

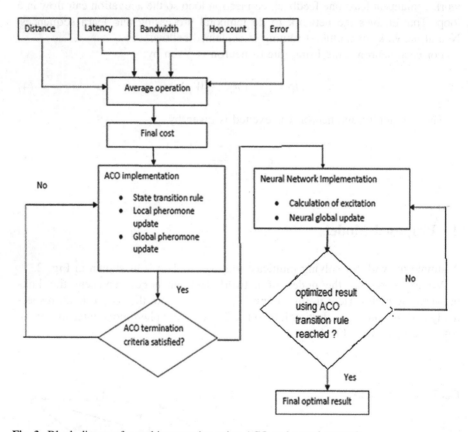

Fig. 3 Block diagram for multicast routing using ACO and neural network

The proposed model is a hybridization of ACO and Neural Network in which around about 60 % convergence is carried out using ACO and then the best states are taken for optimization using neural network. The block diagram shown in Fig. 3 illustrates the model for Multicast routing problem using ACO and Neural Network.

Table 2 chosen values of parameters

Probability	Distance	Bandwidth	Packet loss	Hop count	Delay	Average cost	Pheromone intensity
P(r,v)	0.468	0.399	0.4	0.399	0.4	0.4132	2.42
P(r,s)	0.468	0.399	0.4	0.399	0.4	0.4132	2.42
P(r,u)	0.031	0.1	0.1	0.1	0.101	0.0864	11.5
P(r,t)	0.031	0.1	0.1	0.1	0.101	0.0864	11.5
P(v,u)	0.084	0.1826	0.1826	0.179	0.1827	0.16218	6.20
P(v,s)	0.084	0.1826	0.1826	0.179	0.1827	0.16218	6.20
P(v,t)	0.106	0.138	0.138	0.23	0.1382	0.15004	6.66
P(v,r)	0.767	0.4958	0.495	0.410	0.496	0.53276	1.877
P(s,v)	0.089	0.1826	0.1826	0.179	0.1827	0.16218	6.20
P(s.t)	0.089	0.1826	0.1826	0.179	0.1827	0.16218	6.20
P(s,u)	0.106	0.138	0.138	0.23	0.1382	0.15004	6.66
P(s,r)	0.767	0.4958	0.495	0.410	0.496	0.53276	1.877

5 Implementation

Parameters are considered for calculating average cost function of an undirected network by taking into account the bandwidth, distance, error, latency, and hop count. Real time data has been traced by communicating different websites (www. iitkgp.ac.in, www.google.com, www.bitmesra.ac.in, www.yahoo.com) at different time and normalized to find the cost function as shown in Table 2.

Ant colony optimization is applied on the normalized cost function obtained from the mentioned websites with ACO termination condition of 60 % convergence. The best states are further optimized till termination condition of NN is reached. The termination condition for NN is assumed as 80 % convergence of ants. From the experimental results it is observed that the convergence rate degrades after 80 % convergence without modifying the results.

6 Result and Discussion

Research shows that ACO has slower convergence thus increase the computational time. Although neural network is more complicated than ACO, it is proven to give faster convergence and have dynamic approach. For the analysis of execution time, eleven test networks with varying number of nodes (5–14) and number of links (9–40) are considered and a comparative study has been made among Dijikstra algorithm, Conventional ACO, and ACO NN as shown in Fig. 4. It can be noticed that Dijikstra and ACO gives similar performance. With increasing number of nodes both Dijikstra and ACO tends to take higher execution time while the hybrid

Fig. 4 Execution time of Dijikstra algorithm, conventional ACO, and ACONN

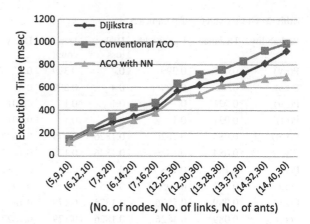

(No. of nodes, No. of links, No. of ants)

ACONN is found to be comparatively faster. It is also observed that the number of ants initialized has no significant impact on the performance of the algorithm if the ants considered are greater than the number of nodes in the network.

7 Conclusion

It can be concluded from this work that hybridizing ACO and neural network results a better technique which is more cost effective than primitive Dijkstra's algorithm and conventional ACO. It not only improves the number of iteration and convergence speed but also reduces the time complexity in various scenario considering all possible constraints like-cost, delay, bandwidth etc. The proposed model can also be applied in various sectors where ACO has been successfully implemented.

References

1. Chen, X., Liu, S., Guan, J., Liu, Q.: Study on QoS multicast routing based on ACO-PSO algorithm. In: International Conference on Intelligent Computation Technology and Automation (ICICTA), vol. 3, pp. 534–537 (2010)
2. Li, C., Cao, C., Li, Y., Yu, Y.: Hybrid of genetic algorithm and particle swarm optimization for multicast QoS routing. In: IEEE International Conference Controls Automation, pp. 2355–2359 (2007)
3. Wang, H., Meng, X., Zhang, M., Li, Y.: Tabu search algorithm for RP selection in PIM-SM multicast routing. Elsevier Comput. Commun. **33**, 35–42 (2009)
4. Wang, H., Meng, X., Li, S., Xu, H.: A tree-based particle swarm optimization for multicast routing. Comput. Netw. **54**(15), 2775–2786 (2010)

5. Zhou, J., Cao, Q., Li, C., Huang, R.: A genetic algorithm based on extended sequence and topology encoding for the multicast protocol in two-tiered WSN. Expert Syst. Appl. **37**(2), 1684–1695 (2010)
6. Wang, H., Xu, H., Yi, S., Shi, Z.: A tree-growth based ant colony algorithm for QoS multicast routing problem. Expert Syst. Appl. **38**, 11787–11795 (2011)
7. Patel. M.K., Kabat, M.R., Tripathy, C.R.: A hybrid ACO/PSO based algorithm for QoS multicast routing problem. Ain Shams Eng. J. **5**(1), 113–120 (2014)
8. Shimamoto, N., Hiramatsu, A., Yamasaki, K.: A dynamic routing control based on a GA. In: Proceedings of the IEEE International Conference on Neural Network, pp. 1123–1128 (1993)
9. Zhang, L., Cai, L., Li, M., Wang, F.: A method for least-cost QoS least-cost multicast routing based on genetic simulated annealing algorithm. Comput. Commun. **31**, 3984–3994 (2008)
10. Frank, A.J., Wittie, L.D., Bernstein, A.J.: Multicast communication on network computers. IEEE Softw. **2**(3), 49–61 (1985)
11. Yen, J.Y.: An algorithm for finding shortest routes from all source nodes to a given destination in general networks. Q. Appl. Math. **27**, 526–530 (1970)
12. Wang, Y., Xie, J.: Ant colony optimization for multicast routing. IEEE APCCAS (2000)
13. Yuan, P., Hai, Y.: An improved ACO algorithm for multicast in ad hoc networks. In: International Conference on Communications and Mobile Computing (2010)
14. Wang, H., Shi, Z., Li, S.: Multicast routing for delay variation bound using a modified ant Colony algorithm. J. Netw. Comput. Appl. (2008)
15. Pan, D.-R., Xue, Y., Zhan, L.-J.: A multicast wireless mesh network (WMN) network routing algorithm with ant colony optimization. In: International Conference on Wavelet Analysis and Pattern Recognition, pp. 744–748 (2008)
16. Singh, G., Kumar, N., Verma, A.K.: Ant colony algorithms in MANETs: a review. J. Netw. Comput. Appl. **35**, 1964–1972 (2012)
17. Wang, H., Xu, H., Yi, S., Shi, Z.: A tree-growth based ant colony algorithm for QoS multicast routing problem. Exp. Syst. Appl. **38**, 11787–11795 (2011)
18. Huang, Y.: Research on QoS multicast tree based on ant colony algorithm. Appl. Mech. Mater. 635–637; 1734–1737 (2014)
19. Liu, W., Wang, L.: Solving the delay constrained multicast routing problem using the transiently chaotic neural network. Advances in Neural Networks. Lecture Notes in Computer Science, vol. 4492, pp. 57–62
20. Pour, H.M., Atlasbaf, Z., Mirzaee, A., Hakkak, M.: A hybrid approach involving artificial neural network and ant colony optimization for direction of arrival estimation. In: Canadian Conference on Electrical and Computer Engineering, pp. 001059–001064 (2008)
21. Dorigo, M., Maniezzo, V., Colorni, A.: The ant system: optimization by a colony of cooperating agents. IEEE Trans. Syst. Man Cybern. Part B **26**, 29–41 (1996)
22. Srivastava, S., Sahana, S.K., Pant, D., Mahanti, P.K.: Hybrid microscopic discrete evolutionary model for traffic signal optimization. J. Next Gener. Inf. Technol. (JNIT) **6**(2), 1–6 (2015)
23. Sahana, S.K., Jain, A., Mahanti, P.K.: Ant colony optimization for train scheduling: an analysis. I. J. Intell. Syst. Appl. **6**(2), 29–36 (2014)
24. Sahana, S.K., Jain, A.: High performance ant colony optimizer (HPACO) for travelling salesman problem (TSP). In: 5th International Conference on ICSI 2014, Hefei, China, October 17–20, 2014, In: Advances in Swarm Intelligence, vol. 8794, Springer International Publishing, Lecture Notes in Computer Science (LNCS), pp. 165–172 (2014)
25. Sahana, S.K., Jain, A.: An improved modular hybrid ant colony approach for solving traveling salesman problem. Int. J. Comput. (JoC) **1**(2), 123–127 (2011)
26. Mehmet Ali, M.K., Kamoun, F.: Neural networks for shortest path computation and routing in computer networks. IEEE Trans. Neural Netw. **4**(6), 941–954 (1993)

Artificial Intelligence (AI) Based Object Classification Using Principal Images

Santosh Kumar Sahoo and B.B. Choudhury

Abstract Now-a-days an object detection and classification is a unique perplexing difficulties. In the meantime the morphology and additional topographies of the defected objects are unlike from normal or defect free object, so it is possible to classify them using such artificial intelligence (AI) based structures. Here an substitute methodology followed by several AI procedures are established to categorize the defective object and defect free object by means of principal image texture topographies of various defective object like a soft drinks or cold drinks bottle and applying the pattern recognition techniques after that the successful accomplishing the image spitting, quality centered parameter abstraction as well as successive sorting of substandard and defect free bottles. Our results validated that Least Square support vector machine, linear kernel and radial function has maximum overall performance in terms of Classification Ratio (CR) is about 96.35 %. Thus, the proposed setup model is proved as a best choice for classification of an object.

Keywords Artificial neural network (ANN) · Support vector machine (SVM) · Radial basis function (RBF) · Least square support vector machine (LSSVM) · Kth nearest neighbor

1 Introduction

In order to find a defect free bottle from a manufacturing unit it is essential to scrutinize the damaged bottle perfectly and rapidly for instigating corrective action. Hence the detection and classification of damaged bottle at the manufacturing end

S.K. Sahoo (✉)
Department of ECE, IGIT, Sarang, Utkal University, Odisha, India
e-mail: santosh.kr.sahoo@gmail.com

B.B. Choudhury
Department of Mechanical Engineering, IGIT, Sarang, Odisha, India
e-mail: bbcigit@gmail.com

© Springer India 2016
H.S. Behera and D.P. Mohapatra (eds.), *Computational Intelligence in Data Mining—Volume 2*, Advances in Intelligent Systems and Computing 411, DOI 10.1007/978-81-322-2731-1_13

143

has endured a challenge. On the other hand it is a sturdy necessity to identify the damaged bottle centered on the training facilitated by means of numerous defective picture or image texture parameters for a rapid analysis of the bottle. The proposed process will be much tranquil and if trained well, then it is suitable for classification of damaged one. This proposed model can identify damaged bottle speedily and consequently the remedial action will be ongoing at a primary phase of the man-ufacturing process as a result the degradation of quality risk can be optimized. An ANN specimen is widely used in numerous fields related to non-linear utility estimation and classification because of its softness and usefulness. Likewise the SVM method is also an intelligent machine learning algorithm having remarkable performance and more benefits over other approaches for resolving recent signal and image processing tasks. At this moment, our proposed research work aims to grow a system to classify damaged bottle by means of AI-centered classifier which will directs several structures of principal images and categorize them into a specific group of bottle like damaged bottle or defect free bottles. Here several pattern recognition techniques are applied using different types of features pull out from 95 bottle images. Here four various machine learning classifiers like KTH NN, ANN, SVM and LSSVM for grouping of defective bottle images. Hence it is perceived that the proposed LSSVM with RBF and kernel are confirmed the higher rate of classification matched to other classifiers used during the analysis.

2 Literature Review

Bruylants et al. [1] investigated regarding optimally compression of volumetric medical Images with three dimensional joint photograph and the Volumetric wavelets and the entropy-coding improves the Compression performance. Cai et al. [2] described the image segmentation by using gradient guided active contours. Nie et al. [3] explained the image segmentation by histogram thresholding and Class Variance Criterion. Li et al. [4] introduced a construction of dual wavelet for Image Edge Detection where they used Sobel and Canny filter for comparison purpose. Sharma et al. [5] described the scheming of complex wavelet transform with directional properties and its application. Alarcon et al. [6] explained the results of Some compression of wavelet transforms and thresholding methods. Wu et al. [7] described the variety of arrhythmia Electro Cardio graphic signal utilized for opti-mizing scheme of quantization with wavelet techniques for ECG data using Genetic Algorithm (GA). Eswara et al. [8] analyzed some wavelet methods using com-pression of image in which both sides are analyzed and designed with lifting based wavelet is considered. Alice et al. [9] presented with a compression of picture with 9/7 wavelet transform using lifting design where the soft pictures are retrieved with no loss. Telagarapu et al. [10] described the image density with cosine transform and wavelet by select suitable technique and best result for Peak Signal to Noise Ratio.

3 Design Methodology

[A] Model Design: The damaged bottle of different size and shape used in a cold drinks company like coca- cola for filling up soft drinks were collected with prior permission of coca cola bottling plant khordha and the image of that bottle is captured by NI smart camera 1762. The major aim of this research is to categorize the damaged bottle and non-damaged bottle by pixel processing and artificial intelligent (AI) methods. The anticipated Pattern recognition structural plan is presented in Fig. 1. The process used 95 types of bottle images, where 65 types of images goes to defective bottle and 30 numbers are non-defective bottle. Firstly the bottle is segmented from the image background by an application of image segmentation techniques. This image segmentation and feature mining are analyzed by means of MATLAB software. After evaluating all these features they are united to form a feature dataset. For improved accuracy and implementation the feature numbers are reduced from 95 to 25 by means of principal component analysis (PCA). During the testing and training the concentrated structures are applied as the input to the classifier. The proposed structure reflected 55 % circumstances as training of the compressed data and others as testing. The individual classifier enactment is presented in terms of Classification Ratio (CR) can be conveyed as:

$$\text{Classification Ratio (CR)} = (tp + tn)/(tp + tn + fp + fn) \qquad (1)$$

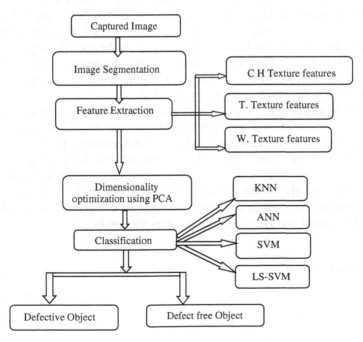

Fig. 1 Schematic structural plan represents the classification of defective object and normal object

Fig. 2 Defective bottles
segmented image

where tp = True positive no., tn = True negative no., fp = False positive no.,
fn = False negative number.

where C.H = Chief histogram based, T. = Tamura, W. = Wavelet based

[b] Image segmentation: An image segmentation is accomplished by segmenting the bottle from image credentials by means of gradient and morphological procedures. This Segmented bottle mostly vary as of the related images by image disparity. An image Gradient can distinguished the discrepancy shown in Fig. 3 which is again modified to a certain level to acquire segmented images. The segmented image for defective is displayed in Fig. 2

[c] Feature extraction: In this process the various features from an image can collected and also reducing the dimensionality. Here the different features are extracted from both defective and non-defective bottle pictures. The Tamura texture features extraction technique is applied to excerpt different features like crudeness, distinction and directionality of 95 bottle images. As Wavelet centered texture features used normally for the fetidness. So here, the 3 level 2D wavelet decomposition methods are used to evaluation of different coefficients of an images in Vertical, Horizontal and Diagonal way are presented in Fig. 4.

Fig. 3 The mask gradient
value

$$\begin{bmatrix} -1 & -2 & -1 \\ 0 & 0 & 0 \\ 1 & 2 & 1 \end{bmatrix}$$

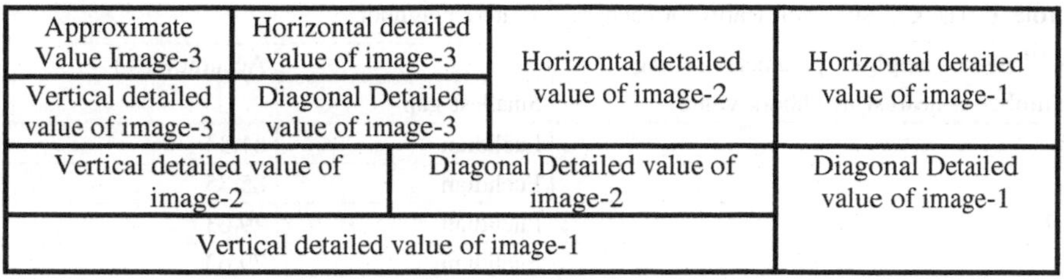

Approximate Value Image-3	Horizontal detailed value of image-3	Horizontal detailed value of image-2	Horizontal detailed value of image-1
Vertical detailed value of image-3	Diagonal Detailed value of image-3		
Vertical detailed value of image-2		Diagonal Detailed value of image-2	Diagonal Detailed value of image-1
Vertical detailed value of image-1			

Fig. 4 Representation diagram displays of three level wavelet transforms

Now Wavelet constants are found in a pixel is stated as:

$$w_\varphi(J_0, M, N) = \frac{1}{\sqrt{m \times n}} \sum_{x=0}^{m-1} \sum_{y=0}^{n-1} F(x, y) \varphi_{J_0, M, N}(X, Y) \qquad (2)$$

$$w_\psi^I(J, M, N) = \frac{1}{\sqrt{m \times n}} \sum_{x=0}^{m-1} \sum_{y=0}^{n-1} F(x, y) \varphi_{J, M, N}^I(X, Y), I = \{h, v, d\} \qquad (3)$$

where $\varphi_{J,M,N}^I(X, Y) = 2^{J/2} \varphi(P^J X - M, P^J Y - N)$ and

$\varphi_{J_0, M, N}(X, Y) = 2^{J/2} \varphi(P^J X - M, P^J Y - N)$ are the scrambled and converted base functions. Where the value of P = 2. Now an energy is intended laterally in all the ways like, Vertical, Horizontal and Diagonal (Fig. 4). As 3 level of decomposition is used so the energy for each wave let coefficient is stated as:

$$e_{cl} = \frac{1}{l \times l} \sum_{I=0}^{l} \sum_{J=0}^{ll} (w_{c,l,I,J})^2 \qquad (4)$$

where $w_{C,I,I,J}$ is the coefficient of wave let at [I, J] location.

As we attained 9 features for a single wavelet coefficient thus for 3 level disintegration of a solo image by using wavelet transform, total 81 features are extracted. The proposed technique is continual for all images, i.e. defective and defect free object class and by gathering all the information related to an image then a data set is formed in order to providing the input variable and the output. A binary value '0' is assigned to represent the output for a non-defective object and '1' for a defective object. And also by using PCA the dimensions of feature datasheet is reduced.

4　Results and Discussions

The reduced features obtained using PCA algorithm are fed to different classifiers to distinguish defective and defect free class. The Schematic graph in Fig. 1 indicates the application of different classifiers to differentiate defective and defect free

Table 1 The Classification Ratio for each K^{TH} nearest neighbor

K^{TH} nearest neighbor parameter setting		Accuracy (%)
Number of nearest neighbors value	Smallest gap	
I	Euclidian	81.34
II	Euclidian	85.85
III	Euclidian	79.63
IV	Euclidian	79.63
V	Euclidian	80.43
VI	Euclidian	80.43
VII	Euclidian	81.85

bottles. In order to accomplish better accuracy in KNN classifier the two parameters like the nearest neighbors (K) value as well as the smallest gap from nearest neighbors as Euclidean are used. After that the Classification Ratio of KNN classifier is calculated.

The Classification Ratio (CR) for unlike adjoining neighbors (K) are conveyed in Table 1 where it is perceived that the highest Classification Ratio 85.85 % at K = II. The CR will reduce if we consider more and more neighbors due to increase in FP and FN. From the Table 1 it is concluded that an accuracy is highest at K = II. Hence, to develop an improved model of KNN by using K = II.

In Fig. 5 the deviation of Mean Squared Error (MSE) w.r.t the number of iteration relatively specifies that the MSE regularly declined while the number of iterations enlarged more than thousand times. The CR is perceived here as 91.78 % shown in Table 2. By considering the classification degree of the analyzed data, results of support vector machine classifier is estimated. A proposed SVM model competent with linear and Kernel RBF functions in-order to acquire improved trial information. Again the trial information can tested by SVM linear and radial basis kernel classifier which adjusted and trained earlier. Now the valuation of trial

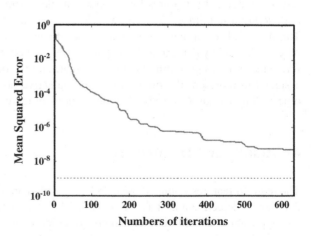

Fig. 5 Deviations of error w.r.t total iterations in multi feed neural network classifier

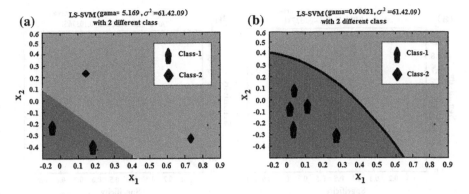

Fig. 6 a Class segmentation with respect to hyper-plane using RBF Kernel LSSVM. **b** Class segregate by linear kernel LSSVM

information is articulated through confusion matrix (CR), conveyed in Table 2 in addition to this the Receiver Operation Characteristics (ROC) outline is exposed in Fig. 7. The performance of LSSVM classifier can estimated in a way related to that of SVM. The Fig. 6a, b represents the distribution of data w.r.t optimum hyper-plane for Least Square—support vector machine along with linear and Radial Basis Function kernel classifier. This graph postulates the deviation of training output or class's w.r.t the most significant features of the exercised data. Figure 6 represents the class segmentation in a better way where the large blue area indicates the class 2 or defect free bottle. Table 2 signifies the presentation of LSSVM classifier in terms of Classification Ratio (CR). Also in Fig. 7 the receiver operating characteristic (ROC) graph is presented. Moreover Table 2 also proves that the LSSVM by kernel radial basis function and linear kernel has maximum precision of 96.35 % during object classification.

By following the Fig. 7 it is concluded that the area under ROC curve for RBF Kernel LSSVM classifier is about 0.96 which one is higher than the region of linear kernel LSSVM classifier as a result the space below the ROC Arc directs the

Table 2 The classifiers performance table

Classifiers	tp	fp	tn	fn	Classification Ratio (CR) in (%)
Multi feed neural network	28	1	11	3	91.78
Linear kernel support vector machine	31	10	2	0	86.84
RBF kernel support vector machine	28	0	12	3	94.06
Least square support vector machine with linear kernel	29	0	12	2	96.35
Least square support vector machine with radial basis function kernel	29	0	12	2	96.35

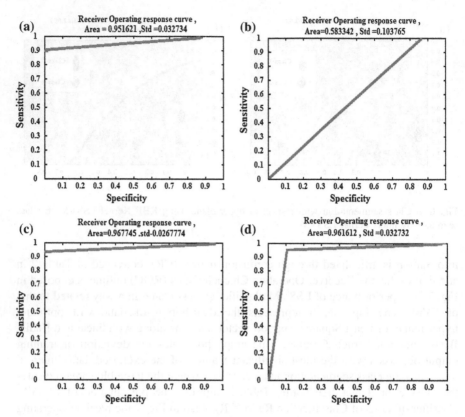

Fig. 7 Receiver operating characteristic strategy for evaluation of testing data. **a** RBF kernel support vector machine response. **b** Linear kernel support vector machine response. **c** Radial basis kernel least square support vector machine response. **d** Linear kernel least square support vector machine response

classifier's preciseness. Hence it is decided that LSSVM with RBF kernel classifier is the best classifier among all used in the present analysis for the classification of defective object and defect free object.

5 Conclusion

By using the morphological way of image analysis which forecast the defective bottles in the manufacturing process in a precise manner. Current research utilizing the pattern recognition is effectively realized to categorize the defective and defect free bottle from refined principal images.

The actual contests are involved for documentation of structures. These structures are effectively categorized as defective and defect free bottle by several artificial intelligence classifiers like K[TH] nearest neighbor, artificial neural network,

Support Vector Machine, Least Square Support Vector machine. The particular LSSVM functional blocks effectively demonstrated as a best analytical tool with advanced Classification Ratio of 96.35 % for analysis.

References

1. Bruylants, T., Munteanu, A.: Wavelet based volumetric medical image compression. Sign. Process. Image Commun. **31**, 112–133 (2015) (Elsevier)
2. Cai, B., Liu, Z., Wang, J., Zhu, Y.: Image segmentation framework using gradient guided active contours. Int. J. Sign. Process. **8**(7), 51–62 (2015)
3. Nie, F., Zhang, P.: Image segmentation based on framework of two-dimensional histogram and class variance criterion. IJSP, Image Process. Pattern Recognit. **8**(7), 79–88 (2015)
4. Li, Y.: A construction of dual wavelet and its applications in image edge detection. IJSP, Image Process. Pattern Recognit. **8**(7), 151–160 (2015)
5. Sharma, N., Agarwal, A., Khyalia, P.K.: Squeezing of color image using dual tree complex wavelet transform based preliminary plan. Signal Image Process. Int. J. (SIPIJ) **5**(3) (2014)
6. Alarcon-Aquino, V., Ramirez-Cortes, J.M., Gomez-Gil, P.: Lossy image compression using discrete wavelet transform and thresholding techniques. The Open Cybern. Syst. J. **7**, 32–38 (2013)
7. Wu, T.C., Hung, K.C., Liu, J.H., Liu, T.K.: Wavelet-based ECG data compression optimization with genetic algorithm. J. Biomed. Sci. Eng. **6,** 746–753 (2013) (Scientific Research)
8. Eswara Reddy, B. Dr., Venkata Narayana, K.: A lossless image compression using traditional and lifting based wavelets. Signal Image Process. Int. J. (SIPIJ) **3**(2) (2012)
9. Alice Blessie, A., Nalini, J., Ramesh, S.C.: Image compression using wavelet transform based on the lifting scheme and its implementation. IJCSI Int. J. Comput. Sci. Issues **8**(3), No. 1 (2011)
10. Telagarapu, P., Naveen, V.J., Prasanthi, A.L., Santhi, G.V.: Image compression using DCT and wavelet transformations. Int. J. Sign. Process. Image Process. Pattern Recognit. **4**(3) (2011)

Signature Verification Using Rough Set Theory Based Feature Selection

Sanghamitra Das and Abhinab Roy

Abstract An offline signature verification system based on feature extraction from signature images is introduced. Varieties of features such as geometric features, topological features and statistical features are extracted from signature images using Gabor filter technique. As all the features are not relevant, only the salient features are selected from the extracted one by a Rough Set Theory based reduct generation technique. Thus only the relevant features of the signatures are retained to reduce the dimension of feature vector so as to reduce the computation time and are used for offline signature verification. The experimental results are expressed using few parameters such as False Rejection Rate (FRR), False Acceptance Rate (FAR).

Keywords Gabor filter · Feature extraction · Rough set theory · Feature selection

1 Introduction

Signature is considered to be one of the most widely accepted parameter for human identification. In this era of digital revolution, signature verification system is almost an 'inevitable application' not only in the field of financial transactions using credit/debit cards, cheque cashing but also for certificate verification, contract paper verification etc. In comparison with other identification technologies like face detection, fingerprint, retina, and so on, signature verification is more advantageous as an identity verification mechanism.

S. Das (✉)
Department of Computer Science and Engineering, Hooghly Engineering
and Technology College, Hooghly 712103, West Bengal, India
e-mail: sanghamitra.das@hetc.ac.in

A. Roy
Department of Computer Science and Technology, IIEST, Shibpur,
Howrah 711103, West Bengal, India
e-mail: abhinabroy92@gmail.com

© Springer India 2016
H.S. Behera and D.P. Mohapatra (eds.), *Computational Intelligence in Data Mining—Volume 2*, Advances in Intelligent Systems and Computing 411, DOI 10.1007/978-81-322-2731-1_14

153

Signatures can be authenticated in two ways; either on-line or off-line. The signature image is scanned at a high resolution before giving it as input to the offline system. For online systems, the signature images are dynamically captured using devices like tablet, stylus, or digitizer.

Offline signature verification system plays a pivotal role in identifying skilled forgery of signature against genuine signature. In case of online signatures, dynamic information makes the signature difficult to forge. On the other hand, offline signatures are easier to imitate for an imposter.

This paper presents a novel set of features for the purpose of verification. Each signature image is preprocessed by some general preprocessing techniques and subsequently 2D-Gabor filter is applied for feature extraction. As all the features are not equally important, only the salient features are selected from the extracted one by a Rough Set Theory (RST) based reduct generation technique and finally few classification techniques are applied using the WEKA tool [1] to measure the effectiveness of the method. Experimental results show that the proposed technique provides better outcome compared to some previous verification techniques [2] using the same GPDS (Digital Signal Processing Group corpus).

The paper has been divided in five sections and accordingly discussed. Section 2 describes Related Work. In Sect. 3 the proposed feature extraction and features selection method is described. The performance of the proposed method on experimental data set is discussed in Sect. 4. Finally conclusion and future perspectives are provided in Sect. 5.

2 Related Work

Offline signature verification is a well known area where different features and classification approaches are used for signature authentication. For instance, in paper [3], grid-based feature extraction method employs information extracted from the signature contour and the verification is done using SVM classifier. Paper [4] uses feature extracted from the local neighborhood of signature image. Cartesian and polar coordinate systems are used to divide the signature into zones and for each zone two separate histogram features are determined: (i) histogram of oriented gradients and (ii) histogram of local binary patterns. Support Vector Machines (SVMs) are used for signature classification. Two feature extraction techniques, Modified Direction Features and the Gradient features are compared in [2]. Both of them have used similar settings for their experiment. Moreover, performance comparison is made between squared Mahalanobis distance classifier and Support Vector Machines by utilizing the Gradient Features. However, in research work [5], two offline signature verification systems are proposed based on local and global approach respectively. The comparison has been done between the systems by employing a huge number of features encoding the orientations of the strokes applying morphology. For an improved signature verification system, features extracted on the basis of boundary of a signature and projections of the same are used in paper [6]. One of the features is obtained from total energy applied by a writer for their signature creation. Another feature highlights the ratio of the distance between key strokes of the signature image and the height/width of the signature image by using information obtained from the horizontal and vertical projections of an image. High pressure points obtained from a signature image (static) are considered as features for offline signature verification system and the extraction of same is depicted in paper [7]. A novel way is presented to calculate High Pressure threshold value from grayscale images. Image holding the high pressure points and the binary image of the primary signature is converted into polar coordinates where density distribution ratio of them is calculated. At the very end, two vectors are calculated to determine how far the points from geometric center of the original signature image. Robustness is tested for simple forgeries using KNN classifier. In [8], the co-occurrence matrix and local binary pattern are representing the features for signature verification and by applying statistical texture features the grey level variations in the image is calculated. For the training of a SVM model, genuine signatures and random forgeries have been used. Besides for the purpose of testing, random and skilled forgeries are utilized. Two key aspects of off-line signature verification are reported in [9]. One of them is feature extraction which produces a new graphometric feature set by considering the curvature of the essential segments of the signature. Here the shape of signature is simulated by applying Bezier curves and features are extracted from these curves. In the second aspect, for the improvement of reliability of classification based on graphometric features, an ensemble of classifiers is used. The graphometric feature set also lessens the false acceptance rate. A feature set used in [10] illustrates the signature contour in accordance with spatial distribution of neighboring black pixels around a candidate pixel. Moreover, through the correlation among signature pixels, a texture feature is also calculated for offline signature verification.

3 Proposed Methodology

Initially, the image quality is improved by removing unwanted information from the collected signature images. Thereafter, binary image is inverted and median filtering is done for noise reduction [11] and edge detection. Morphological thinning operation [12] is applied in the process of removing the selected foreground pixels from binary images. Thinning eliminates the variations of thickness of the signatures, because of age, illness, geographic location etc.

3.1 Feature Extraction

After preprocessing the signatures, some important local and global features are extracted.

3.1.1 Feature Extraction Using Gabor Filters

Gabor filters are bandpass filters. The impulse response of Gabor filter is formed by multiplication between a Gaussian function and a complex oscillation. As per the uncertainty principle Gabor Filters hold the optimal localization property in both the spatial and frequency domain. Moreover, Gabor filters can be described using four parameters namely, standard deviation along x and y directions, frequency of the sinusoidal function, and orientation of the filter. The filters focus on particular range of frequencies. The impulse response of Gabor filter is defined using Eq. (1).

$$G(x,y) = \exp\left(\frac{-1}{2}\left(\left(\frac{x'}{\sigma_x}\right)^2 + \left(\frac{y'}{\sigma_y}\right)^2\right)\right) \exp(i(2\pi f x')) \tag{1}$$

where, σ_x and σ_y are the σ_y Standard Deviations along x and y direction
f frequency of the sinusoidal function
x' $x \cos\theta + y \sin\theta$
y' $-x \sin\theta + y \cos\theta$
θ Orientation of the Gabor filter

The standard deviation along x and y axis determines the size of the Gabor filter mask. The distinct values of θ changes the sensitivity to edge and texture orientation. The variation in f will change the sensitivity to high and low frequencies.

3.1.2 Forming the Gabor Filter Mask

- The convolution mask of a Gabor filter bank is a coordinate plane where the rows are numbered from $-\sigma_x$ to σ_x and the columns from $-\sigma_y$ to σ_y.
- The values at each pixel position in the filter mask is the value of the function G (x, y), where x and y are the x and y coordinate of that pixel position, with the given parameters.
- The Gabor filter mask will contain an imaginary number at each pixel position.
- The complex mask is separated into real and imaginary parts and each part is convolved with the input image.

In proposed method two dimensional Gabor filter with different values of the parameters called Gabor filter bank is applied on input signature image shown in

Fig. 1 Input image to Gabor filter

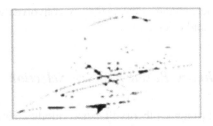

Fig. 2 Signature image after applying Gabor filter bank at f = 0.125, θ = 180°

Fig. 3 Signature image after applying Gabor filter bank at f = 0.25, θ = 60°

Fig. 4 Signature image after applying Gabor filter bank at f = 0.50, θ = 0°

Fig. 1. The pre-processed input binary image is convolved with each Gabor filter bank. Here, seven different orientations (0°, 30°, 60°, 90°, 120°, 150° and 180°) and four different frequencies (0.125, 0.25, 0.5, 0.75) are considered. Standard deviations and energies of the result image at each frequency for seven different orientations are calculated. Hence total 56 features (A_1–A_{56}) are obtained. Figures 2, 3, 4

Fig. 5 Signature image after applying Gabor filter bank at f = 0.50, θ = 150°

and 5 are the four sample signature images that are convolved with Gabor filter bank at different frequencies and orientations.

3.2 Rough Set Theory Based Feature Selection

The various concepts of rough set theory like discernibility matrix, core and attribute dependency are applied together to select the minimum number of important features, called reduct of the signatures. The work applies a reduct generation algorithm [13] that iteratively selects the locally most important feature and finally provides a subset of relevant and important features which are sufficient to represent the signature verification system. The method first separates the core and noncore features and selects one noncore attribute in each iteration until the reduct is found. The heuristic used in this approach is termed as forward attribute selection algorithm. The following algorithm is the rough set based algorithm [13] to find out the important features of signatures where the notations have their usual meanings.

Algorithm: Feature_Selection_From_Signature (DS, CR, NC)
Input: DS = the decision system with C extracted features and D signature class, CR is the set of core features and NC is the non-core feature set obtained using paper [12].
Output: RED = selected important features of signatures
Begin
  Repeat
    R = CR   /*core is considered as initial reduct */
    NC_PREV = NC /* take a copy of initial elements of NC */
    /* Forward selection to give one reduct */

Classifier	R1	R2	R3	R4	R5

    Repeat
      x = highest ranked element of NC
      If (x = ϕ) break /* if no element found in NC */
      If ($\gamma_{R \cup \{x\}}(D) > \gamma_R(D)$) // $\gamma_R(D)$ is the attribute dependency of D on R
      {
          $R = R \cup \{x\}$
          $NC = NC - \{x\}$
      }
    Until ($\gamma_R(D) = \gamma_C(D)$)
    If ($\gamma_R(D) = \gamma_C(D)$) // R is a reduct
        RED = RED $\cup$ R
    If (NC_PREV = NC) then NC = ϕ
  Until (NC is empty)
End.

4 Experimental Results

4.1 Experimental Database

This experiment uses a subset of GPDS-960 corpus. The subset consists of 10 sets where each set consists of 24 genuine signatures and 30 high skilled forgeries. Naturally this experiment involves 240 genuine signatures and 300 high skilled forgeries. The signatures are in "bmp" format, in black and white color and 300 dpi.

4.2 Performance Assessment

The sample configuration of the features selection with their accuracies calculated by the proposed algorithm (FSFS) for different classifiers is summarized in Table 1. The results of our experiments using different features are given in Table 2.

Table 1 Sample reducts and accuracies obtained by forward selection

Classifier	R1	R2	R3	R4	R5
Naïve bayes	95.83	93.33	95.83	94.16	95.83
Logistic	94.16	91.66	95.83	95	95.83
BayesNet	90.83	89.16	92.5	90	91.66
KStar	95	95	95	95	94.16
ClassificationVia Regression	94.16	96.66	93.33	95	97.5
Bagging	88.33	90	91.66	89.16	95
Decorate	90	90.83	89.16	89.16	91.66
MulticlassClassifier	95.83	92.5	95	93.33	94.16
Hyperpipes	88.33	85.83	89.16	86.66	90.83
DTNB	86.66	91.66	90.83	89.16	91.66
FT	97.5	96.66	97.5	97.5	96.66
FilteredClassifier	87.5	87.5	89.16	88.33	87.5
RepTree	89.16	86.66	90.83	91.66	88.33
RandomForest	94.16	90	90.83	92.5	95

Table 2 Experimental results of FAR and FRR using FSFS algorithm for different classifiers

Feature: Gabor filter based features		
Classifier	Feature Selection Algorithm (FSFS)	
	FAR (%)	FRR (%)
FilteredClassifier	13.33	11.67
Bagging	20	10.84
RepTree	20	10
RandomForest	20	7.5

Table 2 gives the results obtained with the selected features by the proposed system on GPDS-960 dataset. We have used classifiers using 10 reference signatures. 24 genuine signatures are used in training purpose and 30 skilled forgeries are used in testing purpose.

5 Conclusion and Future Perspectives

In this paper Gabor filter based feature extraction technique and rough set theory based feature selection algorithm is applied on signature image to retain only the important features for offline signature verification system. Several existing classifiers are applied for performance analysis. The experimental results show that, the accuracies given by various classifiers are comparable with other popular existing methods. Though the FAR and FRR values in Table 2 show that the system can accurately identify the signatures but it is not so helpful in case of forgery detection, which is our main concern as future work.

References

1. http://www.cs.waikato.ac.nz/ml/weka/
2. Nguyen, V., Kawazoey, Y., Wakabayashiy, T., Palz, U., Blumenstein, M.: Performance analysis of the gradient feature and the modified direction feature for off-line signature verification. In: 12th International Conference on Frontiers in Handwriting Recognition, 978-0-7695-4221 (2010)
3. Nguyen, V., Blumenstein, M.: An Application of the 2D Gaussian filter for enhancing feature extraction in off-line signature verification. In: International Conference on Document Analysis and Recognition, pp. 1520–5363 (2011)
4. Yilmaz, M.B., Yanikoglu, B., Tirkaz, C., Kholmatov, A.: Offline signature verification using classifier combination of HOG and LBP features. IEEE, 978-1-4577-1359 (2011)
5. Rekik, Y., Houmani, N., El Yacoubi, M.A., Garcia-Salicetti, S., Dorizzi, B.: A comparison of feature extraction approaches for offline signature verification. IEEE, 978-1-61284-732 (2010)
6. Nguyen, V., Blumenstein, M., Leedham, G.: Global features for the off-line signature verification problem. In: 10th International Conference on Document Analysis and Recognition, 978-0-7695-3725 (2009)
7. Vargas, J.F., Ferrer, M.A., Travieso, C.M., Alonso, J.B.: Off-line signature verification based on high pressure polar distribution. ICFHR (2008)
8. Vargas, J.F., Ferrer, M.A., Travieso, C.M., Alonso, J.B.: Off-line signature verification based on grey level information using texture features. Elsevier, Pattern Recogn. **44**, 375–385 (2011)
9. Bertolini, D., Oliveira, L.S., Justino, E., Sabourin, R.: Reducing forgeries in writer-independent off-line signature verification through ensemble of classifiers. Elsevier, Pattern Recogn. **43**, 387–396 (2010)
10. Kumar, R., Sharma, J.D., Justino, B.: Writer-independent off-line signature verification using surroundedness feature. Elsevier, Pattern Recogn. **33**, 301–308 (2010)
11. Chen, T., Ma, K.K., Chen, L.H.: Tri-state median filter for image denoising. IEEE Trans. Image Process. **8**, 1057–7149 (1999)

12. Jang, B.K., Chin, R.T.: Analysis of thinning algorithms using mathematical morphology. IEEE Trans. Pattern Anal. Mach. Intell. **12**(6), 541–551 (2002)
13. Das, A.K., Chakrabarty, S., Sengupta, S.: Formation of a compact reduct set based on discernibility relation and attribute dependency of rough set theory. ICIP (2012)

Protein Sequence Classification Based on N-Gram and K-Nearest Neighbor Algorithm

Jyotshna Dongardive and Siby Abraham

Abstract The paper proposes classification of protein sequences using K-Nearest Neighbor (KNN) algorithm. Motif extraction method N-gram is used to encode biological sequences into feature vectors. The N-gram generated is represented using Boolean data representation technique. The experiments are conducted on dataset consisting of 717 sequences unequally distributed into seven classes with a sequence identity of 25 %. The number of neighbors in the KNN classifier is varied from 3, 5, 7, 9, 11, 13 and 15. Euclidean distance and Cosine coefficient similarity measures are used for determining nearest neighbors. The experimental results revealed that the procedure with Cosine measure and the number of neighbors as 15 gave the highest accuracy of 84 %. The effectiveness of the proposed method is also shown by comparing the experimental results with those of other related methods on the same dataset.

Keywords Protein · Classification · N-gram · KNN

1 Introduction

It is essential to have knowledge about function of proteins to understand biological processes [1–4]. A large-scale worldwide sequencing projects in recent years reveal new sequences leading to a massive flow of new biological data. However, many

J. Dongardive (✉)
Department of Computer Science, University of Mumbai, Mumbai 400098, India
e-mail: jyotss.d@gmail.com

S. Abraham
Department of Mathematics and Statistics, G. N. Khalsa College University of Mumbai, Mumbai, India
e-mail: sibyam@gmail.com

© Springer India 2016
H.S. Behera and D.P. Mohapatra (eds.), *Computational Intelligence in Data Mining—Volume 2*, Advances in Intelligent Systems and Computing 411, DOI 10.1007/978-81-322-2731-1_15

sequences that are added to the databases are unannotated and await analysis. The experimental methods alone cannot provide functional annotation to these sequences in a reasonable time. So, computer-based methods are needed to predict protein function. These methods identify the function of a protein based on direct sequence similarity analysis or search of conserved sequence motifs. This helps to classify the sequences into a specific protein family or functional class. In the absence of the sequence or structural similarity, the criterion for addition of similar but distantly related proteins into a protein function class becomes increasingly arbitrary. For such instances, function prediction systems that implement machine-learning-based classifiers are effective. There are many machine-learning based classification algorithms. Some examples of these algorithms are K-Nearest Neighbour (KNN) [5], Neural Networks (NN) [6], Support Vector Machines (SVM) [7, 8], Naïve Bayes (NB) [5] and Hidden Markov Models (HMM) [6].

This paper deals with the preprocessing of protein sequences for supervised classification. Motif extraction method N-gram is used to encode biological sequences into feature vectors to enable use of well-known machine-learning classifier KNN.

2 Literature Review

In the past, wide range of computational methods has been developed to classify protein sequences into corresponding classes, sub classes or different super-families. These methods are based on sequence alignments, motifs and machine learning approaches. The method based on sequence alignment uses dynamic programming approach when the sequences do not have enough similarity between them. Needleman and Wunsch [9] developed an algorithm which finds similarity between the protein and DNA sequences by using global alignment. Smith and Waterman [10] used local sequence alignment between protein and DNA sequences to find sequence similarity by looking at the fragments of different lengths in a sequence.

Basic local alignment search tools like BLAST and FASTA are the traditional alignment based methods used for the analysis of both protein and DNA sequences. BLAST [11] uses heuristic algorithms to find similar sequences, while FASTA exploits local sequence alignment to find similar sequence using heuristic search in the database [12]. However, sequence alignment becomes unreliable when the sequences have less than 40 % similarity [13], and unusable for below 20 % similarity [14, 15]. This has generated interest in classification methods that do not make use of alignments. Many computational intelligence approaches using machine learning and pattern recognition have also been developed that do not make use of alignments. A brief review of some of the most recent techniques is presented below.

Jeong et al. [16] introduced a feature extraction method based on position specific scoring matrix (PSSM) to extract features from a protein sequence. They defined four feature sets from PSSM and used four classifiers viz Naïve Bayesian (NB), Support Vector Machine (SVM), Decision Tree (DT) and Random Forest (RF). The maximum classification accuracy obtained was 72.5 %. Mansoori et al. [17] extracted features from a protein sequence using 2 grams and a 2-gram exchange group from the training and test data. A SGERD-based classifier was used to create fuzzy rules. These rules reduced the classification time from 79 to 51 min and the classification accuracy was 96.45 %. Bandyopadhyay used a 1-gram technique for feature encoding [18]. In this, he developed a variable length fuzzy genetic clustering algorithm to find prototypes for each super-family and nearest neighbor algorithm for classification of protein. The classification accuracy of 81.3 % was obtained.

Leslie et al. [19] proposed a technique for SCOP super families which considered subsequences of k-length amino acids (k-spectrum kernel) as a feature vector passed to a support vector machine for classification of protein. The results were compared with SVM-T98, SVM-Fisher and PSI-BLAST. SVM-Fisher produced better results as compared to other approaches.

Caragea et al. [20] explored the feature hashing technique to map high dimensional features to low dimension using hash keys. These high dimensional features were obtained using the k-gram representation. The feature vectors obtained were used to store frequency counts of each k-gram hashed together in the same hash key. The proposed technique reduced the features' size from 222 to 210 and gave classification accuracy of 82.83 %. Yu et al. [21] proposed a k-string dictionary technique to represent protein sequence. To represent each protein sequence properly, Singular value decomposition (SVD) was applied for the factorization of frequency/probability matrix. Using this technique the size of the feature vector was reduced. Yellasiri and Rao [22] proposed a new classification model called Rough Set Classifier for classifying the voluminous protein data based on structural and functional properties of protein. The proposed model was fast and provided 97.7 % accuracy. Saha and Chaki [23] proposed a three phase model for classifying the unknown proteins into known families. In first phase, noisy sequences were removed and thus the input dataset was reduced. In second phase, the necessary features such as molecular weight and isometric point were obtained and then feature ranking algorithm was applied to rank the features. In the final phase, Neighborhood Analysis was used to classify the input sequence into a particular class or family. Jason et al. [24] used a neural network model in their work for classifying the existing protein sequences. The methods used to find the global similarity were 2-gram encoding method and 6-letter exchange group. Minimum description length (MDL) principle was also applied to calculate the significance of motif. This model produced 90–92 % accuracy.

3 Materials and Methods

The section presents the methodology proposed for the classification of protein sequences into their particular class. The different stages starting from collection of data to calculation of performance measures are discussed in the following sub sections.

3.1 Collection of Dataset from Protein Database

The dataset used in the study was constructed by Yu [25], searching oligomeric proteins of each category in the Swiss-Prot database (version 45.4) [26, 27]. It comprises of 717 sequences unequally distributed into seven classes which represent seven quaternary protein structures with a sequence identity of 25 %, the details of which is given in Table 1.

3.2 Feature Extraction: N-Gram Descriptor

The feature extraction from the protein sequence is done using alignment-free encoding technique, known also as N-gram [28]. The N-gram has a predefined length 1, 2, 3…n and is a subsequence composed of N characters, extracted from a larger sequence. This technique helps to extract both global and local features from protein sequences. In this work we have used 3-gram which is obtained by a sliding window of 3 characters on the whole sequence character by character. A subsequence of 3 characters is extracted with each slide and the process is repeated for all the sequences. Finally, only the distinct 3-grams are kept for further analysis. For example, suppose we have a protein sequence VLTINDKGAS; then the protein descriptors of the length 3 amino acid are {VLT, LTI, TIN… GAS}.

Table 1 Experimental dataset

Data source	Identity percentage (%)	Class	Number of sequences	Total sequences
Swiss-prot	25	"Monomer"	208	717
		"Homodimer"	335	
		"Homotrimer"	40	
		"Homotetramer"	95	
		"Homopentamer"	11	
		"Homohexamer"	23	
		"Homooctamer"	5	

3.3 Building Attribute-Value Dataset: Boolean Data Representation

The next step is the construction of the attribute-value table for the 3-gram generated as per Sect. 3.2. To attribute values to features we have used presence/absence of feature approach which, also known as Boolean data representation technique. This representation indicates whether one n-gram is present within a sequence or not based on the rule given below.

$$w_j^i = 1 \text{ if } x_j^i > 0 \text{ and } 0 \text{ else}$$

Thus, our data table as shown in Fig. 1 is Boolean where each line represents a protein sequence and each column represents a 3-grams. The binary value of a case indicates whether the relative 3-grams belong (1) or not (0) to the considered protein sequence.

3.4 K-Nearest Neighbor (KNN) Classification

After generating attribute–value table as mentioned in the above section, the proposed method has used KNN classifier, which is one of the simplest classifiers attempted on protein sequence classification. In this, classification is performed during the test phase by calculating the distance between the vector representation of the unclassified proteins and the vectors representing the training proteins, and finding the nearest neighbors. The unclassified protein is then assigned a function class according to the majority of function classes shared by its nearest neighbors.

The parameters associated with a basic KNN classifier are the number of nearest neighbors to be considered, k. For our sequence based KNN classifier, we have used 3, 5, 7, 9, 11, 13 and 15 neighbors and compared two of the most commonly used measures for determining nearest neighbors for KNN classification viz Euclidean distance and Cosine coefficient similarity. The Euclidean distance

	MPA	PAT	ATS	TSS	SSI	SII	IIT	ITI	TII	IIA	IAV	AVA	VAA	AAC
Seq0	1	1	1	1	1	1	1	1	1	1	1	1	1	1
Seq1	0	0	0	0	0	0	1	1	0	0	1	1	0	0
Seq2	0	1	0	1	0	0	1	0	1	0	0	0	0	0
Seq3	0	0	0	0	1	0	0	0	0	0	0	0	0	0
Seq4	0	1	0	0	0	1	0	0	1	1	0	0	0	0
Seq5	0	0	0	0	0	0	0	0	0	0	0	0	0	0
Seq6	0	0	0	0	1	0	0	0	0	0	0	0	0	1
Seq7	0	0	0	0	0	0	0	0	0	0	0	0	0	1
Seq8	0	0	0	1	0	0	0	0	0	0	0	0	0	0

Fig. 1 Boolean representation of data

measures the planar distance between two n-dimensional feature vectors, p = (p1, p2...pn − 1, pn) and q = (q1, q2...qn − 1, qn) and is defined as:

$$\text{euc}(p, q) = \sqrt{\sum_{i-1}^{n} |pi - qi|^2}$$

The Cosine coefficient is the cosine of the angle between two feature vectors, defined as:

$$\cos(p, q) = \frac{\sum_{i-1}^{n} pi * qi}{\sqrt{\sum_{i=1}^{n}(pi)2} * \sqrt{\sum_{i}^{n}(qi)}}$$

3.5 Performance Evaluation of Classifier

To assess the performance of the classifier, we have used accuracy as the evaluation metric. This measure is calculated based on the number of true positives (TP), false positives (FP), true negative (TN) and false negatives (FN). Accuracy is defined as the proportion of the total number of test proteins that are correctly classified. It is determined using the equation:

$$\text{Accuracy} = (TP + TN)/(TP + FP + FN + TN)$$

3.6 Software

The software used for the experiments is Matlab Version 8.2.0.701 (R2013b). The Bioinformatics Toolbox Version 4.3.1 (R2013b) is used for the implementation of KNN. The computer that was used to perform the experiments for model selection was an Intel(R) Core(TM) 2CPU6300@1.86 GHz.

4 Experimental Results

The work discusses a series of experiments performed on a KNN classifier using the dataset. The parameters associated with a basic KNN classifier are the number of nearest neighbors and the distance measure. In our work we experimented with values 3, 5, 7, 9, 11, 13 and 15 as the number of neighbors. We also used two of the most commonly used measures for determining the nearest neighbors for KNN classification viz Euclidean distance and Cosine coefficient similarity.

4.1 Effect of Parameters of the KNN Classifier

Table 2 shows the effect of the two parameters viz number of neighbors and measure for determining the neighbours. It is seen that the value of the neighbour 15 and cosine measure has obtained the highest accuracy of 84 %.

4.2 Comparison of Results of the Proposed Method with Similar Methods for the Dataset Used

Table 3 shows the comparative results of the same dataset with the ones studied in [25, 29, 30]. We can notice that the worst results were obtained with the AAC

Table 2 Effect of parameters for determining nearest neighbors

Nearest neighbor	Measures for determining nearest neighbors	Accuracy (%)
3	Euclidean	74.9251
3	Cosine	77.3227
5	Euclidean	74.8252
5	Cosine	80.3197
7	Euclidean	80.4196
7	Cosine	82.5175
9	Euclidean	81.4186
9	Cosine	83.7163
11	Euclidean	81.3187
11	Cosine	83.8162
13	Euclidean	82.018
13	Cosine	83.6164
15	Euclidean	81.3187
15	**Cosine**	**84.5155**

Table 3 Comparative results of the proposed method

Methods	Accuracy (%)
Amino acid composition (AAC) and neural networks	41.4
Discriminative descriptors substitution matrix (DDSM) and Naiye Bayesian	59.4
Blast-based	69.6
Functional domain composition (FDC) and neural network	75.2
Discriminative descriptors substitution matrix (DDSM) and neural network	77
Discriminative descriptors substitution matrix (DDSM) and support vector machine	78.9
Discriminative descriptors substitution matrix (DDSM) and C4.5	79.2
N-gram and KNN	**84.51**

method i.e. 41.4 %. Blast arrived at better results, but the accuracy was not very high. The FDC and DDSM methods seemed to be promising since it allowed reaching an accuracy of 75.2 and 79.2 %. But the proposed method, which uses N-gram with KNN, outperforms others as it is quite efficient in terms of accuracy and reached the highest accuracy rate of 84.51 %.

5   Conclusions

This paper proposes a method for classifying protein sequences into structural families using features extracted from motif content method N-gram. Compared with the previous works, protein sequences are converted into feature vectors using global and local features represented by the motif content i.e. N-gram. Having obtained the features, the KNN classifier is used to classify protein sequences into corresponding families based on two parameters-the number of nearest neighbors and measures for determining nearest neighbors of KNN. Comparison of the experimental results of related works on the same dataset reveals that features extracted using the N-gram along with KNN classifier is more effective for protein classification.

References

1. Downward, J.: The ins and outs of signalling. Nature **411**, 759 (2001)
2. Lengeler, J.W.: Metabolic networks: a signal-oriented approach to cellular models. Biol. Chem. **381**, 911 (2000)
3. Siomi, H., Dreyfuss, G.: RNA-binding proteins as regulators of gene expression. Curr. Opin. Genet. Dev. **7**, 345 (1997)
4. Draper, D.E.: Themes in RNA-protein recognition. J. Mol. Biol. **293**, 255 (1999)
5. Webb, A., Copsey, K., Cawley, G.: Statistical Pattern Recognition (2011)
6. Bishop, C.M.: Pattern Recognition and Machine Learning. Springer, New York (2006)
7. Vapnik, V.N.: The Nature of Statistical Learning Theory, 2nd edn. Springer, New York (1999)
8. Burges, C.J.C.: A tutorial on support vector machine for pattern recognition. Data Min. Knowl. Disc. **2**, 121 (1998)
9. Needleman, S.B., Wunsch, C.D.: A general method applicable to the search for similarities in the amino acid sequence of two proteins. J. Mol. Biol. **48**(3), 443–453 (1990)
10. Smith, T.F., Waterman, M.S.: Identification of common molecular subsequences. J. Mol. Biol. **147**, 195–197 (1987)
11. Altschul, S.F., Gish, W., Miller, W., Myers, E.W., Lipman, D.J.: Basic local alignment search tool. J. Mol. Biol. **215**(3), 403–410 (1990)
12. Pearson, W.: Finding protein and nucleotide similarities with FASTA. Current Protocols in Bioinformatics, Chapter 3, unit 3.9 (2004)
13. Wu, C.H., Huang, H., Yeh, L., Barker, W.C.: Protein family classification and functional annotation. Comput. Biol. Chem. **27**(1), 37–47 (2003)
14. Pearson, W.R.: Effective protein sequence comparison. Methods Enzymol. **266**, 227–258 (1996)

15. Pearson, W.R.: Empirical statistical estimates for sequence similarity searches. J. Mol. Biol. **276**(1), 71–84 (1998)
16. Jeong, J.C., Lin, X., Chen, X.: On position-specific scoring matrix for protein function prediction. IEEE/ACM Trans. Comput. Biol. Bioinf. **8**(2), 308–315 (2011)
17. Mansoori, E.G., Zolghadri, M.J., Katebi, S.D.: Protein superfamily classification using fuzzy rule-based classifier. IEEE Trans. Nanobiosci. **8**(1), 92–99 (2009)
18. Bandyopadhyay, S.: An efficient technique for superfamily classification of amino acid sequences: feature extraction, fuzzy clustering and prototype selection. Fuzzy Sets Syst. **152** (1), 5–16 (2005)
19. Leslie, C., Eskin, E., Noble, W.S.: The spectrum kernel: a string kernel for SVM protein classification. In: Proceedings of the Pacific Symposium on Biocomputing, pp. 564–575 (2002)
20. Caragea, C., Silvescu, A., Mitra, P.: Protein sequence classification using feature hashing. In: Proceedings of the IEEE International Conference on Bioinformatics and Biomedicine (BIBM '11), pp. 538–543. Atlanta, GA, USA, Nov 2011
21. Yu, C., He, R.L., S.-T. Yau, S.: Protein sequence comparison based on K-string dictionary. Gene **529**, 250–256 (2013)
22. Ramadevi, Y., Rao, C.R.: Rough set protein classifier. J. Theor. Appl. Inf. Technol. (2009)
23. Suprativ, S., Rituparna, C.: A brief review of data mining application involving protein sequence classification. Advances in Computing and Information Technology (2013)
24. Wang, J.T.L., Ma, Q.H., Shasha, D., Wu, C.H.: Application of neural networks to biological data mining: a case study in protein sequence classification. In: KDD, Boston, MA, USA, pp. 305–309 (2000)
25. Yu, X., Wang, C., Li, Y.: Classification of protein quaternary structure by functional domain composition. BMC Bioinf. **7**, 187–192 (2006)
26. Boeckmann, B., Bairoch, A., Apweiler, R., Blatter, M.C., Estreicher, A., Gasteiger, E., Martin, M.J., Michoud, K., O'Donovan, C., Phan, I., Pilbout, S., Schneider, M.: The SWISS-PROT protein knowledgebase and its supplement TrEMBL in 2003. Nucleic Acids Res. **31**, 365–370 (2003)
27. Bairoch, A., Boeckmann, B., Ferro, S., Gasteiger, E.: Swiss-Prot: juggling between evolution and stability. Brief Bioinform. **5**, 39–55 (2004)
28. Leslie, C., Eskin, E., Noble, W.S.: The spectrum kernel: a string kernel for SVM protein classification. In: Proceedings of the Pacific Symposium on Biocomputing (PSB), pp. 564–575 (2002)
29. Chou, P.Y.: Prediction of protein structural classes from amino acid composition. In: Fasman, G.D. (ed.) Prediction of Protein Structure and the Principles of Protein Conformation. Plenum Press, New York, pp. 549–586 (1989)
30. Saidi, R., Maddouri, M., Nguifo, E.M.: Protein sequence classification by means of feature extraction with substitution matrices. BMC Bioinf. **11**, 175 (2010)

Harmony Search Algorithm Based Optimal Tuning of Interline Power Flow Controller for Congestion Management

Akanksha Mishra and G.V. Nagesh Kumar

Abstract With rapid increase in private power producers to meet the increasing power demand, results in the congestion problem. As the power transfer is increasing the operation of power systems is becoming difficult due to higher scheduled and unscheduled power flows. Interline Power Flow Converter (IPFC) is a most flexible device and effective in reducing the congestion problem. In this paper line utilization factor (LUF) is used for finding the best location to place the IPFC. The Harmony Search (HS) Algorithm is used for proper tuning of IPFC for a multi objective function which reduces active power loss, total voltage deviations, security margin and the capacity of installed IPFC of installed IPFC capacity. Simulation is carried out on IEEE-30 bus test system and the results are presented and analyzed to verify the proposed method.

Keywords Interline power flow converter · Harmony search algorithm · Congestion · Line utilization factor

1 Introduction

There are many challenges in terms of system operation because of the upgrading of generation and transmission systems have not been adequate with the increasing in load. This also affects the security of the system which is measured through the system congestion levels. It is now necessary to use the transmission system effectively for better reliability and stability of the system [1]. In the new competitive electric market, it is now mandatory for the electric utilities to operate such that it makes better utilization of the existing transmission facilities in conjunction with maintaining the security, stability and reliability of the supplied power.

A. Mishra · G.V. Nagesh Kumar (✉)
Department of EEE, GITAM Institute of Technology, GITAM University,
Visakhapatnam, India
e-mail: gundavarapu_kumar@yahoo.com

© Springer India 2016   173
H.S. Behera and D.P. Mohapatra (eds.), *Computational Intelligence in Data Mining—Volume 2*, Advances in Intelligent Systems and Computing 411, DOI 10.1007/978-81-322-2731-1_16

Several authors [2–4] have used index based methods for optimal placement of Flexible AC Transmission Systems (FACTS) devices. Gitizadeh et al. [5] investigated a Simulated Annealing based optimization method for placement of FACTS devices in order to relieve congestion in the transmission lines while increasing static security margin and voltage profile of a given power system. Ye et al. [6] proposed an algorithm for optimal congestion dispatch calculation with UPFC control. A decomposition control method was introduced to solve this optimal power flow problem. Reddy et al. [7] has presented optimal location of FACTS controllers considering branch loading (BL), voltage stability (VS) and loss minimization (LM) as objectives at once using Genetic Algorithm for management of congestion. Mohamed et al. [8] has compared three variants of PSO namely basic PSO, Inertia weight approach PSO and constriction factor approach PSO considering a single objective i.e. to minimize the transmission line loss.

FACTS devices are preferred in modern power systems based on the requirement and are found to deliver good solution [9, 10]. Out of all FACTS devices IPFC is considered to be most flexible, powerful and versatile as it employs at least two VSC's with a common DC link. Hence IPFC has the capability of compensating multi transmission line. FACTS devices like TCSC and SSSC are also placed on the most congested line. However, IPFC is a device connected to multiple transmission lines. In its simplest form it has at least two converters placed on two transmission lines connected to a common bus [11, 12]. Proper placement of IPFC is therefore a subject to be analyzed. Location and tuning of FACTS devices in the power system is one of the important issues and hence optimal placement and tuning of IPFC has been proposed based.

In this paper, Line Utilization Factor has been used for determination of the optimal location of IPFC. It gives an estimate of line overloading in terms of MVA. Once the IPFC has been placed at a proper location, it is very important to correctly tune the IPFC parameters to be able to properly utilize the device for enhancing the system performance to the maximum level. Harmony Search Algorithm is one of the most recent music centered metaheuristic algorithm which is reported to have been successfully implemented in several engineering applications. Implementation of Harmony Search Algorithm is considered to be easier in comparison to many existing algorithms. HS algorithm is a population based algorithm and is considered to have higher efficiency in comparison to other metaheuristic algorithms. The Harmony Search Algorithm can show promising results for tuning of IPFC. A multi objective optimization has been formulated for optimal tuning of IPFC using Harmony Search Algorithm. The multi objective function comprises of reduction of active power loss, minimization of total voltage deviations and minimization of security margin with the usage of minimum value of installed IPFC. Tuning of IPFC for reduction of loss further reduces line congestion. Reduction of Voltage deviation and security margin ensure power quality and system security. The proposed method is implemented and tested on an IEEE 30 bus system with different loading conditions.

2 Proposed Harmony Search Algorithm for Tuning of IPFC

Harmony Search (HS) is a population based metaheuristic algorithm. It is inspired from the musical process of searching for a perfect state of harmony. It was proposed by Zong Woo Geem in 2001. In the HS algorithm, each (musician = decision variable; plays = generates; a note = a value; for finding a best harmony = Global optimum). The pitch of each musical instrument determines the aesthetic quality. Just as the fitness value determines the quality of the decision variables. In the process of music, all players sound pitches within the possible range together to make one harmony. If all the pitches make a good harmony, each player stores in his memory and the possibility of making a good harmony is increased next time. In optimization also the same thing follows, the initial solution is generated randomly from decision variables within the possible range. If the objective function value of these decision variables is good to make a promising solution, then the possibility of making a good solution is increased next time. The parameters of this algorithm are

HMS	the size of the harmonic memory
HMCR	the rate of choosing a value from the harmony memory
PAR	the rate of choosing a neighboring value
δ	the amount between two neighboring values in discrete candidate set
FW (fret width)	the amount of maximum change in pitch adjustment

The Harmony Search Algorithm mainly depends upon three rules namely Harmony Memory Consideration Rule (HMCR), Pitch Adjustment Rate (PAR) and Random initialization rule [13–15]. The Flowchart is given in Appendix.

3 Optimal Tuning of IPFC

A multi-objective function is formulated to find the optimal size of IPFC [8, 12, 16–18]. The function comprises of

1. Minimization of the active power loss.
2. Minimization of total voltage deviations
3. Minimization of security margin
4. Minimization of the value of installed IPFC.

A multi objective function formulated is given in Eq. (1)

$$\text{Min}F = \text{Min} \sum_{i=1\text{to}4} w_i f_i \tag{1}$$

where, w_1, w_2, w_3, w_4 are the weighting factors.

The weights represent the preference of the respective objective functions. In this study, equal preference has been given to each objective. Hence,

$$w_1 + w_2 + w_3 + w_4 = 1 \tag{2}$$

$$w_1 = w_2 = w_3 = w_4 = 0.25 \tag{3}$$

The optimal tuning of IPFC is done taking into consideration the above mentioned multi-objective function using Harmony Search Algorithm.

4 Results and Discussion

An IEEE 30 bus test system is considered, in which bus no. 1 is considered as a slack bus and bus nos. 2, 5, 8, 11, 13 are considered as PV buses while all other buses are load bus as shown in Fig. 1. This system has 41 interconnected lines. The IEEE 30 bus test system load flow is obtained using MATLAB Software [19–21]. A power injection model of IPFC is taken for load flow study. Only load buses are considered for IPFC placement. Equal weights of 0.25 have been considered for all objectives. The results have been analyzed for normal loading. An IPFC with two converters have been taken into consideration.

From the load flow analysis of IEEE 30 bus system, it is established that line 3–4 is the most congested line connected between the load buses with LUF value of 0.8415 p.u. as mentioned in Table 1. Hence it is established that the first converter of the IPFC has to be placed on line 3–4. In the 30 bus system, two lines have been connected to line 3–4, line 4–6 and line 4–12. The LUF values of the lines have

Fig. 1 IEEE 30 bus test system with IPFC installed at line connected between buses 3–4 and 4–12

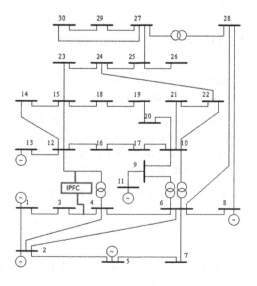

Table 1 LUF values of sample lines

S. no.	From bus	To bus	LUF line 2 (p.u.)
1	4	6	0.7173
2	4	12	0.5284
3	3	4	0.8415

been presented in Table 1. From the values given in Table 1, it is observed that line connected between buses 4–12, has lesser LUF value in comparison to the line 4–6. Thus line 4–12 is a healthiest line connected to line 3–4. Hence lines between buses 3–4 and buses 4–12 have been selected for optimal placement of IPFC.

The parameters of the Harmony Search Algorithm were varied to observe its effect on the various objective functions and the total objective function. The values obtained have been tabulated in Table 2. Effect of variation of HMS and HMCR on objective function value is observed from Fig. 2. Effect of variation of Pitch Adjustment Rate on Objective Function has been shown in Fig. 3. From Figs. 2 and 3 it is clear that minimum objective is obtained at HMS = 1, HMCR = 0.8, and PAR = 0.2. Value of FW is fixed at 0.1. The parameters of Harmony Search Algorithm used for tuning the IPFC have been mentioned in Table 3.

The final values of the IPFC after tuning using Harmony Search Algorithm have been given in Table 4. The LUF values before and after tuning of IPFC have been compared in Fig. 4. It is observed that tuning of IPFC using Harmony Search Algorithm, reduces the congestion in line 3–4 from 0.8415 p.u to 0.833 p.u. System Real and reactive power loss, security margin, voltage deviation, and installed IPFC

Table 2 Effect of variation of Harmony Search algorithm parameters on the objective function

HMS	FW	HMCR	PAR	Security margin	Voltage deviation	Inst. IPFC capacity	Active power loss	F1
1	0.1	0.7	0.35	15.5008	2.38162	1.76E-06	21.6067	9.8748
10	0.1	0.7	0.35	15.5719	2.3822	1.80E-06	21.628	9.8851
50	0.1	0.7	0.35	15.606	2.3822	1.80E-06	21.6368	9.8998
100	0.1	0.7	0.35	15.6061	2.3822	1.80E-06	21.6368	9.8952
1	0.1	0.8	0.35	15.5716	2.3816	1.76E-06	21.6107	9.8961
10	0.1	0.8	0.35	15.4352	2.3822	1.80E-06	21.6054	9.8919
50	0.1	0.8	0.35	15.5633	2.3822	1.80E-06	21.6258	9.8868
100	0.1	0.8	0.35	15.603	2.3816	1.76E-06	21.6161	9.8917
1	0.1	0.9	0.35	15.5603	2.3816	1.76E-06	21.6217	9.8812
10	0.1	0.9	0.35	15.5207	2.3822	1.81E-06	21.6156	9.879
50	0.1	0.9	0.35	15.4779	2.3822	1.80E-06	21.6039	9.8958
100	0.1	0.9	0.35	15.6061	2.3822	1.80E-06	21.6368	9.901
10	0.1	0.8	0.1	15.5091	2.3822	1.80E-06	21.6364	9.89392
10	0.1	0.8	0.2	15.549	2.377	1.80E-06	21.622	9.86334
10	0.1	0.8	0.3	15.6061	2.3822	1.80E-06	21.6368	9.8692
10	0.1	0.8	0.4	15.6061	2.3822	1.80E-06	21.6368	9.8741
10	0.1	0.8	0.5	15.6061	2.3822	1.81E-66	21.64	9.8806

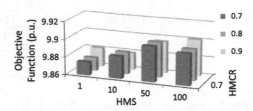

Fig. 2 Effect of variation of HMS and HMCR on objective function

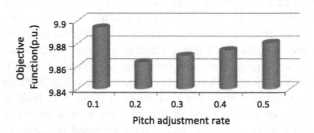

Fig. 3 Effect of variation of PAR on objective function

capacity, without and with optimally tuned IPFC have been mentioned in Table 5. The real and reactive power loss of the system with untuned IPFC is 21.909 MW and 101.334 MVAR respectively. After tuning the IPFC with Harmony Search Algorithm the active and reactive power loss of the system is reduced to 21.619 MW and 100.931 MVAR respectively. Similarly a good improvement in the above mentioned parameters is noticed with optimal tuning of IPFC using Harmony Search Algorithm The Voltage profile of the 30 bus system without and with Harmony tuned IPFC has been compared in Fig. 5. It is observed from Fig. 5 that tuning of IPFC using Harmony Search Algorithm improves the voltage at the buses to a large extent.

Table 3 Harmony Search algorithm parameters for IPFC tuning

Parameters	Values
HMS	1
HMCR	0.8
PAR	0.2
bw	0.1

Table 4 IPFC parameters after tuning by Harmony Search Algorithm

IPFC parameters	Untuned IPFC	Tuning of IPFC using HS
VSe1 (V)	0.0050	0.0013
VSe2 (V)	0.0100	0.0088
Θse1 (Radian)	−159.8295	−21.2163
Θse2 (Radian)	180	180

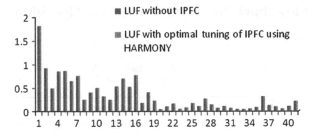

Fig. 4 LUF of lines of 30 bus system without and with optimally tuned IPFC using Harmonic Search Algorithm

Table 5 Comparison of objective function values with untuned IPFC and with HS tuned IPFC

Parameters	Untuned IPFC	Tuning of IPFC using HS
Real power losses (MW)	21.909	21.619
Reactive power loss (MVAR)	101.334	100.931
Voltage deviation of all buses (p.u.)	2.3889	2.3822
Security margin of all lines (p.u.)	18.2714	15.5377
Capacity of installed IPFC (p.u.)	0.000406	1.8025e-006

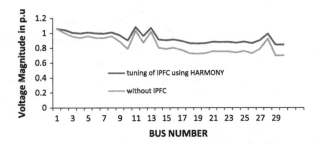

Fig. 5 Voltage profile without and with HS tuned IPFC

5 Conclusion

In this paper, Harmony Search Algorithm has been used for IPFC tuning. A multi objective function comprising of reduction of active power loss, minimization of total voltage deviations, and minimization of security margin with the usage of minimum value of installed IPFC has been considered. The proposed method is implemented for IEEE-30 bus test system. It is observed that placement and tuning of IPFC by the proposed methodology causes an effective reduction in congestion in the lines. Simulation results have demonstrated the effectiveness and accuracy of the Harmony Search algorithm technique to achieve the multiple objectives and to determine the optimal parameters of the IPFC. A reduction in Real power loss, Voltage deviation, security margin has been achieved with much smaller capacity of installed IPFC. Reduction in loss helps in congestion management of the system. Reduction in security margin protects the system against collapse. Lower the capacity of IPFC lower is the cost. Hence the overall system performance has been improved at a minimum cost using Harmony Search Algorithm.

Appendix: Flowchart for Harmony Search Algorithm

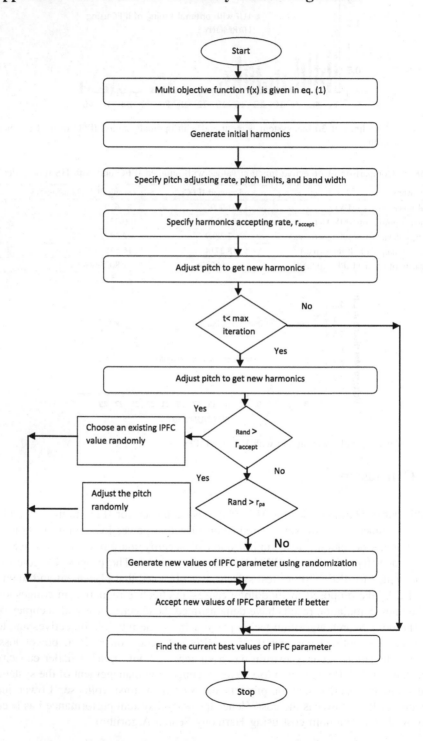

References

1. Zhang, X.P., Yao, L.: A vision of electricity network congestion management with FACTS and HVDC. In: The Third International Conference on Electric Utility Deregulation and Restructuring and Power Technologies. Nanjing, China, IEEE PES, IET, CSEE, pp. 116–121 (2008)
2. Qian, F., Tang, G., He, Z.: Optimal location and capability of FACTS devices in a power system by means of sensitivity analysis and EEAC. In: The Third International Conference on Electric Utility Deregulation and Restructuring and Power Technologies. Nanjing China, IEEE PES IET CSEE, pp. 2100–2104 (2008)
3. Acharya, N., Mithulananthan, N.: Locating series FACTS devices for congestion management in deregulated electricity markets. Electr. Power Syst. Res. **77**(3–4), 352–360 (2007)
4. Mandala, M., Gupta, C.P.: Congestion management by optimal placement of FACTS device, pp. 1–7. New Delhi, India, PEDES and Power India (2010)
5. Gitizadeh, M., Kalantar, M.: A new approach for congestion management via optimal location of FACTS devices in deregulated power systems. In: The Third International Conference on Electric Utility Deregulation and Restructuring and Power Technologies. Nanjing China, IEEE PES, IET CSEE, pp. 1592–1597 (2008)
6. Ye, P., Yang, Y., Wang, T., Sun, F., Zhao, H.: Decomposition control of UPFC for optimal congestion dispatch. In: The Third International Conference on Electric Utility Deregulation and Restructuring and Power Technologies. Nanjing China, IEEE PES IET CSEE (2008)
7. Reddy, S.S., Kumari, M.S, Sydulu, M.: Congestion management in deregulated power system by optimal choice and allocation of FACTS controllers using multi-objective genetic algorithm. In: Transmission and Distribution Conference and Exposition, IEEE PES, pp. 1–7 (2010)
8. Mohamed, K.H., Rama Rao, K.S., Md. Hasan, K.N.: Application of particle swarm optimization and its variants to interline power flow controllers and optimal power flow. In: International Conference on Intelligent and Advanced Systems (ICIAS). Kualalumpur, pp. 1–6 (2010)
9. Kirschner, L., Retzmann, D., Thumm, G.: Benefits of FACTS for power system enhancement. In: Transmission and Distribution Conference and Exhibition: Asia and Pacific IEEE/PES. Dalian, China, pp. 1–7 (2005)
10. Abdel-Moamen, M.A., Padhy, N.P.: Optimal power flow incorporating FACTS devices-bibliography and survey. In: Proceedings of the 2003 IEEE PES Transmission and Distribution Conference and Exposition, pp. 669–676 (2003)
11. Zhang, J.: Optimal power flow control for congestion management by interline power flow controller (IPFC). In: International Conference on Power System Technology 2006. Chongqing, China, pp. 1–6 (2006)
12. Teerthana, S., Yokoyama, A.: An optimal power flow control method of power system using interline power flow controller (IPFC). In: TENCON 2004, IEEE Region Conference, pp. 343–346 (2004)
13. Yang, X.S.: Harmony search as a metaheuristic algorithm. Stud. Comput. Intell. **191**, 1–14 (2009)
14. Pavelski, L.M., Almeida, C.P., Goncalves, R.A.: Harmony search for multi-objective optimization. In: Brazilian Symposium on Neural Networks (SBRN). Curitiba, pp. 220–225 (2012)
15. Ricart, J., Hüttemann, G., Lima, J., Barán, B.: Multiobjective harmony search algorithm proposals. Electr. Notes Theor. Comput. Sci. **281**, 51–67 (2011)
16. Benadid, R., Boudour, M., Abido, M.A.: Optimal placement of FACTS devices for multiobjective voltage stability problem. In: Power Systems Conference and Exposition. Seattle, WA, pp. 1–11 (2009)
17. Obadina, O.O., Berg, G.J.: Determination of voltage stability limit in multimachine power systems. IEEE Trans. Power Syst. **3**(4) (1988)

18. Gitizadeh, M.: Allocation of multi-type FACTS devices using multi-objective genetic algorithm approach for power system reinforcement. Electr. Eng. **92**(6), 227–237 (2010)
19. Hingorani, N.G., Gyugyi, L.: Understanding FACTS: concepts and technology of flexible AC transmission system. IEEE Press (2000)
20. Acha, E., Fuerte-Esquivel, C., Ambriz-Perez, H., Angeles, C.: FACTS: Modelling and Simulation in Power Networks. John Wiley & Sons, USA (2004)
21. Zhang, X.P.: Modeling of the interline power flow controller and the generalized unified power flow controller in Newton power flow. IEEE Proc. Generat. Trans. Distrib. **150**(3), 268–274 (2003)

Investigation of Broadband Characteristics of Circularly Polarized Spidron Fractal Slot Antenna Arrays for Wireless Applications

Valluri Dhana Raj, A.M. Prasad and G.M.V. Prsad

Abstract In this work, a novel broadband circularly polarized spidron fractal slot antenna has been considered. The design and implementation of single Spidron antenna, arrays of 2 × 2 and 4 × 4 are considered along with 4 × 4 array with the modified ground has been considered. HFSS13.0 has been used for the design and simulation, and fabrication FR4 epoxy material has been used. The broadband characteristics are examined simulation, and experimental results were presented. Theoretically a 4 × 4 array antenna offers the gain of is 5 dB and wide band ranging from 12.17 to 14.98 GHz. Its bandwidth is 2813 MHz. Practically a 4 × 4 array antenna operates from 9.9 to 14 GHz. The grid Spidron array performance evolution and comparison with 2 × 2 and 4 × 4 arrays is presented.

1 Introduction

Microstrip antennas are extensively used for the historical few decades because of its distinct options like low-profile, light-weight weight, ease to integrate with active devices [1–8]. In particular, considerable attention for the circularly polarized microstrip antennas has been given in many wireless applications like Satellite communications [9] and Radio-frequency identification readers [10, 11]. Based on some feeds microstrip antennas are classified as single or dual feed. In order to avoid the usage of additional components like 90° hybrid couplers, single feed

V.D. Raj (✉)
JNTUK, Kakinada, India
e-mail: drvalluri7777@gmail.com

A.M. Prasad
Department of ECE, JNTUK, Kakinada, India
e-mail: a_malli65@yahoo.com

G.M.V. Prsad
BVCITS, Amalapuram, India
e-mail: drgmvprasad@gmail.com

© Springer India 2016
H.S. Behera and D.P. Mohapatra (eds.), *Computational Intelligence in Data Mining—Volume 2*, Advances in Intelligent Systems and Computing 411, DOI 10.1007/978-81-322-2731-1_17

183

antennas are much preferred. Many a designs were described for single feed circularly polarized (CP) microstrip antennas, Koch fractal boundary CP microstrip antennas [12], a compact CP patch antenna loaded with a metamaterial [13], and a shorter CP patch antenna using a High permittivity substracte [14]. All of the single feed CP antennas usually described have a narrow bandwidth that is undesired. The influence of the feeding position on the radiation characteristics and resonance behavior were assessed. To increase the information measure performance, square-slot antennas with multiple monopole parts [15] or a halberd-shaped feeder [16] were proposed; these yield 3 dB AR bandwidths of 28.03 and 29.1 %, respectively. a technique based on a metamaterial chiral structure was conjointly applied to boost the performance of the CP antenna [17]. In [18], a pair element technique was used for a 4 × 4 array of spiral slot antennas to widen the resistivity and AR bandwidths. In order to enhance each the axial magnitude relation information measure and purity of the circular polarization, a method for sequent rotation feeding involving the applying of a physical rotation of the diverging element associated an acceptable phase offset to element excitation was introduced [19]. Many 4 × 4 arrays incorporating sequent rotation feeding techniques are developed. Several 4 × 4 arrays incorporating successive rotation feeding techniques are developed [20–22]. A 4 × 4 consecutive revolved patch antenna array fed by a series feed with a 3 dB AR information measure of 12. 4 % and a 2:1 voltage standing wave quantitative relation information measure of 14. 7 % was reported in [20]. In [21], another 4 × 4 successive rotation array of aperture antennas fed by a planar waveguide (CPW) was additionally planned. This array exhibited AN ohmic resistance information measure of 10.6 % and a 3 dB AR information measure of 11 %. During a recent study, a 4 × 4 array of consecutive revolved, stacked CP patches was conferred for L-band applications [22]. By using a successive rotation feeding technique, AN array consisting of 16^2 patches exhibited a broad ohmic resistance information measure of 25th and a 3 dB AR information measure of 100 %. In this, we have a tendency to propose a completely unique broadband CP antenna array is utilizing Spidron pattern slots as diverging components that square measure fed by parallel sequential networks [23]. The only antenna component of the projected array is comprehensively investigated in terms of its style issues and operation principle. In our style, the diverging Spidron pattern slot is excited by one microstrip feeding line to realize broadband CP radiation. 2 sorts of antennas square measure fed by a consecutive revolved feeder. A 2 × 2 and 4 × 4 arrays are designed, made-up and tested.

1.1 Single Spidron Fractal Slot Antenna

The configuration and design parameters for the one Spidron pattern slot are shown in Fig. 1.

From Fig. 1a, it may be seen that the hypotenuse of each right triangle coincides with one in every of the legs of its succeeding, down-scaled triangle. It ought to be

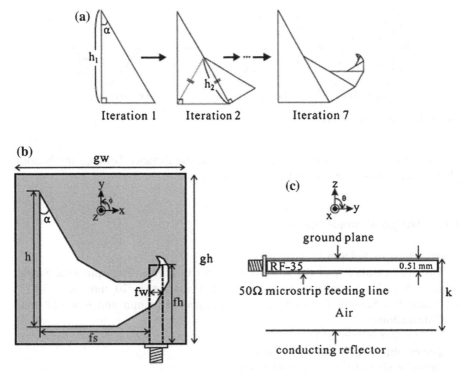

Fig. 1 Configuration of a Spidron fractal slot antenna with a conducting reflector: **a** Geometry of the Spidron fractal, **b** top view of the antenna, and **c** side view of the antenna

noted that α is one in every of the angles opposite the correct angle within the 1st right triangle whereas h is that the length of the leg adjacent to angle α. The scaling issue (δ) is that the magnitude relation between the lengths of the perimeters of 2 successively generated right triangles and were set by

$$\delta = \frac{h_{n+1}}{h_n} = \tan \alpha \text{ for } 0^\circ < \alpha < 45^0 \qquad (1)$$

The anticipated antenna element includes of a Spidron pattern slot etched from the bottom plane, a microstrip feeding line printed on the bottom of the insulator Substrate and a conducting reflector located on the substrate, is shown in Fig. 1b, c. The antenna is fabricated on an FR-4 epoxy substrate having a thickness of 1.6 mm, dielectric constant of 4.4, and loss tangent of 0.02.

The Spidron pattern slot utilized in this design consists of seven iterated reductions of a right-angled triangle. The 50 Ω microstrip feeding line situated at AN offset distance of fs from one facet of the Spidron pattern slot features a dimension and length of fw and fh, respectively. The dimensions of the one antenna

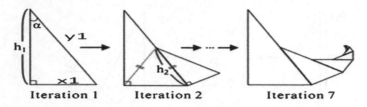

Fig. 2 Implementation of Spidron

area unit gw × gh × k mm³, wherever k is that the distance between the bottom plane and therefore the conducting reflector.

1.1.1 Design of Single Spidron

Given details of Spidron is as follows:

The implementation of the antenna is shown in Fig. 2. The optimized Single Spidron fractal slot antenna have these dimensions, h = 10 mm, α = 30.1°, k = 6 mm, fs = 8.3 mm, fw = 1 mm, fh = 6 mm, gw = 12.8 mm, and gh = 12.8 mm:

Calculations:

Fixed value of α = 30.1°

The height of right angle triangle is h_i.

Opposite side of the right angle triangle is x_i.

The hypotenuse of the right angle triangle is y_i.

a. **Iteration-1**: $h_1 = 10$ mm

$$\cos \alpha = \frac{h_1}{y_1} \tag{2}$$

$$y_1 = \frac{10}{\cos 30.1°} = 11.56 \, \text{mm}$$

$$\sin \alpha = \frac{x_1}{y_1} \tag{3}$$

$$x_1 = 11.56 * \sin 30.1° = 5.8 \, \text{mm}$$

b. **Iteration-2**:

$$h_2 = \frac{y_2}{2} = 11.56 = 5.78 \, \text{mm}$$

$$\cos \alpha = \frac{h_2}{y_2}$$

$$y_2 = \frac{5.78}{\cos 30.1°} = 6.68 \text{ mm}$$

$$\sin \alpha = \frac{x_2}{y_2}$$

$$x_2 = 6.68 * \sin 30.1° = 3.35 \text{ mm}$$

c. **Iteration-3**:

$$h_3 = \frac{y_2}{2} = \frac{6.68}{2} = 3.34 \text{ mm}$$

$$\cos \alpha = \frac{h_3}{y_3}$$

$$y_3 = \frac{3.34}{\cos 30.1°} = 3.86 \text{ mm}$$

$$\sin \alpha = \frac{x_3}{y_3}$$

$$x_3 = 3.86 * \sin 30.1° = 1.94 \text{ mm}$$

d. **Iteration-4**:

$$h_4 = \frac{y_3}{2} = \frac{3.86}{2} = 1.93 \text{ mm}$$

$$\cos \alpha = \frac{h_4}{y_4}$$

$$y_4 = \frac{1.93}{\cos 30.1°} = 2.23 \text{ mm}$$

$$\sin \alpha = \frac{x_4}{y_4}$$

$$x_4 = 2.23 * \sin 30.1° = 1.119 \text{ mm}$$

e. **Iteration-5**:

$$h_5 = \frac{y_4}{2} = \frac{2.23}{2} = 1.115 \text{ mm}$$

$$\cos \alpha = \frac{h_5}{y_5}$$

$$y_5 = \frac{1.115}{\cos 30.1°} = 1.288 \, mm$$

$$\sin \alpha = \frac{x_5}{y_5}$$

$$x_3 = 1.288 * \sin 30.1° = 0.646 \, mm$$

f. **Iteration-6**:

$$h_6 = \frac{y_5}{2} = \frac{1.288}{2} = 0.644 \, mm$$

$$\cos \alpha = \frac{h_6}{y_6}$$

$$y_6 = \frac{0.644}{\cos 30.1°} = 0.744 \, mm$$

$$\sin \alpha = \frac{x_6}{y_6}$$

$$x_6 = 0.744 * \sin 30.1° = 0.373 \, mm$$

g. **Iteration-7**:

$$h_7 = \frac{y_6}{2} = \frac{0.744}{2} = 0.372 \, mm$$

$$\cos \alpha = \frac{h_7}{y_7}$$

$$y_7 = \frac{0.372}{\cos 30.1°} = 0.429 \, mm$$

$$\sin \alpha = \frac{x_7}{y_7}$$

$$x_6 = 0.429 * \sin 30.1° = 0.215 \, mm$$

The corresponding iteration right angle triangles are shown in Fig. 3. Single Spidron dimensions are shown in Table 1.

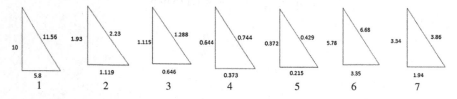

Fig. 3 Various iterations from 1–7 of the right angle triangle

Table 1 Measurements for all iterations

Iteration	Base (x_i in mm)	Height (h_i in mm)	Hypotenuse (y_i in mm)
Iteration-1	5.8	10	11.56
Iteration-2	3.35	5.78	6.68
Iteration-3	1.94	3.34	3.86
Iteration-4	1.119	1.93	2.23
Iteration-5	0.646	1.115	1.288
Iteration-6	0.373	0.644	0.744
Iteration-7	0.215	0.372	0.429

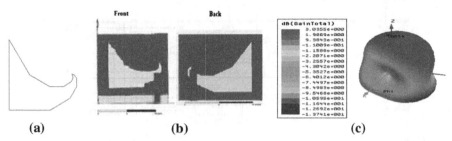

    (a)   (b)   (c)

Fig. 4 Simulated single Spidron antenna. **a** Shape. **b** Front and back. **c** 3D Gain polar plot

After combining all triangles that are corresponding to 7 iterations, we got Spidron shape. The simulation results of single Spidron antenna is shown in Fig. 4.

1.1.2 Design Specifications of Single Spidron Slot Antenna

Operating frequency (f_0) = 12.5 GHz
    Height of the substrate = 1.6 mm
    Dielectric constant (ε_r) = 4.4
    The substrate material used => FR-4 epoxy with double sided copper clad.
    Simulation results:
    According to simulated results, we got the gain of single Spidron slot antenna is 3 dB (Fig. 4).
    Single Spidron slot antenna offers less loss for the frequencies in the range from 6.1 to 7.5 GHz. So we can use this antenna for transmitting and reception of signals that are in its operating range. It offers bandwidth of 1312 MHz. Single Spidron slot antenna has low VSWR (1.42) at 6.75 GHz as shown in Fig. 5.

1.2 2 × 2 Sub Antenna Array and the Feeding Network

After the excellent assessment of the electrical resistance matching and radiation characteristics of one Spidron form slot antenna, our scope extended to the look of a

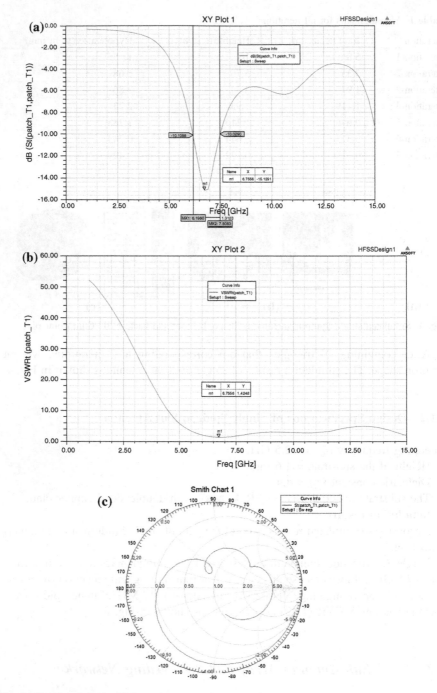

Fig. 5 Simulated single Spidron antenna. **a** Shape. **b** Front and back. **c** 3D Gain polar plot

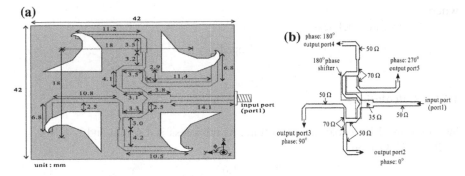

Fig. 6 2 × 2 subarray. **a** Top view. **b** Detailed dimensions of the feeding network

consecutive revolved CP array so as to boost each the antenna gain and information measure. During this section, configurations for a 2 × 2 subarray and 4 × 4 array that have each been integrated with parallel ordered feeding networks square measure conferred thoroughly.

Figure 6 shows the configuration of a 2 × 2 subarray that utilizes the Proposed Spidron fractal slots as its radiating element and a sequential feeding network on a 42 mm × 42 mm square substrate. The distance between each of the Spidron fractal slots along both the x- and y-axes is 18 mm, corresponding to 12.5 GHz. Detailed dimensions are given in Fig. 6a. To attain LHCP operation, the shape slots square measure consecutive turned dextrorotatory by 900. The feeding network here is intended to excite every squeeze flip with consecutive part delays of 900. To implement this network style, 3 T-junction dividers and resistance transformers square measure wont to match the impedances between the 50 Ω input port and four 50 Ω output ports, as shown in Fig. 6b. The middle section of the feeding network is AN equal power divider with 1800 part distinction. Every divided, anti-phased port is then connected to a different T-junction divider to get signals with 900 part distinction in 2 sets of output ports. Thus, the whole network consists of 4 output ports, every ordered part delayed by 900 within the dextrorotatory direction.

1.2.1 Design of a 2 × 2 Subarray Spidron

Lengths of parallel sequential feeding network are given in top view of the 2 × 2 subarray. The width of parallel sequential feeding network is given in terms of impedance (z). So we need to calculate the width in terms of 'mm'. General formula that is used to convert impedance into mm is given in Eq. 4.

$$w_z = \frac{30 * \pi * t_s}{z * \sqrt{\varepsilon_r}} \tag{4}$$

where
   t_s = Substrate thickness

$$z = \text{Impedance}$$

1.2.2 Calculations

We are using FR4-epoxy as a substrate. Its thickness is 1.6 mm and $\varepsilon_r = 4.4$.
If z = 35 Ω then

$$w_{35} = \frac{30 * \pi * 1.6}{35 * \sqrt{4.4}} = 2.054 \, \text{mm}$$

If z = 50 Ω then

$$w_{50} = \frac{30 * \pi * 1.6}{50 * \sqrt{4.4}} = 1.44 \, \text{mm}$$

If z = 70 Ω then

$$w_{70} = \frac{30 * \pi * 1.6}{70 * \sqrt{4.4}} = 1.0269 \, \text{mm}$$

.

1.2.3 Results

Simulation Results

2 × 2 sub-array simulation results are shown in Figs. 7 and 8.

According to simulated results, we got the gain of single Spidron slot antenna is 3.65 dB.

The 2 × 2 subarray Spidron slot antenna offers less loss for the frequencies in the range from 4.95 to 5.69 GHz. So we can use this antenna for transmitting and reception of signals that are in its operating range. It offers bandwidth of 741 MHz The gain of the 2 × 2 subarray is improved by 0.5 dB, but bandwidth is reduced by

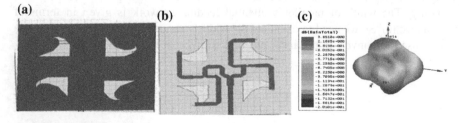

Fig. 7 Simulation of 2 × 2 subarray. **a** Top view. **b** Feeding network. **c** 3D-Gain plot

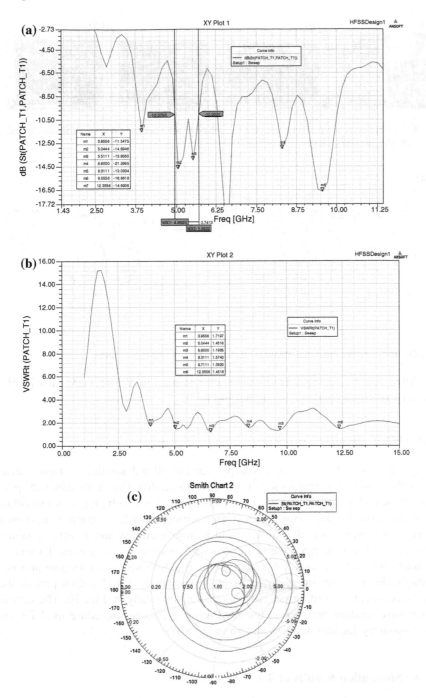

Fig. 8 Simulation of 2 × 2 sub array. **a** Return loss plot. **b** VSWR plot. **c** Smith chart

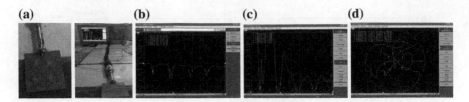

Fig. 9 Experimental setup and results of 2 × 2 subarray. **a** Experimental setup. **b** Return loss plot. **c** VSWR plot. **d** Smith chart

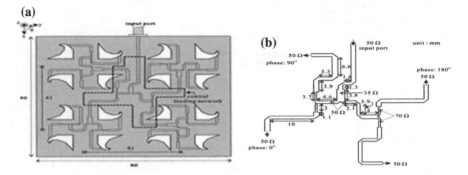

Fig. 10 Top view of the 4 × 4 subarray. **a** Top view. **b** Feeding network

571 MHz. A 2 × 2 subarray Spidron slot antenna has low VSWR (1.32) at 9.7 GHz. The Experimental results were shown in Fig. 9.

Experimental Results of 2 × 2

A high-gain, CP 4 × 4 array developed for broadband satellite communication within the Ku-band is represented during this sub-section. Figure 10 shows the pure mathematics of the 4 × 4 array that features four 2 × 2 sub-arrays as individual divergent parts. The sub-arrays ar organized throughout a 2 × 2 configuration on a sq. substrate having overall dimensions of 80 mm × 80 mm and a center-to-center distance of 41 mm between the sub-arrays on every the x- and y-axes. To service the four consecutive turned 2 × 2 sub-arrays, an extra consecutive feeding network is employed. For this, every subarray input is connected to 1 output port of the central consecutive feeding network, as shown intimately in Fig. 10. The central consecutive feeding network is intended using the same procedure used for the 2 × 2 subarray feeding network shown in Fig. 10b.

1.2.4 Simulation Results of 4 × 4

According to simulated results shown in Figs. 11 and 12, we got the gain of 4 × 4 array Spidron slot antenna is 5 dB. The 4 × 4 array Spidron slot antenna offers less

(a)   **(b)**

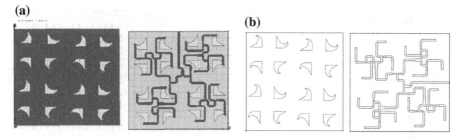

Fig. 11 4 × 4 array. **a** HFSS simulation. **b** CAD model

loss for the frequencies in the range from 12.17 to 14.98 GHz as shown in the Fig. 12. So we can use this antenna for transmitting and reception of signals which are in its operating range. It offers bandwidth of 2813 MHz Gain of 4 × 4 array is improved by 2 dB and bandwidth is increased by 1500 MHz 4 × 4 array Spidron slot antenna has low VSWR (1.08) at 8.46 GHz.

1.2.5 Experimental Results of 4 × 4 Array

The experimental setup and results of 4 × 4 array are shown in Fig. 13.

2 Performance of Spidron Grid Array

The 4 × 4 array has been modified to investigate its radiation characteristics. The 3-dB gain of the array is shown in Fig. 14. The performance characteristics of the Spidron grid array has been verified with the vector network analyzer E5071C and the results were presented in Figs. 15 and 16. It is observed that the grid array has exhibited multi-band rather wide band characteristics.

8.1 Practical results

The simulation and experimental results are compared

2.1 Theoretical Values

The simulation results were obtained using HFSS13.0 presented in Table 2.

2.2 Practical Values

Table 3 shows the comparison of arrays.

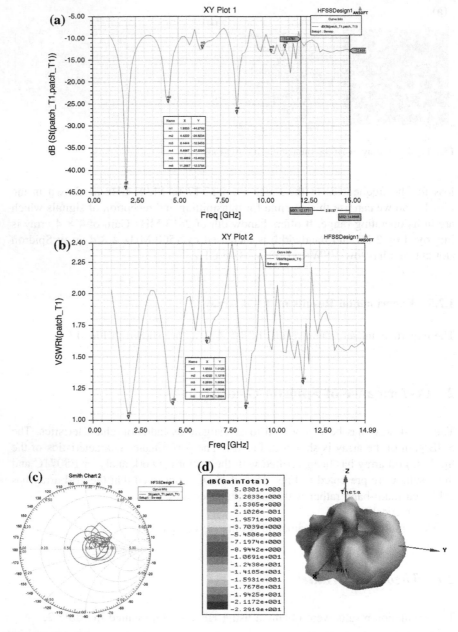

Fig. 12 Simulation results 4 × 4 array. **a** Return loss curve. **b** VSWR plot. **c** Smith chart. **d** 3D-Polar plot

(a)   (b)   (c)   (d)

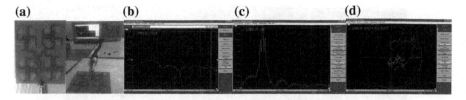

Fig. 13 Experimentation results 4 × 4 array. **a** Experimental setup. **b** Return loss curve. **c** VSWR plot. **d** Smith chart

Table 2 Theoretical values

	Gain (in dB)	Return losses (in dB)	Bandwidth (in MHz)	VSWR
Single Spidron	3.03	−15 dB at 6.7 GHz	1312	1.42
2 × 2 array	3.65	−21 dB at 6.6 GHz	741	1.18
4 × 4 array	5	−44 dB at 1.9 GHz	2813	1.01
4 × 4 grid array	6.4938	−51 dB at 10.4 GHz	9000	1.7

Table 3 Practical values

	Return losses (in dB)	Bandwidth (in MHz)	VSWR
2 × 2 array	−27 dB at 4.65 GHz	300	1.6
4 × 4 array	−31 dB at 10 GHz	4000	1.1
4 × 4 grid	−30 dB at 10.4 GHz	8800	1.771

Fig. 14 Gain plot of 4 × 4 grid array

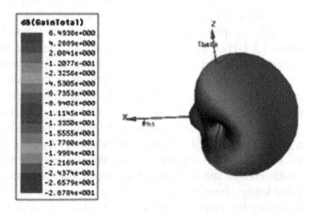

Fig. 15 Experimentation results 4 × 4 grid array

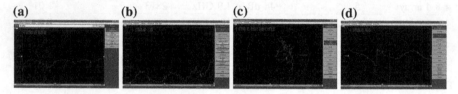

Fig. 16 Experimentation results Spidron 4 × 4 grid array. **a** Return loss. **b** VSWR plot. **c** Smith chart. **d** Phase plot

3 Conclusions

In this work a UWB Spidron slot antenna and the array is considered. A 4 × 4 array provides good characteristics comparatively with single Spidron and 2 × 2 subarray.

Theoretically the fourth iterative wideband circular fractal antenna offers gain of 4.33 dB and offers the wide band ranging from 0.8 to 6.5 GHz, i.e., bandwidth is about 7.3 GHz, but practically this antenna operates from 0.782 to 5.76 GHz and cover the L, S and some range of C-bands in microwave frequency spectrum.

Theoretically a 4 × 4 array antenna offers the gain of is 5 dB and wide band ranging from 12.17 to 14.98 GHz. Its bandwidth is 2813 MHz. practically a 4 × 4 array antenna operates from 9.9 to 14 GHz. The results are indicating that the as the number of elements are increasing the gain as well as bandwidth is increasing from a single element to 4 × 4 array. In the case of Spidron grid array, the improvement in bandwidth not significant and the designed array has exhibited multiband characteristics and is suitable for x-band. The experimental gain can be studied by considering alternate elements in the array with the help of a chamber. Further designing the antenna array can be studied for more compact and multiband features. Using different material the array performance evaluation can be made.

References

1. Pozar, D.M.: Microstrip antennas. Proc. IEEE **80**(1), 79–91 (1992)
2. Asimakis, N.P., Karanasiou, I.S., Uzunoglu, N.K.: Non-invasive microwave radiometric system for intracranial applications: a study using the conformal L-notch microstrip patch antenna. Prog. Electromagn. Res. **117**, 83–101 (2011)
3. Tiang, J.J., Islam, M.T., Misran, N., Mandeep, J.S.: Circular microstrip slot antenna for dual-frequency RFID application. Progr. Electromagn. Res. **120**, 499–512 (2011)
4. Monavar, F.M., Komjani, N.: Bandwidth enhancement of microstrip patch antenna using Jerusalem cross-shaped frequency selective surfaces by invasive weed optimization approach. Progr. Electromagn. Res. **121**, 103–120 (2011)
5. Gujral, M., Li, J.L.-W., Yuan, T., Qiu, C.-W.: Bandwidth improvement of microstrip antenna array using dummy EBG pattern on the feedline. Prog. Electromagn. Res. **127**, 79–92 (2012)
6. Moradi, K., Nikmehr, S.: A dual-band dual-polarized microstrip array antenna for base stations. Progr. Electromagn. Res. **123**, 527–541 (2012)
7. Wei, K.P., Zhang, Z.J., Feng, Z.H.: Design of a dualband omnidirectional planar microstrip antenna array. Progr. Electromagn. Res. **126**, 101–120 (2012)
8. Garcia-Aguilar, A., Inclan-Alonso, J.-M., Vigil-Herrero, L., Fernandez-Gonzalez, J.M., Sierra-Perez, M.: Low-profile dual circularly polarized antenna array for satellite communications in the X band. IEEE Trans. Antennas Propag. **60**(5), 2276–2284 (2012)
9. Wang, X., Zhang, M., Wang, S.-J.: Practicability analysis and application of PBG structures on cylindrical conformal microstrip antenna and array. Progr. Electromagn. Res. **115**, 495–507 (2011)
10. Lau, P.-Y., Yung, K.K.-O., Yung, E.K.-N.: A low-cost printed CP patch antenna for RFID smart bookshelf in the library. IEEE Trans. Ind. Electron. **57**(5), 1583–1589 (2010)
11. Wang, P., Wen, G., Li, J., Huang, Y., Yang, L., Zhang, Q.: Wideband circularly polarized UHF RFID reader antenna with high gain and wide axial ratio beamwidths. Progr. Electromagn. Res. **129**, 365–385 (2012)
12. Rao, P.N., Sarma, N.V.S.N.: Fractal boundary circularly polarised single feed microstrip antenna. Electron. Lett. **44**(12), 713–714 (2008)
13. Dong, Y., Toyao, H., Itoh, T.: Compact circularly-polarized patch antenna loaded with metamaterial structures. IEEE Trans. Antennas Propag. **59**(11), 4329–4333 (2011)
14. Tang, X., Wong, H., Long, Y., Xue, Q., Lau, K.L.: Circularly polarized shorted patch antenna on the high permittivity substrate with wideband. IEEE Trans. Antennas Propag. **60**(3), 1588–1592 (2012)
15. Rezaeieh, S.A., Kartal, M.: A new triple band circularly polarized square slot antenna design with crooked T and F-shape strips for wireless applications. Progr. Electromagn. Res. **121**, 1–18 (2011)
16. Sze, J.-Y., Pan, S.-P.: Design of broadband circularly polarized square slot antenna with a compact size. Progr. Electromagn. Res. **120**, 513–533 (2011)
17. Zarifi, D., Oraizi, H., Soleimani, M.: Improved performance of circularly polarized antenna using semi-planar chiral metamaterial covers. Progr. Electromagn. Res. **123**, 337–354 (2012)
18. Nakano, H., Nakayama, K., Mimaki, H., Yamauchi, J., Hirose, K.: Single-arm spiral slot antenna fed by a triplate transmission line. Electron. Lett. **28**(22), 2088–2090 (1992)
19. Hall, P.S.: Application of sequential feeding to wide bandwidth, circularly polarised microstrip patch arrays. Proc. Inst. Elect. Eng. Microw., Antennas Propag. **136**(5), 390–398 (1989)
20. Evans, H., Gale, P., Sambell, A.: Performance of 4 × 4 sequentially rotated patch antenna array using series feed. Electron. Lett. **39**(6), 493–494 (2003)
21. Soliman, E.A., Brebels, S., Beyne, E., Vandenbosch, G.A.E.: Sequential-rotation arrays of circularly polarized aperture antennas in the MCM-D technology. Microw. Opt. Technol. Lett. **44**(6), 581–585 (2005)

22. Kaffash, S., Kamyab, M.: A sequentially rotated RHCP stacked patch antenna array for INMARSAT-M land applications. In: Proceedings 6th European Conference Antennas Propagation (EUCAP), pp. 1–4. Prague, Czech Republic (2012)
23. Hwang, K.C.: Broadband circularly-polarised Spidron fractal slot antenna. Electron. Lett. **45** (1), 3–4 (2009)

A Fuzzy Logic Based Finite Element Analysis for Structural Design of a 6 Axis Industrial Robot

S. Sahu, B.B. Choudhury, M.K. Muni and B.B. Biswal

Abstract Six axis industrial robots are widely used for carrying out various operations in industry and in the process it is subjected to varying payload conditions. This paper shows a methodology to determine the optimum value of pay load vis-à-vis the design parameters considering the criteria of reducing the material used to build the structure of industrial robot based on the finite element method (FEM). Different loads are applied at gripper and the total deformation is calculated. Finally, the weak area of the robot arm is found out and relative improving suggestions are put forward, which leads to the foundation for the optimized design. In the fuzzy-based method, the weight of each criterion and the rating of each alternative are described by using different membership functions and linguistic terms. By using this technique, the deformation is determined in accordance to load to the robotic gripper. The results of the analysis are presented and it is found that the triangular membership function is the effective one for deformation measurement as its surface plot shows a good agreement with the output result.

Keywords Industrial robot · Fuzzy logic · Finite element

S. Sahu (✉) · B.B. Choudhury · M.K. Muni
Department of Mechanical Engineering, I.G.I.T Sarang, Dhenkanal, Odisha, India
e-mail: supriyaigit24@gmail.com

B.B. Choudhury
e-mail: bbcigit@gmail.com

M.K. Muni
e-mail: manoj1986nitr@gmail.com

B.B. Biswal
Department of Industrial Design, NIT Rourkela, Odisha, India
e-mail: bbbiswal@nitrkl.ac.in

© Springer India 2016
H.S. Behera and D.P. Mohapatra (eds.), *Computational Intelligence in Data Mining—Volume 2*, Advances in Intelligent Systems and Computing 411, DOI 10.1007/978-81-322-2731-1_18

1 Introduction

Robots are the mechanical devices that are controlled by programming. An Aristo robot is a 6 axis articulated robot powered by actuators and can be used to lift parts with great accuracy. These are often used for tasks such as welding, painting, and assembly. Finite element techniques of analysis and simulation of mechanical systems are used to build mathematical models and to analyze the static and dynamic behavior of the structural elements directly on the computer. It is required to correlate with the kinematic model of the joints and it is used to establish the loads and to build a dynamic model to determine the behavior. FEM tools are used for modeling and analysis of the structure of the robot arm.

Hardeman et al. [1] have derived the dynamic equations of motion using finite element method suitable for both simulation and identification. Huang et al. [2] have conducted analog simulation experiment using finite element analysis and established FE model of flexible tactile sensor sensitive unit. Ghiorghe [3] showed a methodology to determine the optimum values for the design parameters considering the criteria of reducing the material used to build the structure of industrial robot, using an structural optimization and topology algorithm. Ristea [4] showed the difference of a composite substance and the conventional aluminum for the design of the robot elements used. Doukas et al. [5] investigated the structural behavior of industrial robots and developed a model, to predict the correctness of robot in definite positions of arm with loading environment by use of Finite Element Method (FEM). Gasparetto et al. [6] considered an useful method for modeling spatial industrial robots of low weight based on the same Rigid Link System approach from an experimental corroboration outlook. Fahmy et al. [7] proposed a neuro fuzzy controller for robotic manipulators by using a learning method to produce the essential inverse modeling rules from input and output values taken and have got the best results compared with conventional PID controller. Chen et al. [8] used a six-axis force/torque sensor as a component for the large manipulator in the space station. In order to obtain the large measurement range of force/torques, an elastic body based on cross-beam with anti-overloading capability is designed, and the size is optimized by using FEA.

2 The Finite Element Model

2.1 The Finite Element Model of the Robot Arm

The Industrial robot consists of subassemblies of motion which are initially developed and assemble all the components as shown in Fig. 1. FEA is a method used to solve engineering problems. The first step of FEM is to build of a model for the complete structure of the robot. The model of the industrial robot was imported from CATIA into the environment of finite element for analysis.

Fig. 1 6 axis industrial robot
3D-model

Although ANSYS Workbench is a very useful platform in the finite element calculation but its function in the aspect of modeling is not strong. Therefore, our objective is to use solid modeling using CATIA V5 software to establish a three-dimensional entity model of robot arm. The data exchange of the model could be imported into ANSYS Workbench through software interface. Before the model was analyzed in ANSYS Workbench, the materials needed used for the structure of the robot arm are chosen. The "Engineer data" module of the ANSYS Workbench was used to select structural steel for the purpose. As the robot is a six degrees- of— freedom structure, it required the fixed constraints on the base surface. The mesh size was taken as 0.01. The FEA analysis is presented in Fig. 2.

For the FEM analysis of the robot structure, we considered gripper loads of 1.25–125 N as input data. The structure is tested under static conditions in order to obtain the total deformations.

2.2 Fuzzy Logic System

This research mainly focuses to predict closeness using the fuzzy methodologies. A fuzzy system is an alternative to classical ideas of set membership and logic and applications at the prominent edge of Artificial Intelligence. This research work emphasizes the foundations of fuzzy systems along with some of the new notable demurrals to its use, with examples haggard from existing research in the field of Artificial Intelligence. Finally, it is established that the practice of fuzzy systems creates a feasible addition to the field of Artificial Intelligence. Here Fuzzy Logic is used for training of the datasets of the data table for the prediction of failure and its

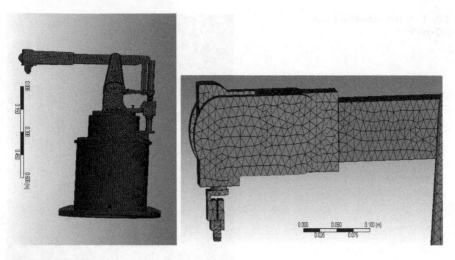

Fig. 2 Finite element mesh of an industrial robot

severity. Basically, the fuzzy logic provides a corollary organization that enables suitable human thinking abilities that machines do not have. It uses fuzzy rules, which is the illustration of knowledge frequently presented with rules comprising "if-then" announcements or different cases having different reality arrangements.

2.3 Fuzzy Controller/Fuzzy IF-THEN Rule

When fuzzy set theory and fuzzy logic are combined with each other and develop fuzzy logic controllers. The mathematical model is controlled by rules of fuzzy logic as an alternative of equations.

The rules are generally in the form of IF-THEN-ELSE statements. The IF part is is known as antecedent and THEN part is consequent. The antecedent is linked with Boolean operator like AND, OR, NOT etc. The final output from the model are converted to a appropriate form by block of fuzzification. All the rules had clear based on applications and the process begins with computation of the rule consequences. The rule consequences take place within the computation unit. Lastly the set of fuzzy is defuzzified to one crisp control action using defuzzification module.

In this research the fuzzy controller has been designed using one type of memberships, i.e. triangular (Fig. 3) membership function. Figures 4 and 5 show the input and output variables used in Mamdani FIS with Triangular membership functions.

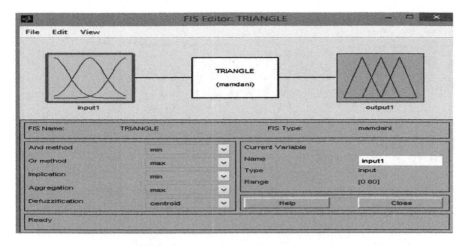

Fig. 3 Fuzzy controller using triangular membership functions

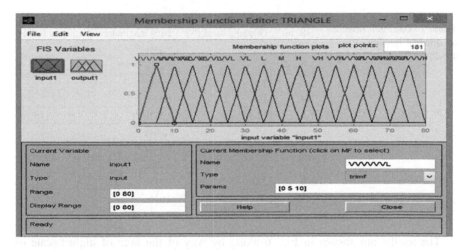

Fig. 4 Input variables of Mamdani FIS with triangular membership function

3 Results and Interpretations

The results obtained in terms of maximum deformation for ten loads out of fifty loads applied are presented in Table 1. The static analysis comprises an assessment of the total deformation for life and damage and safety factor. Figure 6 displays some results of deformations corresponding to the four kinds of typical gripper loads from fifty loads applied. The lowest value of the deformation was found at the bottom of the part (Dark blue colour in Fig. 6) while the maximum is at the top of it near the gipper (Red colour in Fig. 6) and the dialog box contains explicit values.

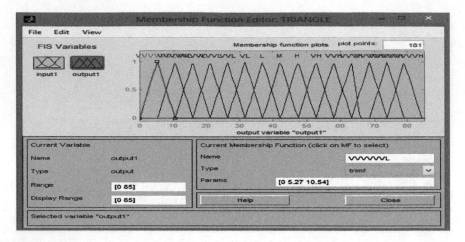

Fig. 5 Output variables of Mamdani FIS with triangular membership function for closeness of the robot

S. no.	Gripping loads (N)	Total maximum deformation (mm)
1	2.5	$2.635e^{-6}$
2	5	$5.270e^{-6}$
3	7.5	$7.90e^{-6}$
4	10	$1.054e^{-5}$
5	12.5	$1.317e^{-5}$
6	15	$1.581e^{-5}$
7	17.5	$1.844e^{-5}$
8	20	$2.106e^{-5}$
9	22.5	$2.317e^{-5}$
10	25	$2.635e^{-5}$

Table 1 Total deformations on different loading conditions

The results are shown in Fig. 6 along by way of the area of higher scale of deformation with different colors. Light blue and dark blue colours indicate low deformations, while the yellow to red indicate high deformations.

From the observation it is found that as the load increases the deformation increases. Since up to force value of 2.5 N there is no red colour found, at indicates that up to 2.5 N is safe load that can be applied to the Robot. From 5 N onwards there is gradual increase in colour red and it is maximum for the force value of 125 N. Figure 7 shows the rule viewer in Mamdani FIS (MAT Lab) with Triangular membership function. It represents the comparison of closeness and capacity for different membership functions. Surface plots of different membership function shown in Fig. 8.

The fuzzy controllers are developed here to predict the deformation of the Industrial Robot. The predicted result from fuzzy controllers for deformation of

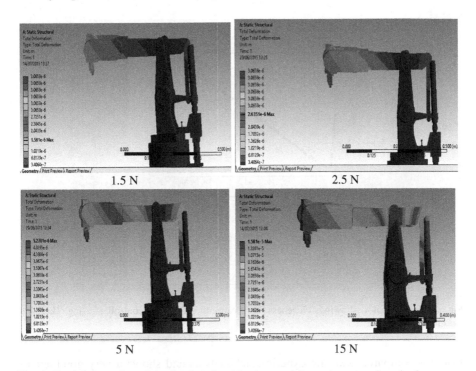

Fig. 6 Total deformation of the 6 axis industrial robot model structure at various loads

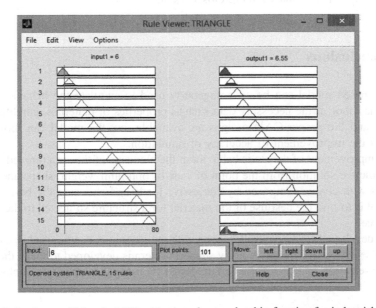

Fig. 7 Rule viewer of Mamdani FIS with triangular membership function for industrial robot

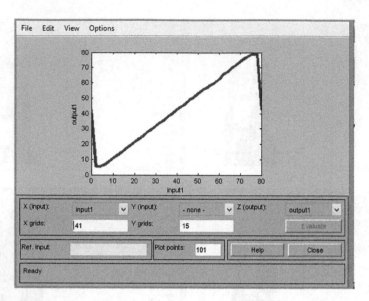

Fig. 8 Surface plot showing closeness and capacity of fuzzy controller using Gaussian membership function

Robot is compared with the experimental analysis and shows a very good agreement. It has been observed that the result of Fuzzy controller using Triangular membership function shows more closer result.

4 Conclusions

The FE based model can be advantageously used to simulate the behavior of an industrial robot, if the necessary data such as properties of materials, conditions of loading and the proper CAD designs are available, but the model can further be used for the improvement of accuracy of simulation.

An improvement of the static behavior of the elements of the structure led to find a constructive solution from the point of view of optimum weight, shape and static stiffness with changes imposed by successive loads and choosing the best option. The goal is to find the best use of the material for a structure subjected to the action of a force.

The developed fuzzy controller along with the technique can be used as a robust tool for selection of Robots. A fuzzy support system is developed to make the most suitable selection of Robot for a certain design by using the capacity and closeness value.

References

1. Hardeman, T., Aarts, R., Jonke, B.: Modelling and identification of robots with joint and drive flexibilities. In: IUTAM Symposium on Vibration control of Nonlinear Mechanisms and structures, pp. 173–182 (2005)
2. Huang, Y., Huang, P., Wang, M., Wang, P., Ge, Y.: The finite element analysis based on ANSYS perssure-sensitive conductive rubber three-dimensional tactile sensor. CCIS **2**, 1195–1202 (2007)
3. Ghiorghe, A.: Optimization design for the structure of an RRR type industrial robot, U.P.b. Sci. Bull D **72**(4) (2010)
4. Ristea, Alin: FEA analysis for frequency behavior of industial robot's mechanical elements, pp. 26–28. Romania, Sinaia (2011)
5. Doukasa, C., Pandremenosa, J., Stavropoulosa, P., Foteinopoulosa, P., Chryssolourisa, G.: On an empirical investigation of the structural behavior of robots. Procedia CIRP **3**, 501–506 (2012)
6. Gasparettol, A., Moosavil, A.K., Boscarioll, P., Giovagnon, M.: Experimental validation of a dynamic model for lightweight robot. Int. J. Adv. Robot. Sy **10**(182), 201 (2013)
7. Fahmy, A.A., Abdel Ghany, A.M.: Neuro-fuzzy inverse model control structure of robotic manipulators utilized for physiotherapy applications. Ain Shams Eng. J. **4**, 805–829 (2013)
8. Chen, D., Song, A., Li, A.: Design and calibration of a six-axis force/torque sensor with large measurement range used for the space manipulator. Procedia Eng. **99**, 1164–1170 (2015)

The Design and Development of Controller Software for ScanDF Receiver Subsystem (Simulator) Using Embedded RTOS

P. Srividya and B. I Neelgar

Abstract In modern receiver control applications achieving real time response is of main importance. The paper is a part of software development for CSM controller system. In this paper a real time application, for controlling SCAN DF receiver simulator will be implemented using POSIX socket APIs on VxWorks RTOS and process the data from the scanDF receiver subsystem. The processed data can be used to develop a database for electronic order of battle. This application will be able to interact with receiver using socket APIs, to exchange command and data packets. TCP/IP based socket APIs will be used to realize the application.

Keywords Electronic warfare (EW) · ScandF receiver · TCP/IP sockets · Client · Server · VxWorks RTOS · Powerpc

1 Introduction

This paper discusses about the development of a real time controller for ScanDF receiver subsystem. The operations of the ScanDF receiver subsystem include the interception of the signal and report the frequency, direction of arrival, amplitude, time of interception and signal bandwidth between the frequency ranges of 20–1000 MHz. The simulator is responsible for the acceptance of the commands from the user and displays the result. To perform such operations a real time application

Technologies Used—VxWorks RTOS, Development IDE Wind River Workbench 2.5, Programming Language 'C' on LINUX Operating System, ChampAV IV Board.

P. Srividya (✉) · B. I Neelgar
GMR Institute of Technology, Razam, India
e-mail: Srividya.pidaparthi@gmail.com

B. I Neelgar
e-mail: hod_ece@gmrit.org

© Springer India 2016
H.S. Behera and D.P. Mohapatra (eds.), *Computational Intelligence in Data Mining—Volume 2*, Advances in Intelligent Systems and Computing 411, DOI 10.1007/978-81-322-2731-1_19

like VxWorks is essential. VxWorks is a hard Real Time Operating System. This application is implemented using POSIX Socket APIs that are used to create a Client-Server environment. This application will interact with operator and should be able to send and receive the commands and data packets.

Four main functional units of this subsystem are operator interface unit, VxWorks RTOS, ScanDF receiver subsystem and CHAMPAV IV Board. The operator unit, the embedded system with VxWorks RTOS based ChampAV IV Hardware and ScanDF Subsystem units are connected through LAN where the data transfer is done using the client server model.

2 Electronic Warfare

The swift expansion of modern electronically controlled, directed, commanded weapons in military science field is called Electronic Warfare (EW) [1]. The actual idea of EW is to utilize the enemy's electromagnetic emissions in all parts of the electromagnetic spectrum in order to provide intelligence on intentions and capabilities, and to use counter measures to contradict effective use of communication and weapon systems while defending own electromagnetic spectrum. The new idea is that EW is an important part of an overall military strategy which concentrates on neutralization of an enemy's Command, Control and Communication system while maintaining own system.

EW is structured into the three major categories: Electronic Support Measures (ESM), Electronic Counter Measure (ECM), and Electronic Counter Counter Measures (ECCM). ESM involves actions taken to search for, intercept, locate, and instantly identify the radiated electromagnetic energy for the intention of immediate threat recognition and the strategic service of forces. [1] ECM is set of actions taken to avert or lessen the enemy's effective use of the electromagnetic spectrum. ECCM combines both ESM and ECM technologies [1].

3 TCP/IP Protocol Based Socket Programming

A protocol is a common language that the server system and the client systems both understand. TCP is fine for transporting data across a network. Socket is a method of communication between computers (clients and server) using standard UNIX file descriptors [2]. A Socket is used in client server application frameworks. Client Process is the process which usually makes a request for information. After getting the response this process may conclude or may do some other processing. Server Process is the process which obtains a request from the clients. After getting a request from the client, this server process will do essential processing, congregate the information and will send it to the requestor client [3].

Client process involves the following steps: [3]

- Creation of socket using socket () call.
- Connecting the socket to the address of the server using connect () call.
- write () and read () the data to the socket and from the socket.

Server process involves the following steps: [3]

- Creation of socket using socket () call.
- Bind the socket to an address using the bind () call.
- Listen for connections and Accept a connection with the listen () and accept () system call.
- read () and Write () the data from the socket and to the socket.

4 Actual System

The actual system used for Commanding and controlling in DLRL is Communication Support Measure (CSM) Controller subsystem (Fig. 1). The CSM System functional block diagram is shown in Fig. 2. The ScanDF Subsystem (SDF), the Monitoring receiver and Analysis Subsystem (MAS), wideband Surveillance Subsystem (WS) and CSM Controller are on an internal bus realized through LAN. Subsystem are controlled and coordinated by the CSM Controller. Intelligence received from these subsystems are processed and formatted into data files and sent to further echelons.

5 Functional Blocks for ScanDF System

Through the operator interface/ command control system the operator will give the command to the system. The operator interface and ChampAV IV are connected through LAN and works on Client Server model where the image of VxWorks is

Fig. 1 Actual block diagram

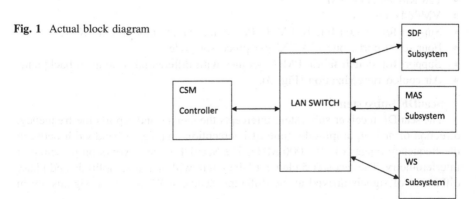

Fig. 2 Functional blocks of
ScanDF system

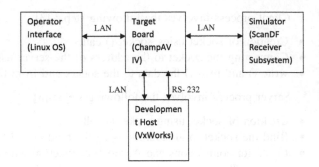

dumped onto ChampAV IV board. The board is connected to the Simulator and the Client-Server model is again implemented. Here VxWorks RTOS is used to display the results that are received from the board. VxWorks RTOS and Champ AV IV board are connected through LAN and RS-232 cable. The simulated results from the Simulator are displayed on Operator interface through VxWorks.

Operator Interface System:

The Operator interface system or command control system based on Linux OS is used to give the commands to the Simulator. The response to the command will be again displayed on the command controller. Operator interface unit is a system through which the operator gives command.

Host System:

VxWorks RTOS [4] is the host system. The host is a Windows XP PC with Wind River 2.5 installed on it.

Target System:

Target is CHAMP AV 4 [5] POWERPc Board with

- Quad PowerPC 7447A CPUs operating at 1.0 GHz
  64 Kbyte L1 and 512 Kbyte (7447A) L2 internal caches operating at core processor speed
- 256 Mbytes DDR-25O SDRAM with ECC per processor (1 GByte total)
- 256 Mbytes Flash memory with write protection jumper
- 128 Kbytes NVRAM
- VME64x interface
- Support for two 64-bit, 100 MHz PCI-X mezzanine modules (PMC-X)
- Four serial ports, one EIA-232 per processor node
- Support for switch fabric PMC modules with differential routing to backplane
- Air-cooled ruggedization (Fig. 3).

ScanDF Subsystem:

The scanDF receiver subsystem intercepts the signals and reports the frequency, direction of arrival, amplitude, time of interception and signal bandwidth between the frequency ranges of 20–1000 MHz. The ScanDF system works on principle of interferometry. The receiver 5 element DF system will find the amplitude and phase difference of signals arrived at the different receivers. When all the signals are in

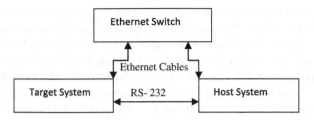

Fig. 3 Host target connectivity

inphase then the interception will starts. The receiver data is send to VxWorks and then to command control system.

6 Application Level Algorithm

- Set the configuration parameters of ScanDF subsystem as per IRS (Interface Requirement System).
- The target system (embedded) will forward the commands to the simulator/ScanDF subsystem.
- The ScanDF subsystem/Simulator will send the intercepted data to the target system.
- The target system will forward the data to command control/ operator interface system for the display of received data.

6.1 Working of Subsystems

Once the client server model is implemented between the systems, the user has to configure the parameters as attenuations (rf and if) of signal, threshold level of signal, integration factor for the system. Attenuation of signal is to a reduce strength during transmission to a particular level, for example, if rf attenuation is to set then the user has to select the attenuation level in between 0db and 30db so that the signal will be attenuated. The threshold level is the point that must be exceeded to begin producing the interception; here the threshold level must be in between -120 Hz and 40 Hz. The integration factor is to define the number of times the interception is needed. After the configuration of parameters is set, the user from client (operator interface) has to fix the resolution bandwidth so that the simulator will check the frequencies of signals with a difference of resolution bandwidth. When the resolution bandwidth is also set then the client will pass the start analysis command to the embedded system which acts as both server and client for the subsystem. Here VxWorks is used to prioritize the tasks between monitoring of signal and Scanning the scenario. The image that is developed in VxWorks will be dumped on to ChampAV IV board. The image for ScanDF subsystem will be sent to one of the processors in quad POWERPc board through which the commands from client (operator interface) will be sent to ScanDF simulator (server).

The ScanDF system will continuously scan the scenario. The ScanDF subsystem will do both the scanning of the signal and finding the direction of arrival of the signals at a same time for a given frequency. The embedded system will filter and process the data for the intercepting needs. For example it will do band filtering and AOI filtering. AOI filtering means it filters the data in an area by intercepting the area in sectors. Until the stop analysis command is given by the user it will scan the scenario and process the data packets; if analysis by user is without DF (Direction Finding) mode then only the amplitude packet data that is about the amplitude packet frequency, Power level, time of interception and noise level of the signal will be sent by the simulator through embedded system, If the analysis selected by the user is with DF mode then the simulator will send the data about direction packet data that is about the direction packet frequency, power level, noise level, direction of arrival and the time of interception of the signal.

7 Results

The screenshots as in Figs. 4, 5, 6 and 7 indicate the results of the data that is intercepted by the ScanDF Subsystem. The data from Subsystem is sent to the embedded system (target) for further process and it will be forwarded to operator interface for display and for the use of information at the higher echelons. After the client (Command controller) got the acceptance from the Embedded Server the setting of parameters is displayed on the screenshot. When the start analysis command is given by the operator the simulated result from the server (Simulator) will be displayed on the screen of Command control are shown in Fig. 4. After the Client is accepted by the server (Simulator), when the start analysis command is read by the server the simulation will starts and the data is send to client, the process is shown in Fig. 5. Figure 4 represents the display on client process; here mode value is given as 1 which means only Scan mode. Figure 6 represents the display on client process; here mode value is given as 2 which means ScanDF mode. The Fig. 8 points to the outputs displayed on the screen of VxWorks RTOS after the VxWorks image is dumped and burned onto ChampAV IV board.

The screenshots as in Figs. 4 and 6 indicate the results of the data that is intercepted by the ScanDF Subsystem. The data from Subsystem is send to the embedded system (target) for further process; it will be forwarded to operator interface for display and for the use of information at the higher echelons. After the client (Command controller) get acceptance from the Embedded Server the setting of parameters is displayed on the screenshots when the start analysis command is given by the operator the simulated result from the server (Simulator) will be displayed on the screen of Command control are shown in Figs. 4 and 6 as per the

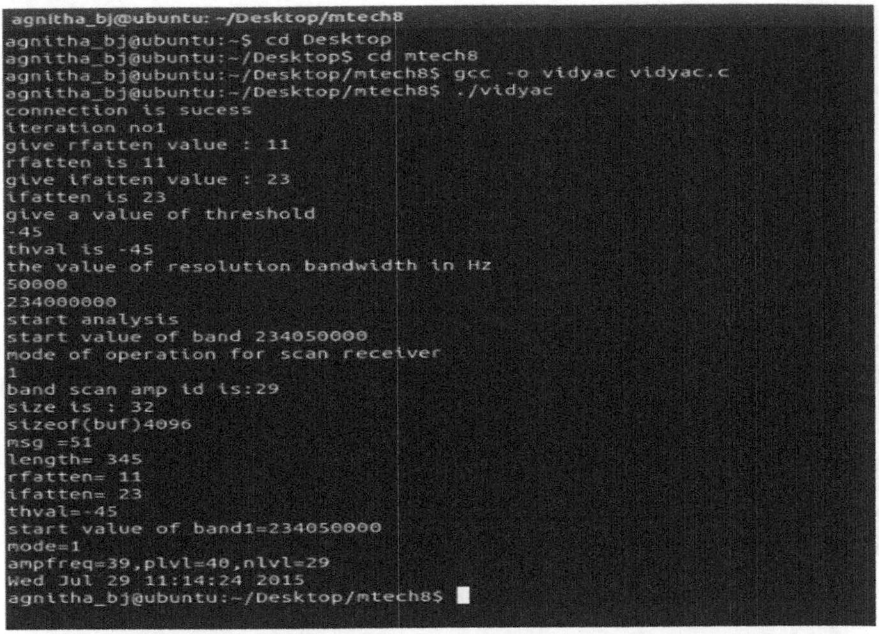

```
agnitha_bj@ubuntu: ~/Desktop/mtech8
agnitha_bj@ubuntu:~$ cd Desktop
agnitha_bj@ubuntu:~/Desktop$ cd mtech8
agnitha_bj@ubuntu:~/Desktop/mtech8$ gcc -o vidyac vidyac.c
agnitha_bj@ubuntu:~/Desktop/mtech8$ ./vidyac
connection is sucess
iteration no1
give rfatten value : 11
rfatten is 11
give ifatten value : 23
ifatten is 23
give a value of threshold
-45
thval is -45
the value of resolution bandwidth in Hz
50000
234000000
start analysis
start value of band 234050000
mode of operation for scan receiver
1
band scan amp id is:29
size is : 32
sizeof(buf)4096
msg =51
length= 345
rfatten= 11
ifatten= 23
thval=-45
start value of band1=234050000
mode=1
ampfreq=39,plvl=40,nlvl=29
Wed Jul 29 11:14:24 2015
agnitha_bj@ubuntu:~/Desktop/mtech8$
```

Fig. 4 Input command and outputs on command controller

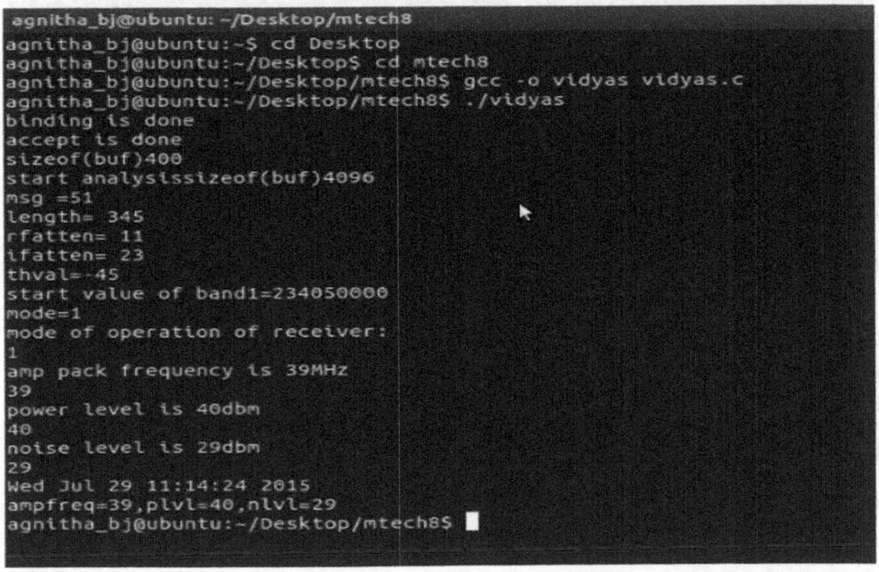

```
agnitha_bj@ubuntu: ~/Desktop/mtech8
agnitha_bj@ubuntu:~$ cd Desktop
agnitha_bj@ubuntu:~/Desktop$ cd mtech8
agnitha_bj@ubuntu:~/Desktop/mtech8$ gcc -o vidyas vidyas.c
agnitha_bj@ubuntu:~/Desktop/mtech8$ ./vidyas
binding is done
accept is done
sizeof(buf)400
start analysissizeof(buf)4096
msg =51
length= 345
rfatten= 11
ifatten= 23
thval=-45
start value of band1=234050000
mode=1
mode of operation of receiver:
1
amp pack frequency is 39MHz
39
power level is 40dbm
40
noise level is 29dbm
29
Wed Jul 29 11:14:24 2015
ampfreq=39,plvl=40,nlvl=29
agnitha_bj@ubuntu:~/Desktop/mtech8$
```

Fig. 5 Simulation of ScanDF system for the given scan command

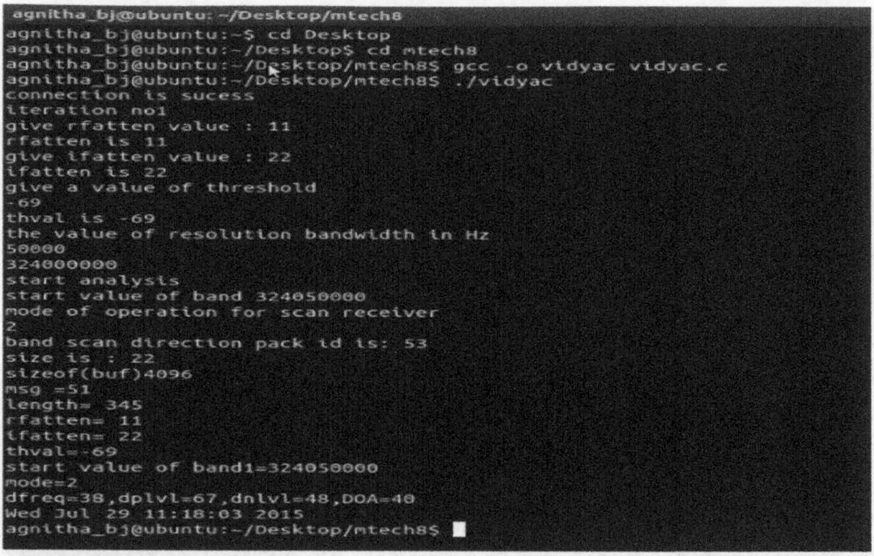

Fig. 6 Input command and outputs on command controller

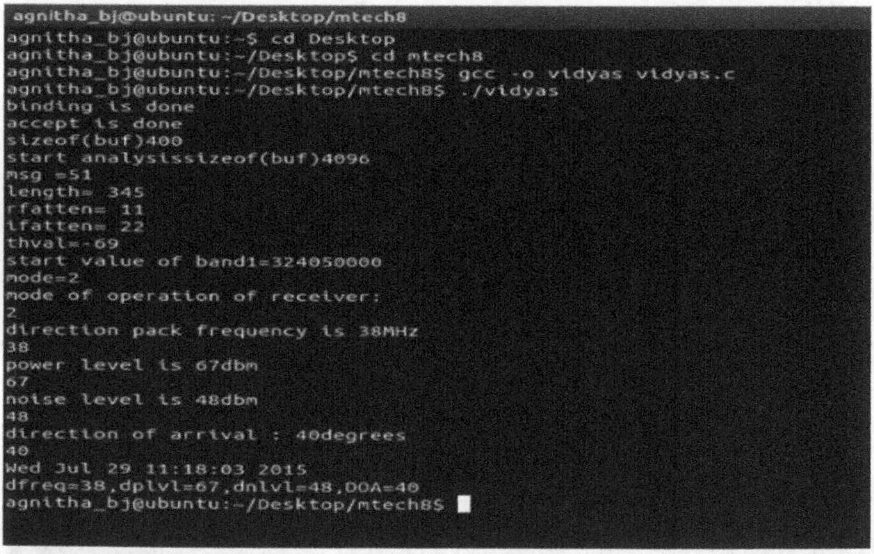

Fig. 7 Simulation of system for the given ScanDF command

given mode of operation. After the Client got acceptance by the server (Simulator), when the start analysis command is read by the server the simulation will starts and the data is send to client, the process is shown in Figs. 5 and 7.

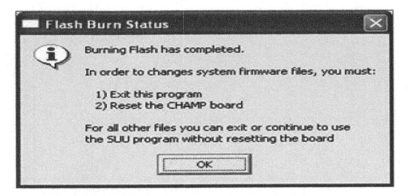

Fig. 8 The output displayed on VxWorks after burning the image on ChampAV IV

8 Conclusion

The paper is a part of software development for CSM controller system. The main functionality of the CSM software is to command and control various receiver subsystems. The software acts as server for various subsystems and provides necessary data for display. In The basic concept of electronic warfare is to exploit, intercept, reduce or prevent hostile use of electromagnetic spectrum and action which retains friendly use of electromagnetic spectrum. The threat warning function is many times coupled with a defensive capability in the form of a self protection jammer in combination with decoys which can divert weapons from the defended target. It is to keep track of the enemy even in the peacetime because unfortunately some very important signals may be passed this might lead to war. In modern receiver control applications achieving real time response is of prime importance. In this project a real time application for controlling SCAN receiver, will be implemented using POSIX socket APIs on VxWorks RTOS. This application must be able to interact with receiver using socket APIs, to exchange command and data packets. TCP/IP based socket APIs will be used to realize the application.

References

1. Curtis Schleher, D.: Introduction to electronic warfare, 1 Dec 1986
2. Stevens, R.: UNIX network programming, 3rd edn., vol. 1
3. Gaughan, K.: Client-server programming with TCP/IP sockets, March 22 (2003)
4. Alameda: Windriver systems Inc, Tornado User's guide, CA: windriver systems Inc. (1999)
5. Compact CHAMP-AV IV QUAD POWERPCTM (SCP-424) user manual
6. Poisel, R.: Introduction to communication electronic warfare systems, 1st Feb 2002

Problem Solving Process for Euler Method by Using Object Oriented Design Method

P. Rajarajeswari, A. Ramamohanreddy and D. Vasumathi

Abstract Mathematical techniques are easily understood by using Numerical analysis. This type of analysis is used for solving complex problems to do complicated calculations with the help of Computers. Numerical methods are used for finding the approximate solution to solve the mathematical problems. Modular design process increases debugging and testing process of a program. Additional tasks are performed by using new modules. In this paper object oriented design method is applied for Euler method.

Keywords Problem solving process · Euler method · Object oriented design method · Numerical method

1 Introduction

In the current year's numerical methods plays an important role for solving the problems in an engineering discipline. In this method Solutions are found for solving the problems in a direct manner. Analytical solutions are suitable for problem solving process. Various analytical solutions are used for giving different alternatives to do problem solving process. Numerical methods are acting as a powerful tool for understanding mathematical techniques. Numerical methods are acting as an efficient vehicle for computer learning process. Mathematical techniques are easily

P. Rajarajeswari (✉)
Department of CSE, MITS, Madanapalle, Andhra Pradesh, India
e-mail: perepicse@gmail.com

A. Ramamohanreddy
Department of CSE, S.V. University, Tirupathi, India
e-mail: ramamohansvu@yahoo.com

D. Vasumathi
Department of CSE, JNTU, Hyderabad, India
e-mail: rochan44@gmail.com

© Springer India 2016
H.S. Behera and D.P. Mohapatra (eds.), *Computational Intelligence in Data Mining—Volume 2*, Advances in Intelligent Systems and Computing 411, DOI 10.1007/978-81-322-2731-1_20

understood by Numerical methods. Object-oriented design programming method is defined by using objects and classes. In this design programming low level features are shared with high level programming languages. Variables are used for storing the information in the form of data types, integers. Functions are methods, procedure etc. In this method abstract classes are defined only by the methods. Concrete classes are used for doing the implementation of methods. Concrete classes are inherited from the Abstract classes. Modular programming reduces the complexity of design process. Object-oriented design programming is performed with the help of computer programming. Object-oriented design approach is applied for finding the solution of Euler method in a quick manner. This paper presents the design process for finding the solution of Euler method by using object-oriented design method. Description of Mathematical model is performed with the help of ordinary differential equations that are given in Sect. 2. In Sect. 3 object–oriented design approach is used for doing Modular programming. Object-oriented design approach for Euler method is given in the Sect. 4. Conclusions are given in the Sect. 5.

2 Mathematical Model

Mathematical model represents the features of system behavior in terms of mathematical formulas, equations, etc.

Dependent variable is a function of variable independent, variable functions, parameters, forcing functions.

Numerical methods are used for finding the approximate solution to solve the various mathematical problems that are formulated in Engineering and Science field by using Ordinary differential equations. Numerical method provides approximate values for finding the solution of problem.

2.1 Problem Solving Process in an Engineering Field

Engineering problems are solved by using various techniques or methods. Mathematical modeling provides solution for problem in an accurate manner. Find out the solution for the problem is given Fig. 1 [1]. Mathematical formulas, expressions, rules are used for giving solutions for the problems.

2.2 Ordinary Differential Equations

An ordinary differential equation has more number of derivatives with an unknown function.

Fig. 1 Problem solving
process in an engineering field

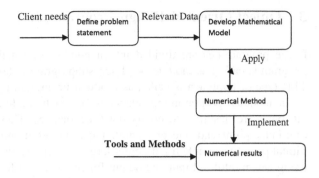

This type of equations are used to solve the problems in two types. One is initial value problem and second one is boundary value problem. At first point, Initial-value problem is defined with the specifications of all the conditions. The conditions are specified at initial point and final point. All the conditions are specified at t = 0 or x = 0. Here t or x as an independent variables [2].

2.2.1 First Order Degree Differential Equation

Differential equations are used for solving algebraic expressions. These expressions are solved by using certain techniques and fit in certain pattern. The methods are focused on the same initial problem. We will be given a differential equation with the derivative a function of the dependent and independent variable and an initial condition [3]. Function of (x, y) is taken as derivative of y with respective to x at the point of (x_0, y_0).

The solution which we call a function $y(x)$ to such an equation can be pictured graphically. The point (x_0, y_0) must be on the graph. The function $y(x)$ would also satisfy the differential equation if you plugged $y(x)$ in for y. Here $y'(x)$ is defined as **function** of (x, y(x)).

2.3 Euler Method

This method is used for solving first-order differential equations. It is mainly suitable for quick programming. Forward, Backward, Modified are the three types for Euler method [3].

3 Object Oriented Design Programming

Large programmes are divided into a number of small programmes. Each subprogram is called as module [4]. Each subprogram is tested in a separate manner. This type of approach is called as modular programming.

In modular programming attributes for each module are independent. These attributes are designed to do special functions [5]. These types of functions have one entry point and one exit point. Different programmers are working on individual parts for completing different large projects. Additional task is performed by using new modules. Input and output functions are performed by using INPUT and OUTPUT statements

Structure gives a framework for declaring a User defined Complex-data type. In C++ programming define the class for providing framework to user defined data type [4]. Programmers are able to build their own data type by using Class and Structure mechanism. The basic data types in standard programming are integers, floating point numbers, boo lean flags, and arrays etc.

- **Examples**: Person, Dog, PDE solver.
  Classes are composed of data and methods. Classes are considered as a collection of related data with functions.
  Classes–inheritance: Properties are inherited from super class to subclass. New class is derived from a base class.
  Suppose we want to write a class for a '**Scholar**'.
  Inheritance gets around this
  Class Scholar inherits from person:
  Data:
  N Num Papers
  Methods:
  Publish Paper ()
  Get Num Papers ()
- **Abstract classes**: Abstract classes are classes that contain a pure virtual method [6]. Abstract classes are defined only interfaces but not focused on implementation.
- Class Abstract bird b:
  Data:
  b is hungry
  Methods:
  Eat () (set b is hungry to false)
  Make Noise () (Abstract method, implementation is not given)
  Abstract classes cannot be instantiated. A subclass must be written which implements the abstract method.

4 UML Diagrams for Euler Method

UML diagrams consist of static diagrams and dynamic diagrams. Class diagram, object diagram are static diagrams. Sequence and collaboration diagram are interaction diagrams. It shows the behavioral characteristics of the system. Use case and actors are the elements in Use case diagram. Each use case shows functions of software system in Fig. 2.

Here analyzer and numerical analyzer are the actors. Problem Analyzer identifies needs of clients and defines the problem. Find out the solution of the problem and derives the solution for problem is performed by analyzer. Numerical analyzer applies the numerical method for the problem and derives the solution for problem.

Sequence diagram
Sequence diagram shows the time order of messages. All actions are shown in sequential order [7] (Fig. 3).

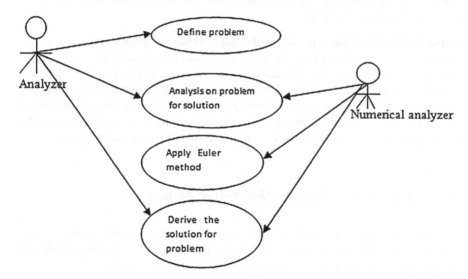

Fig. 2 Use case diagram for Euler method

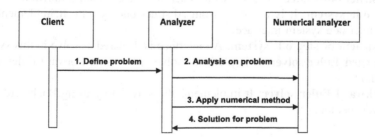

Fig. 3 Sequence diagram for Euler method

5 Applying Object-Oriented Design Approach for Euler Method

In this approach Abstract ode system is an Abstract class. Here My Ode system is inherited from Abstract Ode system. My Ode system is called as Concrete class. Object-oriented design implementation method is called as solver. Here abstract one step ode solver is an abstract class. Forward Euler solver and Backward Euler solver are inherited from abstract one step ode solver class [8].

Object-Oriented design approach for Euler method:
The following steps are followed for describing the Euler method with the help of Object-oriented design approach.

Abstract ode system is considered as one class. In this class Member variable is the dimension of the vector **y** and Abstract method is Evaluate the function $\mathbf{f(t, y)}$. My ode system is inherited from an abstract ode system that is considered as subclass of the system. Implementation of this method is Evaluate one particular choice of the function $\mathbf{f(t, y)}$.

Solver method for Euler method with Object-Oriented design approach: In this method abstract one step ode solver is consider as an one class. In this class abstract method is defined as Solve (abstract ode system, t_0, t_1, initial condition). Forward Euler solver and Backward Euler solver are the two subclasses for Abstract one step ode solver. These two are inherited from abstract one step ode system. Implemented method is Solve () for both the methods. Solve () is the function used for implementing Forward Euler solve method and Backward Euler solver method.

5.1 Structure for Abstract Ode Solver System

Abstract ode system, Abstract one step ode system are the abstract classes. My ode system is concrete class for abstract class [9]. Forward Euler solver and Backward Euler solver system are concrete classes of abstract class (Fig. 4).

Participants for Abstract ode solver system

- **Abstract ode system**: It declares an interface for evaluating function.
- **My ode system**: It is inherited from Abstract ode system. It implements the Abstract ode system interface.
- **Abstract one step ode system**: An interface is declared as solve in this system.
- **Forward Euler solver**: It implements solve () by using Forward Euler solver method.
- **Backward Euler solver**: It implements the solve () by using Backward Euler solver method.

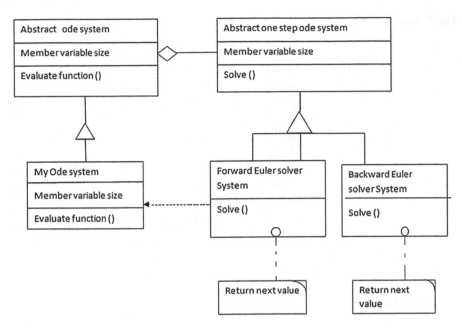

Fig. 4 Design structure for abstract ode solver system

6 Numerical Example

Find out velocity of sky driver and plot the solution for t ≤ 20 s after the sky driver jumps from the air plane. $M = 70$ **kg,** $c = 0.27$ **kg/m h = 0.1 s**

Euler method is one of the methods for Numerical Method. This method is used for solving First-order ODEs. This is particularly suitable for Simple and Quick programming method. Euler method has three types Forward Euler, Modified Euler, Backward Euler Method. These methods are used for providing numerical solutions for initial-value problems. We want to solve the function f (t, y) is so difficult in a practical way. It takes number of steps for solving the functions. To cope up this problem computer program is written for solving Euler method with the help of pseudo-code. We have applied Forward Euler method for finding velocity of sky driver. We provide a solution for Forward Euler Method is given in Fig. 5. Here solution is found by using MATLAB [10, 11].

Fig. 5 Numerical example

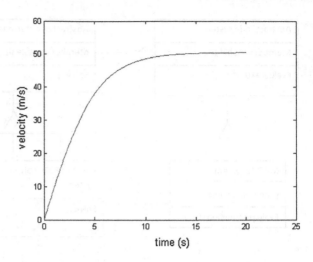

7 Conclusions

We presented Euler method by using first order differential equations. We presented design structure for Abstract ode solver system. We applied object oriented design approach for Euler method. We presented problem solving process for Euler method with the help of object-oriented design approach.

References

1. Van Loan, C.F.: Introduction to Scientific Computing, 2nd edn. Prentice Hall, Upper Saddle River (1997)
2. Coombes, K.R., Hunt, B.R., Lipsman, R.L., Osborn, J.E., Stuck, G.J.: Differential Equations with MATLAB. 3rd edn. ISBN: 978-1-118-37680-5 (2012)
3. Shampine, L.F.: Numerical Solution of Ordinary Equations. Chapman and Hall, New York (1994)
4. Subburaj, R.: Object Oriented Programming with C++, ANSI/ISO Standard. ISBN:81-259-1450-1
5. Forsythe, G., Malcolm, M., Moler, C.: Computer Methods for Mathematical Computations. Prentice-Hall, New Jersey (1977)
6. Pathmanathan, P.: Numerical methods and object oriented design (2011)
7. Booch, G., Rumbaugh, J., Jacobson, I.: The Unified Modeling Language User Guide, 1st edn. Addision-Wesley, 20 October 1999, ISBN:0-201-57168-4
8. Gilat, A.: MATLAB: An Introduction with Applications, 3rd edn. Wiley (2004)
9. Booch, G.: Design pattern, elements of reusable object-oriented software, Pearson Education, ISBN-978-81-317-0007-5
10. Nakamura, S.: Numerical analysis and graphic visualization with Matlab, 2nd edn. (2011)
11. Moler, C.B.: Numerical Computing with MATLAB. Siam (2004)

Intelligent TCSC for Enhancement of Voltage Stability and Power Oscillation Damping of an Off Grid

Asit Mohanty, Meera Viswavandya, Sthitapragyan Mohanty
and Pragyan Paramita

Abstract This paper discusses the enhancement of voltage stability and damping in isolated hybrid system. Fuzzified compensation has been carried out with TCSC for proper enhancement of stability and reactive power. The wind energy conversion system with linear approximation is experimented with different loading conditions. Overall compensation has been achieved with TCSC Controller for different loadings and uncertain wind power inputs. Both proportional and integral constants are properly tuned and the optimized results are achieved with Fuzzy controller. The results vindicate the performance of the proposed compensator and desired results are achieved.

Keywords TCSC · Voltage stability · Fuzzy control · WECS · Reactive power compensation

1 Introduction

Due to the uncertain nature of renewable energy sources, DG based hybrid systems are implemented quite often where two or more energy sources are combined to deliver power to the remote station. DG based hybrid system has a clear advantage than other, where the combined output power of different renewable abolish the

A. Mohanty (✉) · M. Viswavandya
Department of Electrical Engineering, CET, Bhubaneswar, India
e-mail: asithimansu@gmail.com

M. Viswavandya
e-mail: mviswavandya@rediffmail.com

S. Mohanty
Department of Computer Science, CET, Bhubaneswar, India
e-mail: msthitapragyan@gmail.com

P. Paramita
Department of Mechanical Engineering, VSSUT, Burla, India

© Springer India 2016
H.S. Behera and D.P. Mohapatra (eds.), *Computational Intelligence
in Data Mining—Volume 2*, Advances in Intelligent Systems
and Computing 411, DOI 10.1007/978-81-322-2731-1_21

shortcomings [1, 2]. Wind Diesel system is the most accepted form of hybrid energy system that proves to be economical and perform in a excellent way in the remote areas [3–5]. Induction generator based wind turbine performs robustly and is quite rugged in nature. It becomes essential to provide reactive power for smooth running of induction generator and the requirement varies with non uniform wind and load inputs. The role of synchronous generator is very crucial which provides the much needed reactive power to the induction generator. It is quite evident from the fact that the reactive power need of turbine is not met by synchronous generator fully. Many research works have discussed the use of capacitor banks to achieve reactive power compensation [6, 7]. But the unpredictable wind power input and unexpected load change have necessitated to go for custom power devices.

TCSC [7–10] helps in increasing line power transfer and enhancing system stability. The main components of TCSC are by pass inductor, capacitor bank and bidirectional thyristors which not only compensates the reactive power of the system but enhances stability. Furthermore the device improves the system damping. The importance of reactive power in a power system can be well understood from the fact that misbalance in the management of reactive power makes the system unstable. TCSCs with PI controller have been widely used in the past and the controller's performance improves a lot with Fuzzy logic based TCSC controller. The basic aim of the controller is to tune proportional and integral constant values and to increase stability margin.

Firing angle of Thyristor Controlled Series Capacitor (TCSC) changes and invariably controls the reactance according to system algorithm. Variation of Thyristor firing angle and the conduction angle the reactance given to the hybrid power system changes. Controllers like proportional integral controllers and proportional integral derivative controllers have been experimented in the past but due to their under performances, either they are discarded or used in limited spheres. This paper simplifies the control problem to a great extent and improves the transient stability with small load change like (1–5 %) by implementing fuzzy based compensator. The system mathematical model is the transfer function models of TCSC with wind energy conversion system. The soft computing approach with implementation of fuzzification to adjust different gains is mentioned in one section. Finally the results of simulation of the mentioned hybrid system with different loadings and variable wind power input are shown in the last section followed by the conclusion.

2 System Configuraion and Its Mathematical Modelling

Under shown off grid system in Fig. 1. is having parameters mentioned in Table 2 and is considered with IEEE excitation 1 as in Fig. 2.

From the transfer function block with all parameters mentioned, the reactive power management equation is given by

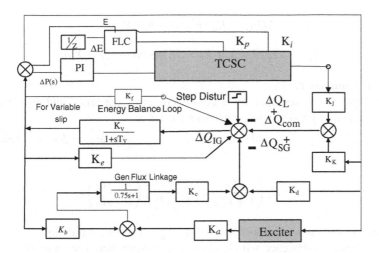

Fig. 1 Transfer function model with fuzzy logic based TCSC controller

Fig. 2 Excitation system 1

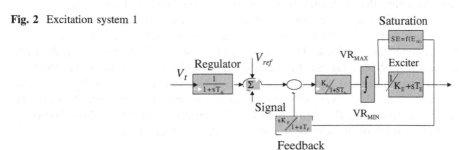

$$\Delta Q_{SG} + \Delta Q_{TCSC} = \Delta Q_L + \Delta Q_{IG} \qquad (1)$$

From the power management equation, change in load modifies the power management equation and affects the reactive power delivery capability of other elements. The system voltage deviates proportionally with the change in value of the reactive power related to other elements.

$\Delta Q_{SG} + \Delta Q_{TCSC} - \Delta Q_L - \Delta Q_{IG}$ influences the system output voltage.

$$\Delta V(S) = \frac{K_V}{1 + ST_V} [\Delta Q_{SG}(S) + \Delta Q_{COM}(S) - \Delta Q_L(S) - \Delta Q_{IG}(S)] \qquad (2)$$

Reactive power delivered by synchronous generator is $Q_{SG} = \dfrac{(E_q' V\cos\delta - V^2)}{X'd}$ for small modification

$$\Delta Q_{SG} = \frac{V\cos\delta}{X'd\Delta E'q} + \frac{E'q\cos\delta - 2V}{X'd\Delta V} \qquad (3)$$

$$\Delta Q_{SG} = K_a \Delta E' q(s) + K_b \Delta V(s) \qquad (4)$$

$$K_a = V cos\delta \ / \ X'd \ K_b = (E'qcos\delta - 2V)/X'd$$

TCSC Controller:
TCSC can be represented by a controlled series reactance and it is variable inductance acting parallel with capacitive reactance.

$X_{TCSC}(\alpha) = \dfrac{X_c X_L(\alpha)}{X_L(\alpha) - X_c}$ and $X_L(\alpha) = X_L \dfrac{\pi}{\pi - 2\alpha - \sin\alpha}$ are inductive reactance values.

$$u = K_P \left(\frac{sT_W}{1 + sT_W} \right) \left(\frac{1 + sT_1}{1 + sT_2} \right) \left(\frac{1 + sT_3}{1 + sT_4} \right) y$$

From the equation α represents delay angular value. Modelling of the said controller has been designed with lead-lag and PI controller. The main modelling of TCSC compensator uses a conventional lead-lag structure and with Proportional Integral controller structure. If we study the conventional lead lag structure of TCSC the input signal remains as speed deviation associated with the generator. The block diagram with a gain block, washout block and two stage lead-lag blocks is mentioned in Fig. 3a, b. The advanced PI Controller abolishes certain demerits associated with lead-lag structure, as the wash out block gives valuable contribution as high pass filter and phase compensation delivers the much needed phase—lead characteristics which eventually compensates any phase lag remaining between the input and the output. K_P and K_I values of PI Controller have been adjusted properly and tuned properly to achieve the much needed compensation.

3 Fuzzy Logic Controller (FLC)

Fuzzy controllers are largely used in system control application. Being a mathematical system it studies the analogue input variables in terms of logical values. The main motive behind this particular controller is to study the system behaviour using set of rules. For each and every controller input a variable has been assigned and

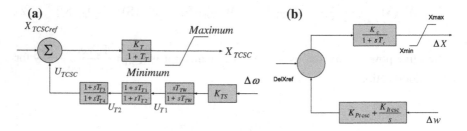

Fig. 3 a–b Block diagram of TCSC with lead-lag and PI controllers

Fig. 4 Block diagram of FLC
(fuzzy logic controller)

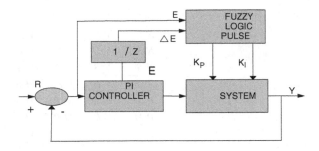

with the help of membership function the process of fuzzification has been carried out. Centre of gravity method has been used in this case for defuzzification process. In case of the TCSC the auto tuned fuzzy controller improves the isolated hybrid system's stability and damping. The output-scaling factor goes for self adjustment and improves with the updating factor (α) which is found out with inputs (E) and (ΔE) of controlled variables. The Fuzzy controller is shown below in Fig. 4.

4 Fuzzy Logic Controller(FLC)

The Fuzzy based PI controller has been designed according to these formulas

$$U(s) = K_P E(s) + K_I \int E(s)$$

$$E.K_P + K_I \int E = U$$

The isolated hybrid power system shows improvement with changing value of K_P and K_I of PI controller. $(K_P$ and $K_I)$ have been modulated with fuzzy inference engine which gives a proper mapping having inputs as error and change in error. The outputs are K_P and K_I which are found out after tuning the gains for a particular values. This is shown in Table 1 and in the FLC block diagram in Fig. 4.

$$\mu P_I = \min[\mu(E), \mu(\Delta E)]$$
$$\mu I_I = \min[\mu(E), \mu(\Delta E)]$$

The input variables such as error (E) and change in error (ΔE) are fed to the controller block and the two output variables are K_P and K_I of PI controller. The

Table 1 Rules of K_P K_I

$E/\Delta E$	NL	NM	NS	Z	PS	PM	PL
NL	VL	VL	VB	VB	MB	M	M
NM	VL	VL	VB	MB	MB	M	MS
NS	VB	VB	VB	MB	M	M	MS
Z	VB	VB	MB	M	MS	VS	VS
PS	MB	MB	M	MS	MS	VS	VS
PM	VB	MB	M	MS	VS	VS	Z
PL	M	MS	VS	VS	VS	Z	Z

controller has been designed with seven linguistic variables and they are having Triangular membership. The input membership functions remain as error (E) and change in error (ΔE) and output membership functions are K_P and K_I. For defuzzification centre of gravity method is applied.

5 Dynamic Responses of Wecs

The complete modeling and simulation have been found out with MATLAB environment taking the fuzzy controller in the WECS. TCSC performance and improvement are tested for improving reactive power and voltage stability of standalone hybrid system. For change in load a step load change of 1 % and for variable wind power inputs, the simulink model has been simulated. Small change in reactive power of diesel generator ΔQ_{SG}, induction generator (ΔQ_{IG}), variation in firing angle ($\Delta Q\alpha$), variation in terminal voltage (ΔV) are noted down and results are compared to find out detailed analysis. The results are shown and are studied using traditional Controller and proposed Fuzzy PI Controller. During minute study from the simulated results from MATLAB, it is accessed that soft computing based controllers deliver the improved performance with increase in size of Diesel generator and Induction generator for good control of terminal voltage. Main parameters like settling time and peak overshoots in FLC based TCSC is found minimum when compared with regular PI Controller and lead lag compensator. From the simulation the performance of fuzzy controller based compensator is justified which is better than the PI Controller. Though Improvement is not large, it is noticeable (Fig. 5).

6 Conclusion

Transfer function small signal models of wind energy conversion system with Fuzzy based TCSC Compensator are modeled and simulated in MATLAB environment. The system stability as well as damping is improved through reactive

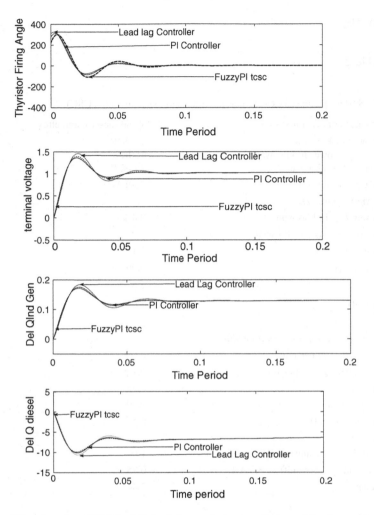

Fig. 5 a–d Results using TCSC for 1 % load change during transient condition (comparison with fuzzy controller)

power management by TCSC with soft computing approach. The compensator's inputs with rules have been identified with different input datas. With TCSC the damping of power oscillation has been adequately improved. The system becomes more robust and achieves transient stability despite frequent load changes and uncertain wind power inputs.

Appendix

See Table 2.

Table 2 System parameters (wind energy conversion system with TCSC)

Wind energy conversion system	Wind-diesel parametres
Wind power in kilo watt	150KW
Diesel energy power in kilo watt	150KW
Load demand in kilo watt	250KW
Base power in kilo watt	250KW
Diesel engine parametres	
Real power P_{SG} in kilo watt	0.4 kw
Reactive power Q_{SG} in kilo watt	0.2
E_q in (per unit)	1.113
E'_q in (per unit)	0.96
V (per unit)	1.0
X_d (per unit)	1.0
T'_{do} (s)	5
Induction generator based wind turbine	
Real power P_{IG}, in per unit	0.6
Reactive power Q_{IG} in per unit	0.189
P_{IN} per unit	0.75
$r_1 = r_2$ (in per unit)	0.19
$X_1 = X_2$ (in per unit)	0.56
Load demand	
Active load P_L (per unit) in KW	1.0
Reactive load Q_L (per unit) in KVAR	0.75
α in radian	2.44

References

1. Kaldellis, J., et al. Autonomous energy system for remote island based on renewable energy sources. In: Proceeding of EWEC 99, Nice
2. Hingorani, N.G., Gyugyi, L.: Understanding FACTS, Concepts and Technology of Flexible AC Transmission System. IEEE Power Engineering Society, NewYork (2000)
3. Murthy, S.S., Malik, O.P., Tondon, A.K.: Analysis of self excited induction generator. IEE Proc. Gener. Transm. Distrib. **129** (1982)
4. Padiyar, K.R.: FACTS Controlling in Power Transmission system and Distribution. New Age International Publishers (2008)
5. Tondon, A.K., Murthy, S.S., Berg, G.J.: Steady state analysis of capacitors excited induction generators. IEEE Trans. Power Apparatus Syst. **PAS-103**, 612–618 (1984)
6. Padiyar, K.R., Verma, R.K.: Damping torque analysis of static VAR system controller. IEEE Trans. Power Syst. **6**(2), 458–465 (1991)

7. Salman, H., Biswarup, D., Pant, V.: A self-tuning fuzzy PI controller for TCSC to improve power system stability. Elect. Power Syst. Res. **78**, 1726–1735 (2008)
8. Magaji, N., Mustafa, M.W.: Optimal thyristor control series capacitor neuro-controller for damping oscillations. J. Comput. Sci. **5**(12), 980–987 (2009)
9. Mok, T.K., Yixin, N., Felix, F.W.: A study of fuzzy logic based damping controller for the UPFC. In: IEEE Proceedings of the 5th International Conference on Advances in Power System Control, Operation and Management (APSCOM), Hong Kong, pp. 290–294 (2000)
10. Wei, L., Chang, X.: Application of hybrid fuzzy logic proportional plus conventional integral-derivative controller to combustion control of stoker-fired boilers. Fuzzy Sets Syst. **111**(2), 267–284 (2000)

Weighted Matrix Based Reversible Data Hiding Scheme Using Image Interpolation

Jana Biswapati, Giri Debasis and Mondal Shyamal Kumar

Abstract In this paper, we introduce a new high payload reversible data hiding scheme using weighted matrix. First, we enlarged the size of original image by interpolation. Then we perform modular sum of entry-wise multiplication with original image block and predefine weighted matrix. We subtract the modular sum from secret data and store the positional value by modifying three least significant bits of interleaved pixel. The original pixels are not affected during data embedding which assures reversibility. The proposed scheme provides average embedding payload 2.97 bits per pixel (bpp) and PSNR 37.97 dB. We compare our scheme with other state-of-the-art methods and obtain reasonably better results. Finally, we have tested our scheme through some attacks and security analysis which gives promising results.

Keywords Reversible data hiding · Weighted matrix · Image interpolation · RS analysis · Relative entropy · Steganography

J. Biswapati (✉)
Department of Computer Science, Vidyasagar University,
Midnapore 721102, West Bengal, India
e-mail: biswapati.jana@mail.vidyasagar.ac.in

G. Debasis
Department of Computer Science and Engineering, Haldia Institute
of Technology, Haldia 721657, West Bengal, India
e-mail: debasisgiri@hotmail.com

M.S. Kumar
Department of Applied Mathematics with Oceanology
and Computer Programming, Vidyasagar University, Midnapore 721102
West Bengal, India
e-mail: shyamal260180@yahoo.com

© Springer India 2016
H.S. Behera and D.P. Mohapatra (eds.), *Computational Intelligence in Data Mining—Volume 2*, Advances in Intelligent Systems and Computing 411, DOI 10.1007/978-81-322-2731-1_22

239

1 Introduction

The modern secret writing is tweaked the cover work in such a way that a secret message can be encoded within them. There are a number of ways to conceal information within cover work. The most usual methodologies are based on the least significant bits (LSBs) substitution, masking, filtering, [1, 2] and the modulus operation [3–5]. Data hiding using block are commonly used to increase visual quality or to achieve reversibility [6, 7]. Reversible Data Hiding (RDH) proposed by Ni et al. [8] which are based on histogram shifting with zero or minimum change of the pixel gray values. Jung and Yoo [9] first proposed data hiding through image interpolation using neighbor mean interpolation with payload 2.28 bpp. Tseng et al. [10] proposed a data hiding scheme using weighted matrix and key matrix for binary image which can hide only 2 bits in a (3 × 3) block of pixels. Fan et al. [11] presented efficient scheme using weighted matrix for gray scale images which can hide 4 bits in a (3 × 3) block. In both the aforesaid matrix based schemes, only one modular sum of entry wise multiplication of weighted matrix W is performed with a (3 × 3) block of pixels in the original image. Only one embedding operation is performed with a single block and only four bits data embed within the block. High data embedding capacity and reversibility is still the important research issues in data hiding through Weighted matrix. No researcher considers RDH through weighted matrix. We present a very high payload weighted matrix based reversible scheme of data hiding. Here, twelve time modular sum of entry wise multiplication operations are performed with each and every block to hide 48 bits secret data within each block. Also the scheme achieves reversibility, good PSNR and payload.

Motivation

- Our motivation is to enhance the embedding capacity and achieve reversibility in steganography. Data embedding using weighted matrix was not reversible. We propose weighted matrix based reversible data embedding scheme.
- Secure data hiding is another important aim of our proposed scheme. To achieve security, we modify weighted matrix W for each new block. For ith block i = 1, 2, ..., we modify the weighted matrix W_{i+1} as $W_{i+1} = (W_i \times \kappa - 1)\ mod\ 9$, where $gcd\ (\kappa, 9) = 1$ and κ is a variable which has six values 2, 4, 5, 6, 7, 8. The weighted matrix is changed using any value of κ for each new block.
- Another motivation is high data embedding capacity. Weighted matrix based data hiding scheme hide only 4 bits secret data within a single block. Performing twelve times sum of entry-wise multiplication on a single pixel block we hide 48 bits secret data within a single pixel block of stego image.

The organization of paper is as follows: Propose data hiding scheme is discussed in Sect. 2. Comparisons with experimental results are discussed in Sect. 3. Steganalysis and Steganographic attacks are presented in Sect. 4. Conclusions and summary are given in Sect. 5.

2 Propose Scheme

Let us consider an original image I, with height M and width N, then we increase the size in double by interpolation and it becomes cover image C of height $(2 \times M - 1)$ and width $(2 \times N - 1)$ as follows:

$$
\begin{cases}
C(i,j) = I(p,q) \\
\quad \{where\, p = 1\ldots M, q = 1\ldots N, \\
\qquad i = 1,3,\ldots(2 \times M - 1), j = 1,3,\ldots(2 \times N - 1)\} \\
C(i,j) = \dfrac{(C(i,j-1) + C(i,j+1))}{2} \\
\quad \{where\, (i\,mod\,2) = 0, (j\,mod\,2) \neq 0, \\
\qquad i = 1\ldots(2 \times M - 1), j = 1\ldots(2 \times N - 1)\} \\
C(i,j) = \dfrac{(C(i-1,j-1) + C(i-1,j+1) + C(i+1,j-1) + C(i+1,j+1))}{4} \\
\quad \{where\, (i\,mod\,2) = 0, (j\,mod\,2) = 0\}
\end{cases}
$$

$$(1)$$

We partition original image into (3×3) pixel block $B_{(3\times3)}$ and cover image into (5×5) pixel block $C_{(5\times5)}$.

$$
\begin{cases}
inpo = \dfrac{(S(i,j-1) + S(i,j+1))}{2} \\
\quad where\, (i\,mod\,2 \neq 0\, and\, (j\,mod\,2) = 0; \\
inpo = \dfrac{(S(i-1,j) + S(i+1,j))}{2} \\
\quad where\, (i\,mod\,2) = 0\, and\, (j\,mod\,2) \neq 0; \\
\qquad i = 1\ldots(2 \times M - 1)\, and\, j = 1\ldots(2 \times N - 1); \\
inpo = \dfrac{(S(i-1,j-1) + S(i-1,j+1) + S(i+1,j-1) + S(i+1,j+1))}{4} \\
\quad where\, (i\,mod\,2) = 0\, and\, (j\,mod\,2) = 0;
\end{cases}
$$

$$(2)$$

Embedding:
Consider the weighted matrix W of size (3×3). Then we perform modular sum of entry wise multiplication (*val*) of original image block $B_{(3\times3)}$ with Weighted matrix W. We calculate data embedding position by subtracting the modular sum (*val*) from secret data unit (D) that is $pos = D - val$. We check the sign of calculated position value

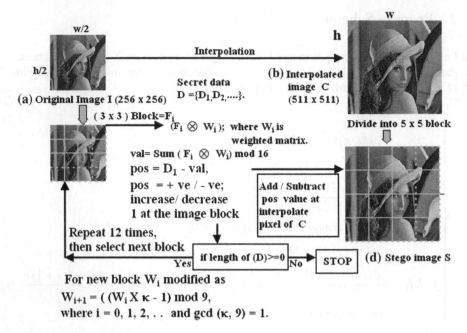

Fig. 1 Schematic diagram of embedding process

(*pos*). If the sign of *pos* is positive/negative then we increase/decrease the desired pixel value by one unit at the desired position of $B_{(3\times3)}$ pixel block. At the same time, we stored the embedding position of original image that is *pos* in the interleaved pixel of the cover image block $C_{(5\times5)}$. We apply twelve time sum of entry-wise multiplication operations and each time we increase/decrease pixel value at $B_{(3\times3)}$ and keep the *pos* in the interleaved pixel $C_{(5\times5)}$ to acquire high data embedding capacity. Our propose scheme can hide $(12 \times 4) = 48$ bits secret information within a single block. After completion of data embedding in a particular $B_{(3\times3)}$ block we update weighted matrix for next block using $W_{i+1} = (W_i \times \kappa - 1) \bmod 9$, where $gcd(\kappa, 9) = 1$. The schematic diagram for the data embedding is shown in Fig. 1.

Extraction

During data extraction, we extract the original pixel from stego image by simply collecting the pixel from odd row and odd column from stego image. The scheme is reversible because after extraction the secret data, the cover image is successfully extracted without any distortion. Now to extract the secret message, we consider $BI_{(3\times3)}$ of size (3×3) from original image and $BC_{(5\times5)}$ of size (5×5) from stego image. Then we calculate interpolation value (*inpo*) using Eq. (2) and calculate position value (*pos*) by subtracting the stego pixel value from (*inpo*) that is $pos = inpo - current\ value$. Now we check the sign of *pos* for realizing increment or decrement. If $(pos \leq 0)$ then $d = 1$ else $d = -1$. We update $BI_{(3\times3)}$ using $BI_{(3\times3)} = BI(3 \times 3) + d$ at the desired position *pos* of weighted matrix. Finally, we extracted 4 bits secret data in each entry wise multiplication operation using

$d_i = (BI_{(3\times3)} \otimes W_i)(mod\ 16)$, where $i = 0, 1, 2, \ldots, 12$. The entry wise multiplication operation is performed on the same block for twelve times and corresponding 48 bits secret data are extracted.

Overflow and Underflow

Overflow situation may occur when we update some pixel by pos value that may exceed the maximum pixel intensity value. For example, consider the pixel pair (254, 255). After interpolation at middle it becomes (254, 254, and 255). If the *pos* value 8 is added with 254 it becomes 262 which are greater than 255, this situation is called overflow situation and it may occur during data embedding. When a *pos* is subtracted from a pixel then there may be a chance to occur underflow. For example, consider the pixel pair (2, 4) then after interpolation it becomes (2, 3, 4). Consider a *pos* is equal to 4. If we subtract 4 from 3 then it is negative which is not a valid pixel value, this is an underflow situation. To overcome the overflow situation we adjust the pixel value by 247. Since, maximum *pos* value is 8, at the time of interpolation we adjust the pixel 247 because $255 - 8 = 247$. To handle the underflow situation, fix the value of interpolated pixel is 8.

$$IP = \begin{cases} 247\ if\ IP > 247 \\ 8\quad if\ IP < 8 \end{cases} \quad \text{where IP is interpolated pixel} \tag{3}$$

3 Experimental Results and Comparison

The developed scheme is tested using gray scale images shown in Fig. 2a. Our developed algorithms: data embedding and extraction are implemented in MATLAB Version 7.6.0.324 (R2008a). Here, Mean Square Error (MSE) is

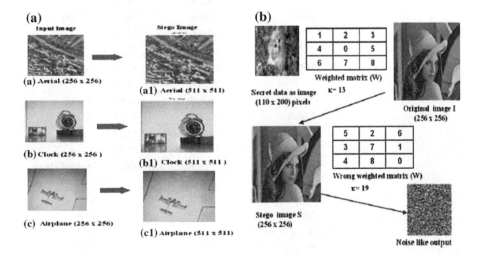

Fig. 2 a Input image and stego image. **b** Result of brute force attacks

calculated to measure the impairment and the quality is measured by the Peak Signal to Noise Ratio (PSNR). The Eq. (4) is used to calculate MSE.

$$MSE = \frac{\sum_{i=1}^{M} \sum_{j=1}^{N} [X(i,j) - Y(i,j)]^2}{(M \times N)}, \tag{4}$$

where M is the number of pixels in the row and N is the number of pixels in the column of the image. The cover image pixels are symbolized by $X(i, j)$ and stego image pixels are symbolized by $Y(i, j)$. The formula of PSNR is as follows:

$$PSNR = 10 \log_{10} \frac{255^2}{MSE} \tag{5}$$

The analysis in terms of PSNR of enlarged image and stego image gives good results which are shown in Table 1 (top). Higher the values of PSNR between two images indicate the better of stego image where as low PSNR demonstrates the opposite. To calculate payload in terms of bits per pixel (bpp) the following equation are used

Table 1 PSNR of stego image with data hiding capacity (left) and comparison with existing scheme (right)

PSNR (dB) with capacity (bits)			
Image I	Capacity (bits)	PSNR	Avg. PSNR
Lena	2,60,096	40.85	37.97
	5,60,000	37.24	
	7,76,224	35.80	
Moon surface	2,60,096	40.87	37.97
	5,60,000	37.24	
	7,76,224	35.81	
Arial	2,60,096	40.83	37.96
	5,60,000	37.24	
	7,76,224	35.80	
Airplane	2,60,096	40.82	37.96
	5,60,000	37.24	
	7,76,224	35.81	
Clock	2,60,096	40.85	37.96
	5,60,000	37.23	
	7,76,224	35.80	

Scheme	Average PSNR (dB)	Payload (bpp)
Ni et al. [8]	30.88	1.11
Jung et al. [9]	33.24	0.96
Lee et al. [12]	33.79	1.59
Tang et al. [13]	33.85	1.79
Proposed scheme	37.97	2.97

$$B = \frac{\left(\lfloor \frac{M+1}{2} - 1 \rfloor\right) \times \left(\lfloor \frac{N+1}{2} - 1 \rfloor\right) \times 48}{(2M - 1)(2N - 1)} \tag{6}$$

Here, M = 256, N = 256, and payload B = 2.97 (bpp).

We compare our experimental results with other existing scheme listed in Table 1 (bottom). The PSNR of Ni et al. [8] scheme is 30.88 dB which is 7.09 dB less than our scheme and payload (bpp) 1.11 unit which is 1.86 unit less that our scheme. The both PSNR and payload of our proposed scheme are higher than other existing schemes shown in Table 1 (bottom). The PSNR is 37.97 dB in our scheme which is more than the existing matrix based on interpolation based schemes which guarantees good visual quality. The payload is very high compared to other existing schemes and it is 2.97 (bpp).

4   Steganalysis and Attacks

Steganalysis is the art of discovering whether or not a secret message exists in a suspected image. Steganalysis does not however consider the successful extraction of the message. Now days, steganographic systems do not achieve perfect security. Here we analyze our scheme by RS analysis and find result of relative entropy.

RS Analysis
We analyze our stego image by the RS analysis. When the value of RS analysis is closed to zero, the scheme is secure. It is observed from Table 2 (left), that the values R_M equals to R_{-M} and S_M equals to S_{-M}. Thus rule $R_M \approx R_{-M}$ and $S_M \approx S_{-M}$ are met. So, the scheme is robust against RS attack. The proposed scheme is also assessed based on statistical distortion analysis by some image parameters like standard deviation (SD) and correlation coefficient (CC) to check the impact on image after data embedding are summarized in Table 2 (right). We achieve good results which indicate our propose scheme is secure against statistical analysis.

Brute Force Attacks
Our scheme has been tested to measure the sustainability against malicious steganographic attacks. Figure 2b shows the revelation example where with wrong key and wrong weighted matrix is used to revels the hidden message. If the original image, stego image and extraction algorithm are known to the adversary still it is not possible to reveal the hidden data without knowing correct weighted matrix and correct secret key. However, we tested our scheme against brute force attack that attempts all possible permutation to reveal the hidden message. Maximum possible weighted matrix to embed r bits data length in each block are $(2^{r-1} + 1)!$. We are using *(M × N)* original matrix and partition (3 × 3) blocks. Total numbers of blocks are *(M/3 × N/3)* and each block used a modified weighted matrix. Consider r bit message need to embed. The possible combinations of r bit are 2^r. As per

Table 2 RS value of stego image with data hiding capacity (top) and standard deviation (SD) and correlation coefficient (CC) (bottom)

Image	Data	RS value of stego image				
		R_M	R_{-M}	S_M	S_{-M}	RS value
Lena	260,096	22,838	22,763	11,743	11,701	0.0034
	560,000	22,356	22,119	15,041	15,058	0.0068
	776,224	21,788	21,641	17,258	17,366	0.0065
Airplane	260,096	33,190	33,666	8359	8190	0.0155
	560,000	28,070	28,319	14,078	13,951	0.0089
	776,224	23,202	23,765	18,698	16,231	0.0246
Clock	260,096	29,089	20,541	9901	9843	0.0131
	560,000	24,499	24,522	15,302	15,323	0.0011
	776,224	22,998	22,738	18,032	18,298	0.0128

Image	Standard deviation		Correlation coefficient
	Image I	Stego image S	I and S
Lena	47.82	47.35	0.9823
Moon surface	27.19	27.47	0.9895
Arial	45.08	44.12	0.9686
Airplane	32.04	32.30	0.9924
Clock	57.30	57.42	0.9978

requirement of weight matrix W will be $(1, 2, \ldots, 2^{r-1})$. Number of values within W is 2^{r+1}. All possible combination of W will be $((2^{r+1})-2^r)!$. The secret key are 128 bits length, so, the number of required trials to reveal the hidden message is $((2^{r-1} + 1)!)$ $(M/3 \times N/3)$. In our scheme, for (256×256) image with r = 4, number of trails will be $(3,62,880)^{7,225}$ which are computationally infeasible for current computers. The proposed scheme achieves stronger robustness against several attacks.

5 Conclusion

A very high capacity reversible data hiding method using weighted matrix is proposed in this paper. We modify the interpolation technique where average value is inserted between zooming and incorporate weighted matrix for data embedding. We simple modify weighted matrix using secret key • to enhanced security in data hiding. Also weighted matrix based data hiding was not reversible, here, in this scheme, we introduce reversible data hiding in case of weight-matrix based data hiding scheme. In this scheme we achieved good PSNR and high capacity data embedding as 2.97 bpp. Also we tested our scheme using RS analysis; calculate some statistical analysis data on stego image like Relative entropy, Standard Deviation and Correlation Coefficient which gives promising result. We test our scheme by several steganographic attacks with wrong weighted matrix and wrong shared secret key, we observed that the scheme is secure and robust against such attacks.

References

1. Chan, C., Cheng, L.: Hiding data in images by simple LSB substitution. Pattern Recogn. **37** (3), 474–496 (2004)
2. Wang, R., Lin, C., Lin, J.: Image hiding by optimal LSB substitution and genetic algorithm. Pattern Recogn. **34**(3), 671–683 (2001)
3. Chang, C., Chan, C., Fan, Y.: Image hiding scheme with modulus function and dynamic programming. Pattern Recogn. **39**(6), 1155–1167 (2006)
4. Thien, C., Lin, J.: A simple and high-hiding capacity method for hiding digit by-digit data in images based on modulus function. Pattern Recogn. **36**(12), 2875–2881 (2003)
5. Wang, S.J.: Steganography of capacity required using modulo operator for embedding secret image. Appl. Math. Comput. **164**(1), 99–116 (2005)
6. Lien, B.K., Lin, Y.: High-capacity reversible data hiding by maximum-span pairing. Multimedia Tools Appl. **52**, 499–511 (2011)
7. Pan, J.S., Luo, H., Lu, Z.M.: A lossless watermarking scheme for halftone image authentication. Int. J. Comput. Sci. Netw. Secur. **6**(2b), 147–151 (2006)
8. Ni, Z., Shi, Y.Q., Ansari, N., Su, W.: Reversible data hiding. IEEE Trans. Circuits Syst. Video Technol. **16**(3), 354–362 (2006)

9. Jung, K., Yoo, K.: Data hiding method using image interpolation. Comput. Stand. Interfaces **31**, 465–470 (2009)
10. Tseng, Y.C., Chen, Y.C., Pan, H.K.: A secure data hiding scheme for binary images. IEEE Trans. Commun. **50**(8), 1227–1231 (2002)
11. Fan, L., Gao, T., Cao, Y.: Improving the embedding efficiency of weight matrix-based steganography for grayscale images. Comput. Electr. Eng. **39**, 873–881 (2013)
12. Lee, C., Huang, Y.: An efficient image interpolation increasing payload in reversible data hiding. Expert Syst. Appl. **39**, 6712–6719 (2012)
13. Tang, M., Hu, J., Song, W.: A high capacity image steganography using multi-layer embedding, Elsevier Science. Optik Int. J. Light Electron Opt. **125**(15), 3972–3976 (2014)

Enhanced iLEACH Using ACO Based Intercluster Data Aggregation

Jaspreet Kaur and Vinay Chopra

Abstract In this paper, an Inter-cluster Ant Colony Optimization algorithm has been used that relies upon ACO algorithm for routing of data packets in the network and an attempt has been made to minimize the efforts wasted in transferring the redundant data sent by the sensors which lie in the close proximity of each other in a densely deployed network. ACO is being widely used in optimizing the network routing protocols. This work has focused on evaluating the performance of iLEACH protocol. The overall goal is to find the effectiveness of the iLEACH when ACO inter-cluster data aggregation is applied on it.

Keywords ACO · iLEACH · WSN

1 Introduction

A WSN consists of a huge number of nodes which may be tightly or arbitrarily deployed in an area in which they have interest. There is Base Stations (BS) situated to sensing area. The base station having major function in WSN as sink send queries to nodes while nodes sense the asked queries and send the sensed information in a joint way reverse to Base station. So the collection of information and send only relevant data to customer via internet is done by Base station.

J. Kaur (✉) · V. Chopra
DAVIET, Punjab, India
e-mail: er.jasdhillon@gmail.com

V. Chopra
e-mail: vinaychopra222@yahoo.co.in

© Springer India 2016   249
H.S. Behera and D.P. Mohapatra (eds.), *Computational Intelligence
in Data Mining—Volume 2*, Advances in Intelligent Systems
and Computing 411, DOI 10.1007/978-81-322-2731-1_23

As we have seen in LEACH [1, 2] after the first node dies the network stability certainly decreased, so spare nodes recover this problem. So work has been done to enhance network stability. So main focus was given by implementing different parameters illumination of those node to be cluster head. For example randomize selection of a cluster of a node as cluster head which suited at edge of network. So energy consumption of these nodes very rapidly and due to this network stability time reduced. So in last decade many new improved techniques have been proposed, they have increased the network stability and lifetime.

2 Literature Review

Liu et al. [3] this paper have explained at new methodology in which reduction of energy load among all the nodes has been presented an improved algorithm LEACH-D based on LEACH. Lu et al. [4] have been concentrated mainly on the nodes those are away from base station and have been elected as cluster head, these node's energy have fallen very rapidly, so to overcome this a new model has been proposed in which three factors have been discussed. Melese and Gao [1] have explained the energy consumption of sensor nodes in Wireless sensor network. Main effort has been done for balancing the energy consumption across the network so that survival time of all nodes can increase. Yektaparast et al. [5] have proposed a technique in which they have divided the clusters into equal parts, called as cell. Every cluster divided into 7 cells. Elbhir et al. [6] has shown that the clustering technique known as a developed distributed energy-efficient clustering scheme for heterogeneous wireless sensor networks has been proposed. Kumar et al. [7] has discussed that the two tier cluster based data aggregation (TTCDA) algorithm for minimizing the cost of communication and computation where the nodes are randomly distributed. As this algorithm helps in minimizing the transmission of the number of data packets to the base station, so this is power and bandwidth efficient. Javaid et al. [8] has discussed that the enhanced developed distributed energy efficient clustering scheme(EDDEEC) for heterogeneous networks. This protocol is adaptive power aware. The probability of sensor nodes for becoming a cluster head in an efficient manner has been altered dynamically in order to distribute same amount of power between sensor nodes. Beiranvand et al. [9] have proposed an enhancement in LEACH named it i-LEACH [5], An Improvement has been done by taking into consideration basically three factors; Residual Energy in nodes,

Distance from base station and number of neighbouring nodes. Chen and Wang [10] have explained an improved model in WSN which has been based on heterogeneous energy of nodes for same initial energy and multiple hop data transmission among cluster heads is proposed.

3 Proposed Methodology

3.1 Algorithm

Begin AcoiLeach (X,Y,BS.X,BS.Y,n,P,E_0)

1. Deploy Sensor nodes
   For i= 1:n
   W(i).xd= rand* X;
   W(i).yd = rand * Y;
   W(i).G = 0;

   W(i).E = E_0;
   W(i).type = 'NCH';
   End

2. While r ≤ MAXTIME
   2.a check for epoch
   If r % (1/P)== 0
   For l= 1:n
   W(i).G = 0;
   End
   2.b check for dead nodes

   $$dead = \begin{cases} 1 & if\ W(i).E \leq 0 \\ 0 & otherwise \end{cases}$$

   dead= dead +1;
   end

$$CH = \begin{cases} 1 & if\ rand \leq 1 - \frac{P}{P*(r\%p)}\ and\ W(i).G \leq 0 \\ 0 & otherwise \end{cases}$$

```
CH = CH +1;
W(i) type= 'CH';
W(i).G = 1/p -1;
2.d Apply ACO
Set d, B & dist_array = G;
city list = C
trial = .5;

m=ch_tabu;
Eta=1./dist_array;
no_ants=ones(ch_tabu,ch_tabu);

ant_sol=zeros(m,ch_tabu);
NC=1;

Rbst=zeros(max_iterations,ch_tabu);
Lbst=inf.*ones(max_iterations,1);
Lavg=zeros(max_iterations,1);
while NC<=max_iterations %
   randloc=[];
   for i=1:(ceil(m/ch_tabu))
     randloc=[randloc,randperm(ch_tabu)];
   end
   antsl(:,1)=(randloc(1,1:m))';
   for j=2:ch_tabu
     for i=1:m
        visit_tour=ant_sol(i,1:(j-1)); %
        J=zeros(1,(ch_tabu-j+1));%
        P=J;
        Jc=1;
        for k=1:ch_tabu
           if length(find(visit_tour==k))==0
              J(Jc)=k;
              Jc=Jc+1;
           end
        end
        for k=1:length(J)
%pheromen as
           P(k)=(no_ants(visit_tour(end),J(k))^Alpha)*(Eta(visit_tour(end),J(k))^Beta);
        end
        P=P/(sum(P));
        %
        Pcum=cumsum(P);
        Select=find(Pcum>=rand);
        to_visit=J(Select(1));
        antsl(i,j)=to_visit;
```

```
        end
    end

Pos=find(Length_bst==min(Length_bst));
brsol=Root_bst(Pos(1),:)
plen=Length_bst(Pos(1))
```

3. Update consumed energy

```
            min_dis;
        if (min_dis>=do)
            W(i).E=W(i).E- ( tx_energy*(4000) + multipath*4000*( min_dis
    *min_dis * min_dis * min_dis));
        end
        if (min_dis<do)
W(i).E=W(i).E- ( tx_energy*(4000) + free_space*4000*( min_dis * min_dis));

        End
Return network life time (r);
```

4 Results and Discussions

In order to implement the proposed algorithm, and implementation has been done in MATLAB. Table 1 shows the parameters used in the implementation along with their values.

Table 1 WSNs characteristics

Parameter	Value
Area (x, y)	100,100
Base station (x, y)	150,150
Nodes (n)	100
Probability (p)	0.1
Initial energy	0.1 J
transmitter_energy	$50 * 10^{-9}$ J/bit
receiver_energy	$50 * 10^{-9}$ J/bit
Free space (amplifier)	$10 * 10^{-13}$ J/bit/m^2
Multipath (amplifier)	$0.0013 * 10^{-13}$ J/bit/m^2
Effective data aggregation	$5 * 10^{-9}$ J/bit/signal
Maximum lifetime	4000
Data packet size	4000 KB

Table 2 First node dead
evaluation

Initial energy	iLEACH	ACO_iLEACH
0.01	25	46
0.02	43	91
0.03	81	138
0.04	111	168
0.05	121	226
0.06	137	273
0.07	180	320
0.08	183	358
0.09	211	394
0.1	264	460

4.1 First Node Dead

Table 2 shows the first node dead evaluation, in the table it is clearly shown that the ACO_iLEACH performs better as compared to the iLEACH.

Figure 1 is showing the comparison of ACO_iLEACH and the iLEACH. X-axis is representing initial energy. Y-axis is representing the number of rounds. It has been clearly shown that the overall numbers of rounds in case of ACO_iLEACH are quite more than that of the iLEACH.

Fig. 1 First node dead
analysis

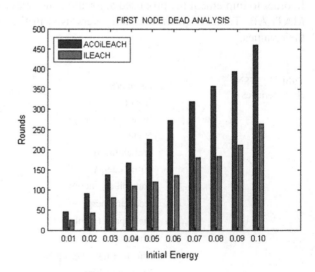

Table 3 Average energy evaluation

Initial energy	iLEACH	ACO_iLEACH
0.01	0.0346	0.0408
0.02	0.1385	0.1631
0.03	0.3087	0.3703
0.04	0.5541	0.6601
0.05	0.8692	1.0339
0.06	1.2236	1.4859
0.07	1.6627	2.0241
0.08	2.1944	2.6345
0.09	2.7732	3.3584
0.1	3.3956	4.1356

4.2 Average Energy

Table 3 shows the average energy, it is clearly shown that the average energy of ACO_iLEACH is better as compared to the iLEACH.

Figure 2 is showing X-axis representing initial energy and Y-axis representing the energy in joules. It has been clearly shown that the overall energy in case of ACO_iLEACH is quite more than that of the iLEACH. Thus ACO_iLEACH outperforms over the iLEACH.

Fig. 2 Average energy evaluation

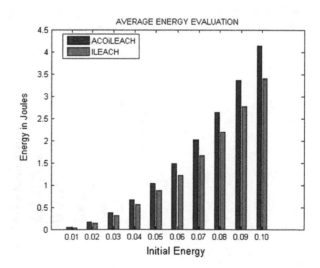

Table 4 Throughput evaluation

Initial energy	iLEACH	ACO_iLEACH
0.01	0.7354	0.9451
0.02	1.4210	1.8569
0.03	2.0715	2.7737
0.04	2.8286	3.6456
0.05	3.5042	4.5724
0.06	4.0349	5.6290
0.07	4.7887	6.4443
0.08	5.4526	7.2729
0.09	6.1265	8.3328
0.1	6.6556	9.1847

4.3 Throughput

Table 4 shows the Throughput evaluation. In the table, it is clearly shown that the Throughput OF ACO_iLEACH is better as compared to the iLEACH.

Figure 3 is showing X-axis representing initial energy and Y-axis representing the number of packets. It has been clearly shown that the overall number of packets in case of ACO_iLEACH are quite more than that of the iLEACH. Thus ACO_iLEACH

Fig. 3 Throughput evaluation

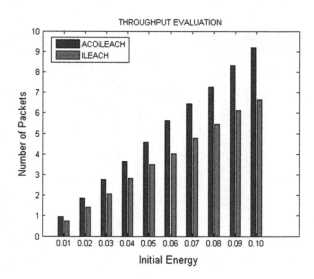

5 Conclusion and Future Scope

The proposed algorithm has reduced the energy consumption problem and also aggregates and transmits the data in efficient manner by using the ant colony optimization. Root cluster head communicates aggregated information to sink and cluster heads communicate data to the root CH directly or indirectly based upon the developed root by ant colony optimization. The proposed algorithm has been designed and simulated in the MATLAB tool. The comparative analysis has shown that the proposed ant colony optimization based iLEACH outperforms over the available protocols in terms of the various quality metrics.

This research work has not considered the use of any compressive sensing technique to enhance the results further, so in near future we will use different compressive sensing techniques to enhance the results further. Also this work is limited to homogeneous wireless sensor networks only so in near future we will consider up to 5 level of heterogeneity i.e. normal, intermediate, advance, super and ultra super nodes.

References

1. Melese, D.G., Xiong, H., Gao Q.: Consumed energy as a factor for cluster head selection in wireless sensor networks. IEEE 6th International Conference on Wireless Communications Networking and Mobile Computing (WiCOM), Sept 2010, pp. 1–4
2. Mao, Y., Chengfa, L., Guihai, C., Wu, J.: EECS: Energy efficient clustering scheme in wireless sensor networks. In: Proceedings of the 24th IEEE International Performance, Computing and Communications Conference (IPCCC), Phoenix, AZ, USA, pp. 535–540, 7–9 April
3. Liu, Y., Luo, Z., Xu, K., Chen, L.: A Reliable clustering algorithm base on LEACH protocol in wireless mobile sensor networks. In: International Conference on Mechanical and Electrical Technology, pp. 692–696, Sept 2010
4. Lu, Y., Zhang D., Chen1 Y., Liu, X., Zong, P.: Improvement of LEACH in wireless sensor network based on balanced energy strategy. In: IEEE Proceeding of International Conference on Information and Automation Shenyang, China, June 2012, pp. 111–115. IEEE (2013)
5. Yektaparast, A., Nabavi, F.H., Sarmast, A.: An improvement on LEACH protocol (Cell-LEACH). In: IEEE 14th International Conference on Advanced Communication Technology, pp. 992–996, Feb 2012
6. Elbhiri, B., Saadane, R., Aboutajdine, D.: Stochastic distributed energy-efficient clustering (SDEEC) for heterogeneous wireless sensor networks. ICGST-CNIR J. 9(2), 11–17 (2009)
7. Kumar, D., Aseri, T.C., Patel, R.B.: EEHC: energy efficient heterogeneous clustered scheme for wireless sensor networks. Comput. Commun. 32
8. Javaid, N., Qureshi, T.N., Khan, A.H., Iqbal, A., Akhtar, E., Ishfaq, M.: EDDEEC: enhanced developed distributed energy-efficient clustering for heterogeneous wireless sensor networks. In: International Workshop on Body Area Sensor Networks (BASNet-2013) in Conjunction with 4th International Conference on Ambient Systems, Networks and Technologies, Halifax, Nova Scotia, Canada (2013)

9. Beiranvand, Z., Patooghy, A. and Fazeli M.: I-LEACH: An efficient routing algorithm to improve performance and to reduce energy consumption in wireless sensor networks. In: IEEE 5th International Conference on Information and Knowledge Technology, pp. 13–18, May 2013

10. Chen, G., Zhang, X., Yu, J., Wang, M.: An improved LEACH algorithm based on heterogeneous energy of nodes in wireless sensor networks. In: IEEE International Conference on Computing, Measurement, Control and Sensor Network, pp. 101–104, July 2012

Evaluating Seed Germination Monitoring System by Application of Wireless Sensor Networks: A Survey

Priyanka Upadhyay, Rajesh, Naveen Garg and Abhishek Singh

Abstract In this grand era of technology the wireless sensor networks outstand to serve multidisciplinary fields of technology whether it be technology, agriculture, healthcare, logistics monitoring etc. Agriculture field has recently adopted the wireless sensor technology to enhance the production and monitoring of crops in a more efficient manner. This paper presents the various applications of wireless sensor networks adopted for monitoring and measuring various parameters for crop and seed germination for effective yield generation. In this paper various parameters are monitored by using wireless sensor networks. Also in this paper a survey of most used sensor nodes and the features that they comprise of are depicted to analyze their usage in the various domains. This paper discusses the advantages of sensor nodes such as cheaper cost, compact size and how they help in managing the agricultural application yielding to best quality crops. The work in implementing wireless sensor nodes is ongoing and a survey in various dimensions is discussed in this paper.

Keywords Wireless sensor network · Agriculture · Seed germination · Germination rate

P. Upadhyay (✉) · N. Garg · A. Singh
Department of Information Technology, Amity School of Engineering and Technology, Amity University Noida, Uttar Pradesh, India
e-mail: priyanka.upadhyay0991@gmail.com

N. Garg
e-mail: er.gargnaveen@gmail.com

A. Singh
e-mail: singhabhishek.0815@gmail.com

Rajesh
Agrionics Division, CSIR, Chandigarh, India
e-mail: calltorajesh@gmail.com

© Springer India 2016
H.S. Behera and D.P. Mohapatra (eds.), *Computational Intelligence in Data Mining—Volume 2*, Advances in Intelligent Systems and Computing 411, DOI 10.1007/978-81-322-2731-1_24

1   Introduction

The enhanced research field of wireless sensor network has enabled the environment monitoring on a large scale with applications in all dimensions of technology.
Further the advancement in microelectronics enables the development of sensors.
The wireless sensor network provides a mechanism that organise and control the
sensor nodes that sense data, process data, communicate and control the monitoring
task. The main motive to use the wireless sensor network is to reduce the human
interference and farmland environment. Basic research in the sensor network
focuses primarily on issues relating to routing, data dissemination, security and data
aggregation. The features of wireless sensor network such as portability, cheap cost,
and wireless monitoring make this technology highly scalable. The advancement in
this technology highlight the adaptability of wireless sensor network towards
agriculture application of electronics. There are various factors that affects the yield
of a crop. Analysing the cause affecting the crop manually is quite impossible. To
speed up the process of detection of any harm caused to the seed germination or
yield of crop we need to apply the wireless sensor networks technology which also
will help in controlling temperature, water content, and moisture requirement
accordingly to the crop or seed. This paper tries to enlighten the development of a
real time information wireless sensor network that can provide the farmers a solid
base to look after their crop yield in a much better way. This paper discusses the
setup of wireless sensor node that can control the yield of seed germination by
sensing the factors affecting its quality. Also a comparison amongst various features
of the wireless sensor nodes widely used in agricultural field for measuring the
factors such as humidity, soil moisture, water, temperature etc. are depicted below.

2   Related Works

There is an ongoing research to enhance the technique used for measuring physical
parameter required for seed germination. We use wireless sensor network for a
variety of purposes in various applications. With respect to agriculture we use it for
monitoring crops, automating the agriculture process. The rapid development and
wide adaptability of wireless sensor networks in agriculture has improved the quality
of the crops produced. Many researchers have worked on real time monitoring
systems. "Low Power Hybrid Wireless Network for Monitoring Infant Incubation"
published by shin et al. in 2004 designed a wireless network to monitor the infant
incubators using (IR) infrared and RF modules [1]. This system was designed to
monitor the temperature of infant incubators and sending related data on related of
host computer lab view software has been used for plotting data using computer.
Wall and king contributed in exploring design for sensor to measure soil moisture
and also proposed distributed sensor network architecture for site specific irrigation

automation [2]. Robert and Sharon discussed Evaluating Seed Germination in various environmental condition by applying computer aided image analysis [3]. The CAIA technique has shortly contributed in development of seed imbibitions monitoring. This technique need to be strictly integrated for high accuracy to describe the performance of seed sample. The system of image analysis consists of two sets of components

A. The system: Petridish, thermostatic cell, time controlled tran illuminator.
B. Computer imaging unit: CCD camera, software packages capable of capturing time lapsed seed images [3].

Yousry and Ian Moss discussed the Seed Germination Mathematical Representation and Parameters Extraction [4]. They found that Germination yield of a seed is characterized by three parameters germination rate, capacity, and time of germination lag. This paper describes a mathematical method which describes cumulative seed germination where the description of four parameter hill function is given. This function has four parameters that allow direct and indirect interpreting of the seed germination behaviour. Another related work is that of measuring soil moisture content by H.R. Bogena, J.A. Huisman stating that the moisture in soil partitions the water and the energy fluxes which provides the atmosphere for precipitation and controls the ground water recharge pattern. Yet, the soil moisture has a lot of importance still no operational way has been discovered to measure moisture content. In this project they aim to develop a sensor network for soil moisture to monitor soil water content and the changes taking place in high spatial scale. They have discussed a Mesh topology Zigbee based wireless sensor network [5].

A project aimed to design and develop a network integrated of sensors to manage the agricultural water of sensors to manage the agricultural water and its management in the semiarid parts of India was started, named as The COMMON- Sense project. The network of this project records the state moisture and watermark contents of soil periodically. Various projects such as Netsens have designed a new system for monitoring called as vine sense which is based on wireless sensor network [6]. Thiruparan and Saleh discuss a real time monitoring system for measuring remote humidity and temperature for studying after ripening process in seeds. They discuss the designing and prototype and testing of the remote monitoring system which is used for the study of seed germination under various geographical conditions. They discover the optimal conditions suitable for the after ripening seeds, their storage and germination. They discuss the development of the remote monitoring system which can precisely measure and monitor the relative temperature and humidityof the closed containers used for seed ripening. The challenge occurs is that the incubator does not allow WiFi signals to permeate through it which hinders the proper communication of the system [7]. They have tried to monitor the closed containers by the real time monitors to measure the temperature and humidity. Another system discussed was designed by Wen-Tsai Sung and Ming-Han Tsai

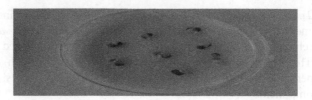

Fig. 1 Healthy mung seed lot

which describes their work "Multi-Sensor Wireless Signal Aggregation for Environmental System via Multi-bit Data Fusion." They discuss the system which uses the ZigBee Protocol, where sensor nodes send data to Zigbee Motherboard which further collects data from all surrounding nodes and compiles data to send to the user through USB. This system advantages in cheapness, compact size. Figure 1, we show a block diagram representing a system that works on the wireless sensor nodes to enhance their working to measure the temperature, humidity, soil moisture or various factors that affect the seed storage, germination, and its after ripening monitoring system. The temperature and humidity are the main factoring which determines the rate of germination of dormant seeds and not in the field, the stress of water often limits germination. In Orobanche aegyptiaca seeds they were germinated at optimum temperature for the seed germination with water and oxygen potential and freely vary the water temperature and the water uptake by the seeds available progress has three stages of germination of seeds in the population. During the first stage, the dry seeds take surrounding moisture and water absorption as indoor humidity, and when to plant size and increase start opening the second step is then initiated couple after beginning of seed germination and third stage Initiate [8]. Temperature is one of the most important environmental factors affecting seed germination of grass. Reliable prediction of the optimum temperature for seed germination is essential for determining the appropriate regions and timing of favorable growing grass seed. In this study, a model assisted by the neural network of double quintic-artificial-propagation (BP-ANN-QE) was developed to improve the prediction of the optimum temperature for seed germination. This model BP-ANN-QE was used to determine optimal times and suitable for sowing Cynodon dactylon cultivars three regions. Facts seed germination temperature of these schemes were used to construct correlation models germination temperature to estimate the germination percentage with confidence intervals. These results indicate that our model BP-ANN-QE has better performance than the rest of the models compared. In addition, data from national networks of temperature generated from the average monthly temperature for 25 years fit into these roles and they were able to assign the germination percentage of these [9]. Dactylon cultivars in China nationwide, and suggested optimal growing regions and times for them.

3   Proposed Methodology

In the proposed model we take a petri dish where we keep the sample seeds and use the germinator machine as shown in Fig. 1, to test the optimum conditions under which a seed germinated. The lighting and humidity conditions are also to be controlled during this experiment. We use distilled water and soaking paper to keep the seeds in a petri dish. Once we get the optimum conditions for the seed germination we note the readings and use them further to control the seed germination by using the wireless sensor nodes placed in a particular area and try to control the environmental factors such as the temperature, humidity and the water content. As once we know the levels of these factors we can prevent our seeds from ripening and can make them germinate in healthy seed lot. The field consist of sensor nodes placed at different levels where they sense and give the readings and the data is analyzed by the system where we have pre modelled the optimum conditions for the seed germination. The aim is to obtain the maximum seed germination rate in a particular seed type. Figure 1 shows a sample seed lot of healthy germinated seeds. Figure 2 shows the schematic diagram of Seed monitoring and evaluation wireless sensor network system. The comparative study of most used node platforms is shown in Fig. 3.

As per the demand of the application we make use of nodes from the Crossbow Berkeley which include MicaZ [10], Mica2 [10, 11], TmoteSky by Shockfish and others. Other platforms used from other companies are the Sensinode, TNOde [12], etc. The adaptability of these nodes depends upon the feature of the node as discussed in Table 1.

Figure 4 depicts the temperature variation and seed size changes observed at particular time. These changes observed help us to identify the germination rate of the seed and find out a seed lot is healthy or not.

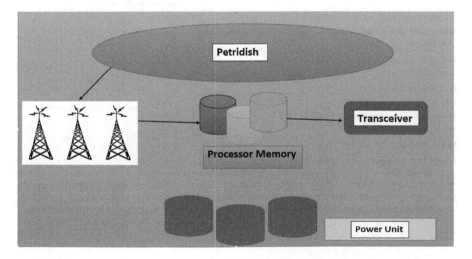

Fig. 2 Seed monitoring and evaluation wireless sensor network system

Fig. 3 Description of widely used node platforms

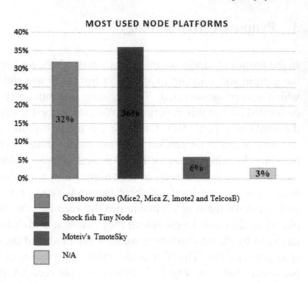

MOST USED NODE PLATFORMS

- Crossbow motes (Mice2, Mica Z, Imote2 and TelcosB)
- Shock fish Tiny Node
- Moteiv's TmoteSky
- N/A

Table 1 Representing various nodes use as a sensor [9]

Tmote sky	MICA Z	MICA 2
TI MSP430F1611 microcontroller at up to 8 MHz	>1 yr battery life on AA batteries	>1 yr battery life on AA batteries
10 k SRAM, 48 k Flash + 1024 k serial storage	2.4 GHz, IEEE 802.15.4 compliant radio	433, 868/916, or 310 MHz multi-channel radio transceiver
250 kbps 2.4 GHz Chipcon CC2420 IEEE 802.15.4 Wireless Transceiver	Light, temperature, RH, pressure, acceleration/seismic, acoustic, magnetic, GPS other sensors	Light, temperature, RH,, acceleration/seismic, acoustic, magnetic, GPS, and other sensors, barometric pressure
On-board humidity, temperature sensors and light sensors	N/A	N/A
Ultra-low current consumption	N/A	N/A
Fast wakeup from sleep (<6 μs)	N/A	N/A
Programming and interface via USB	N/A	N/A
Serial ID chip	N/A	N/A
16-pin expansion port	N/A	N/A
32 × 80 mm	58 × 32 × 7 mm	N/A

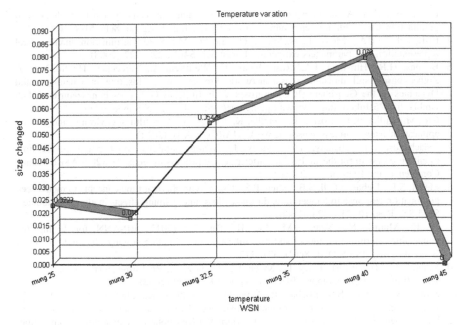

Fig. 4 Temperature variation versus size variation in mung seed

4 Future Work and Conclusion

This paper describes the application of wireless sensor network in the field of agriculture to monitor the seed germination under various geographical condition. The various type of sensor nodes are used accordingly to the features they deliver as per suitability of the application. The paper also describes how the monitoring of seed germination can help us to control the seeds to effectively germinate and in a healthy way. Further the wireless sensor network could also be applied to develop a user interface that collects all the variations and represents it on a graphical system on the basis of which we can monitor and control the seed germination process to enable higher quality storage and produce good yield of crops.

References

1. Shin, D.I., Shin, K.H., Kim, I.K., Park, K.S., Lee, T.S., Kim, S.I., Lim, K.S., Huh, S.J.: Low-power hybrid wireless network for monitoring infant incubators. Med. Eng. Phys. **27**, 713–716 (2005)
2. Wall, R.W., King, B.A.: Incorporating plug and play technology into measurement and control systems for irrigation management. In: Paper Presented at the ASAE/CSAE Annual International Meeting, Ottawa, Canada, p. 042189, Aug. 2004

3. DELL AQUILA: Application of a computer–aided image analysis system to evaluate seed germination under different environmental conditions. Ital. J. Agron. **8**(1), 51–62 (2008)
4. El-Kassaby, Y.A., Moss, I., Kolotelo, D., Stoehr, M.: Seed Germination mathematical representation and parameters extraction. Forest Sci. **54**(2), 220–227 (2008)
5. Bogena, H.R., Huisman, J.A., Rosen-baum, U., Weuthen, A., Vereecken, H.: SoilNet—A Zigbee based soil moisture sensor network. Project Group, Institute of Chemistry and Dynamics of the Geosphere (ICG), Agrosphere Institute, ICG 4, Forschungszentrum Jülich GmbH, 52425 Jülich, http://www.fz-juelich.de/icg/icg-4/index.php.index=739
6. Giri, M., Kulkarni, P., Doshi, A., Yendhe, K., Raskar, S.: Agricultural environmental sensing application using wireless sensor network. IJARCET **3**(3), (2014)
7. Balachandran, T., Sbenaty, S.M., Walck, J.: Remote humidity and temperature real-time monitoring system for the study of the after-ripening process in seeds. In: 120th ASEE Annual Conference and Exposition (2013)
8. Meyer, S.E., Debaene Gill, S.B., Allen, P.S.: Using hydrothermal time concepts to model seed germination response to temperature, dormancy loss, and priming effects in elymuselymoides. Seed Sci. Res. **10**, 213–223, (2000)
9. Pi, E., Mantri, N., Ngai, S.M., Lu, H., Du, L.: (2013) BP-ANN for Fitting the temperature-Germination model and its application in predicting sowing time and region for Bermuda grass. PLoS ONE **8**(12), 371–377
10. Vellidis, G., Tucker, M., Perry, C., Kvien, C., Bednarz, C.: Real-time smart sensor array for scheduling irrigation. Comput. Electron. Agric. **61**, 44–50 (2008)
11. Kumar, S., Iyengar, S.S., Lochan, R., Wiggins, U., Sekhon, K., Chakraborty, P., Dora, R.: Application of sensor networks for monitoring of rice plants: a case study. Science 1–7, (2009)
12. Langendoen, K., Baggio, A., Visser, O.: Murphy loves potatoes experiences from a pilot sensor network deployment in precision agriculture. In: Proceedings of the 20th IEEE International Parallel Distributed Processing Symposium, pp. 1–8 (2006)

A FLANN Based Non-linear System Identification for Classification and Parameter Optimization Using Tournament Selective Harmony Search

Bighnaraj Naik, Janmenjoy Nayak and H.S. Behera

Abstract In this paper, an enhanced version of Harmony Search (HS), called Tournament Selective Harmony Search (TSHS) is used to obtain an optimal set of weights for Functional Link Artificial Neural Network (FLANN) with Gradient Descent Learning (GDL) for the task of classification in data mining. The TSHS performs better than HS and Improved HS (IHS) by avoiding random selection of harmonies for their improvisation by introducing tournament selection strategy. This approach of TSHS to acquire optimal harmony in a population of harmony memory is adopted to find out optimal set of weights for FLANN model. The proposed TSHS-GDL-FLANN is compared with other alternatives by examining on various benchmark datasets from UCI Machine Learning repository. In order to get statistical correctness of results, the proposed method is analyzed by using ANOVA statistical test under null-hypothesis.

Keywords Harmony search · Improved harmony search · Tournament selective harmony search · Functional link artificial neural network · System identification · Classification

1 Introduction

Data collection and applications data mining process is became most promising in the area of web, business management, e-commerce, remote sensors, microarrays gene expression, scientific simulations, production control & engineering design,

B. Naik (✉) · J. Nayak · H.S. Behera
Department of Computer Science Engineering and Information Technology,
Veer Surendra Sai University of Technology, Burla, Sambalpur 768018
Odisha, India
e-mail: mailtobnaik@gmail.com

J. Nayak
e-mail: mailforjnayak@gmail.com

H.S. Behera
e-mail: mailtohsbehera@gmail.com

© Springer India 2016   267
H.S. Behera and D.P. Mohapatra (eds.), *Computational Intelligence in Data Mining—Volume 2*, Advances in Intelligent Systems and Computing 411, DOI 10.1007/978-81-322-2731-1_25

transactions and stocks & bioinformatics etc. Since last few years, System identification has played important roles in advancement of theories and algorithms for software development and industrial applications. System identification is a process of observing input–output data in order to design the mathematical models for dynamic systems. In data mining, the Classification problem can be viewed as a process of system identification that can be used to assign important classes to unknown patterns, by observing relationship between previous input patterns and corresponding target values. Due to non-linear nature of real world data, it is often difficult to determine the optimal model for classification data. Although various classical system identification approaches proposed by many researchers (Akaike's criterion (AIC) [1] or cross validation (CV); parametric prediction error methods (PEMs) [2, 3]), but some weakness related to these classical approaches have been recently reported [4–6]. As a result, a novel approach to system identification has been proposed by cross fertilizing with the machine learning concepts [4] and found to be successful in model selection and parameters optimization.

Various solutions to non-linear data classification problem are suggested by many researchers [7–11] but for the first time, Zhang et al. [11] realized that artificial neural network models are alternative to various conventional classification methods. The artificial neural networks (ANNs) are capable of generating complex mapping between input and the output space and thus they can form arbitrarily complex nonlinear decision boundaries. As compared to higher order neural network [12, 13], classical neural networks (Example: MLP) are suffering in slow convergence and unable to automatically decide the optimal model of prediction for classification. In the last few years, to overcome the limitations of conventional ANNs, some researchers have focused on HONN models for better performance.

In this paper, we have proposed a functional link higher order ANN (FLANN) based system identification model with gradient descent learning for non-linear classification data and a novel Tournament Selective harmony Search based approach for its parameter(weights) optimization. Various hybrids FLANN based models which are used for classification task can be found in [14–22] and other recent developments works can be obtained from [23–33]. Some researchers have also put interest on other HONN based classification models [34–40]. Remaining of this paper is structured as follow: Basic concepts are introduced in Sect. 2, Proposed scheme in Sect. 3. Experimental results are presented in Sect. 4 and Statistical comparisons are provided in Sect. 5 followed by Conclusions in Sect. 6.

2 Basic Concepts

2.1 Functional Link Artificial Neural Network

Basically, In FLANN, the input pattern goes through a transformation (known as functional expansion) of input data by which dimension of input space increases artificially. Then the extended input data are used to train the feed forward network.

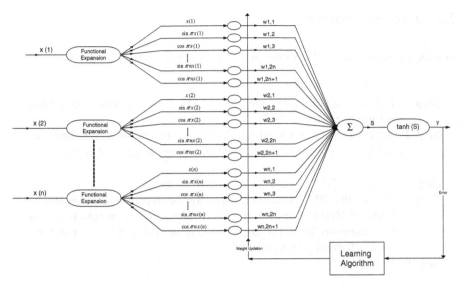

Fig. 1 Functional link artificial neural network architecture

Mathematical functions, such as sine, cosine, log, etc. are used in functional expansion to transform an original input pattern to it extended version. The above functionally expanded values 'φ' is the input to FLANN classifier for training phase. Prior to this, weights of FLANN must be set randomly. Based on net output and given target value, error of FLANN is calculated and a suitable learning method is adapted to adjust weight values of FLANN. An efficient learning method, known as gradient descent learning has been presented in the next sub-section. Figure 1 depicts the basic architecture of FLANN. The details on FLANN and the related implementation can be realized in [20].

2.2 Gradient Descent Learning

Gradient descent learning [41] is the commonly used training methods in which weights are changed in such a way that network error is declined as rapidly as possible. The learning of FLANN model using Gradient descent method with error of the network is successfully used in various recent related works [20, 33].

2.3 Harmony Search

In general, basic steps of Harmony Search [42] can be expressed as follows:

Step 1 Initialize a harmony memory (HM) with randomly generated solution vectors

Step 2 Repeat Steps 3 and 4 until no further significant growth in fitness of solution vector is noticed or the maximum number of iterations is reached

Step 3 Improvise HM to get New Harmony Memory

Step 4 Update the HM based on comparison between solution vectors of HM and NHM in terms of fitness. If any harmony in HM is less fit than harmony in NHM, then harmony in HM will be excluded by adding harmony from NHM

Step 5 Exit

Initially HM is initialized with random solution vectors and gradually, solution vectors in HM improved by using step-3 of harmony search procedure known as HM improvisation step. This step is entirely controlled by parameters: Harmony Memory Consideration Rate (HMCR), Pitch Adjustment Rate (PAR) and Bandwidth (BW).The HMCR and PAR determines Memory Consideration Probability (MCP), Pitch Adjustment Probability (PAP) and Random Probability (RP) as follows:

$$MCP = HMCR * (1 - PAR) * 100$$
$$PAP = HMCR * PAR * 100$$
$$RP = 100 - MCP - PAP$$

Basically, Improvisation of HM is governed by these parameters (MCP, PAP, and RP). In HS, the bw and PAR is fixed and pitch adjustment is done according to Eq. 1.

$$HM_i(t+1) = \begin{cases} HM_i(t+1) = HM_j(t) - rand(1) * bw & if rand(1) < 0.5 \\ HM_i(t+1) = HM_j(t) + rand(1) * bw & if rand(1) > 0.5 \end{cases} \quad (1)$$

In Eq. 1, $HM_i(t+1)$ is the next ith harmony at time $t + 1$ and $HM_j(t)$ is the jth randomly selected harmony for pitch adjustment at time t.

2.4 Tournament Selective Harmony Search

In Tournament selective harmony search (TSHS) [43], a fixed number of harmonies participates in a tournament (usually the size of tournament is less than size of population) and the winner of the tournament (win) is selected based on their fitness values. All the winner harmonies are improvised in step-3 through improvisation process based on MCP, PAP and RP, and included in New Harmony memory. In this strategy, the harmonies with superior fitness have a more possibility of being chosen, thereby eliminating random selection of harmonies for improvisation process. However, by using parameter 't_s' (In Eq. 2), the selection probability can be is easily adjusted. The participant harmonies (t_p) are randomly selected by using Eq. 2.

$$t_p = ceil(HMS * rand(1, t_s))$$ (2)

In Eq. 2, the 'rand' generates a random vector of length 'ts' drawn from a uniform distribution.

The TSHS follows the concepts of Improved Harmony Search [44], in which the bw and PAR are not fixed and these value changes according to HS iterations (Eq. 3 and Eq. 4).

$$bw(iter) = bw_{max} \times exp\left(\frac{ln\frac{bw_{min}}{bw_{max}}}{N} \times iter\right)$$ (3)

In Eq. 3, $bw(iter)$ is the bandwidth in particular iteration 'iter', 'bw_{min}' and 'bw_{max}' are the minimum and maximum bandwidth respectively and N is the number of solution vector in the population.

$$PAR(iter) = PAR_{min} + \frac{PAR_{max} - PAR_{min}}{N} \times iter$$ (4)

In Eq. 4, $PAR(iter)$ is the pitch adjustment rate in particular iteration 'iter'. The 'PAR_{min}' and 'PAR_{max}' are the minimum and maximum pitch adjustment rate and N is the number of solution vector in the population.

3 Proposed Scheme

In this section, the proposed scheme is presented by using problem solving strategy of TSHS in order to design an efficient learning method to get optimal set of weights for FLANN. The proposed method can be realized by following the execution of Algorithms 1–4.

`Algorithm – 1 Tournament Selective Harmony Search-GDL-FLANN (TSHS-GDL-FLANN) Procedure

---

% **HMS** : *Harmony Memory Size (Number of Harmonies)* , **HMCR** : *Harmony Memory Consideration Rate,* **PAR** : *Pitch Adjustment Rate.*

% *Randomly initialize a harmony memory (HM) with size HMS.*

HM = -1 + (1 - -1).*rand (HMS, LHM); % *LHM is the length of each weight-sets (harmonies).*

% *Initialization of HMS, HMCR and PAR .*

  HMS=40;
  HMCR=0.9;
  PAR_{min}=0.01;
  PAR_{max}=0.9;

% *Tournament Selective Harmony Search iteration*

Iter = 0;

While (1)

- % *Changing PAR and bw dynamically with iterations.*
  bw_{min}=zeros(1,L)+0.0001;
  u1=max(HM);
  UB=max(u1);
  L1=min(HM);
  LB=min(L1);
  bw_{max}=zeros(1,L)+(1/((UB-LB)));
  *$PAR=PAR_{min}+((PAR_{max}-PAR_{min})/N)*iter;*   % *Changes of PAR in Iterations.*
  *$bw=bw_{max} * exp(((log(bw_{min}/bw_{max}))/N)*iter);*   % *Changes of bw in Iterations.*

- % *Compute MCP (memory consideration probability), PAP (pitch adjustment probability) and RP (randomization probability).*
  MCP=HMCR*(1-PAR)*100;
  PAP=HMCR*PAR*100;
  RP=100-MCP-PAP;

- *Improvisation of Harmony*
  NHM = Improvisation-of-HM (HM, HMS, MCP , PAP, RP, bw);

- *Updation of HM*
  HM=Update-HM (HM,NHM);

- % *Check for Termination Criteria.*
  if **(iter >= MAX_ITERATION)**
          break;
  end if
  iter=iter+1;

EndWhile

Algorithm – 2 Improvisation-of-HM Procedure

Function NHM = Improvisation-of-HM (HM, MCP , PAP, RP, bw)

for i=1:1:HMS

Select tournament participants 'tp' (weight-sets) from HM. Let TSHM is the set of tournament participants (weight-sets) selected.

Perform tournament among participants in TSHM.

Select winner of the tournament winnerweight-set.

 r=rand(1)*100;

if(1<=r && r<=MCP)

> Select winnerweight-set with memory consideration probability (MCP) which serve as New Harmony memory (NHM).

if(MCP+1<=r && r<=MCP+PAP)

> Select winner$_{weight-set}$ with a probability of PAP for the pitch adjustment to improve quality of winnerweight-set which serves as New Harmony memory (NHM). The appropriate PAP serves this purpose better. This is achieved by using equation-1,3,4.

if(MCP+PAP+1<=r && r<=MCP+PAP+RP)

> Select winner$_{weight-set}$ with random probability (RP) which serve as New Harmony Memory (NHM). In this phase, winnerweight-set is added to NHM by suitably adding or subtracting a random value on it.

Endfor

end

Algorithm – 3 Update-HM Procedure

Function HM= **Update-HM** (HM,NHM)

- *Evaluate all the harmonies (weight-sets) in HM and NHM by using Algorithm-4 (fitfromtrain procedure).*
- *If the new harmony (weight-set) in NHM is better than the harmony in the HM, then add the new harmony into the HM by excluding the worst harmony from the HM.*

end

Algorithm – 4 fitfromtrain Procedure

Function F=<u>fitfromtrain</u> (φ, w, t, μ)

- *Compute S = φ. W*
- *Compute Y = tanh(S);*
- *Compute error e = t - y;*
- *If $\varphi = (\varphi_1, \varphi_2 ... \varphi_L)$, $e = (e_1, e_2 ... e_L)$ and $\delta = (\delta_1, \delta_2 ... \delta_L)$ are vector which represent set of functional expansion, set of error and set of error tern respectively then weight factor of w 'ΔW' is Computed as follow $\Delta W_q = \left(\frac{\sum_{i=1}^{L} 2 \times \mu \times \varphi_i \times \delta_i}{L}\right)$.*
- *Compute error term $\delta(k) = \left(\frac{1-y_k^2}{2}\right) \times e(k)$, for k=1,2...L where L is the number of pattern.*
- *Compute root mean square error (RMSE) by using equation (6) from target value and output.*
- *Find out fitness F=1/RMSE of the network instance of FLANN model. (equation (7))*

. *end*

---

4 Experimental Results

In this section, the parameters setting used for simulation and simulation results are presented. All the models are implemented in MATLAB R2009a and after obtaining results, some statistical analysis has been carried out by using SPSS 16.0 statistical tool. The benchmark datasets (Table 1) used for evaluation of models are originated from UCI machine learning repository [45] and processed by KEEL software [46]. The detail descriptions about all these dataset can be obtained at 'http://archive.ics.uci.edu/ml/' and 'http://keel.es/'.

4.1 Parameter Setting Used for Simulation

4.1.1 FLANN Parameter

During the learning of the FLANN model, the gradient descent learning method is used by setting 'μ' to 0.13. The value of 'μ' is obtained by testing the models in the range 0–3. Each value in the input pattern is expanded to 11 number of functionally expanded input values by setting n = 5. (As FLANN model suggests to generate 2n + 1 number of functionally expanded input values for a single value in the input pattern).

4.1.2 Harmony Search Parameter

Harmony Memory Size (HMS): 40; Harmony Memory Consideration Rate (HMCR): 0.9; Pitch Adjustment Rate (PAR): 0.3; Bandwidth (Bw): 0.0001.

4.1.3 Improved Harmony Search Parameter

Harmony Memory Size (HMS): 40; Harmony Memory Consideration Rate (HMCR): 0.9; Pitch Adjustment Rate (PAR): $PAR_{min} = 0.01$, $PAR_{max} = 0.9$; Bandwidth (Bw): $Bw_{min} = 0.0001$, $Bw_{max} = \frac{1}{20 \times (UB - LB)}$.

4.1.4 Tournament Selective Harmony Search Parameter

Harmony Memory Size (HMS): 40; Harmony Memory Consideration Rate (HMCR): 0.9; Pitch Adjustment Rate (PAR): $PAR_{min} = 0.01$, $PAR_{max} = 0.9$; Bandwidth (Bw): $Bw_{min} = 0.0001$, $Bw_{max} = \frac{1}{20 \times (UB - LB)}$; Tournament size (ts): 20.

4.2 Results Obtained

In this section, the root means square errors (RMSEs) (Eq. 5) of all the models are obtained for various benchmark datasets. RMSEs of all the models are presented in Table 2 and based on this; all the models are compared (Figs. 2, 3, 4, 5, 6 and 7).

The Root Mean Square Error (RMSE) of predicted output values $\hat{y}_i$ of a target variable y_i is computed for n different predictions as follows:

$$RMSE = \sqrt{\frac{\sum_{i=1}^{n} (y_i - \hat{y}_i)^2}{n}} \tag{5}$$

The fitness of the models are computed from RMSEs as follow:

$$F_{W_i} = 1/RMSE_i \tag{6}$$

In Eq. 6, W_i is the ith weight-sets in the population, $RMSE_i$ is the root mean square error of ith weight-set and F_{W_i} is the fitness of ith weight-set W_i.

In this study, the performance of GA, PSO, HS, IHS and TSHS are analyzed in order to know the improvement of harmonies (weight-sets) in the population by these algorithms in different generation. The changes in fitness of weight-sets in

Table 1 Data Set Information

Dataset	Number of pattern	Number of features (excluding class label)	Number of classes
Pima	768	08	02
Dermatology	256	34	06
New Thyroid	215	05	03
Monk 2	256	06	02
Hayesroth	160	04	03
Bupa	345	06	02

Table 2 Comparison of results among various models

Datasets	Maximum fitness obtained by various hybrid models				
	GA-FLANN	PSO-FLANN	HS-FLANN	IHS-FLANN	TSHS-FLANN
Pima	2.216361	2.216474	2.217241	2.217241	2.217241
Dermatology	1.888967	1.981951	2.267907	3.459351	3.459351
New thyroid	2.651259	2.675806	2.675806	3.093559	2.675806
Monk 2	2.024735	2.033361	2.037116	2.075259	3.526605
Hayesroth	1.888252	1.869618	1.895258	2.045696	2.055045
Bupa	1.54692	1.547116	1.548419	1.86651	1.865743

different generations are observed in all the datasets (Table 1) and the Figs. 2, 3, 4, 5, 6 and 7 demonstrates the improvements of fitness of weight-sets in the population.

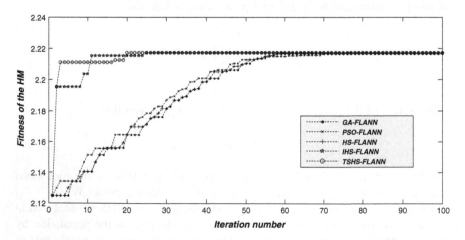

Fig. 2 Improvements in fitness of population in different iterations observed in Pima dataset

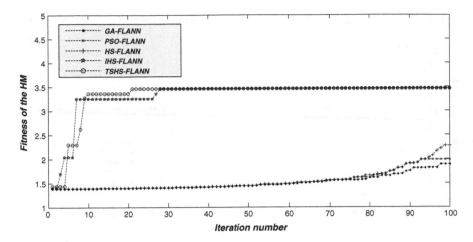

Fig. 3 Improvements in fitness of population in different iterations observed in dermatology dataset

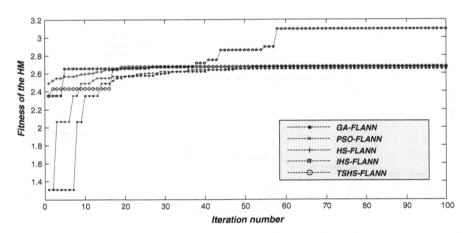

Fig. 4 Improvements in fitness of population in different iterations observed in new thyroid dataset

5 Statistical Comparisons

In this section, the statistical comparison of models over multiple data sets [47] is presented to argue the projected method is statistically better and significantly different from other alternative classifiers. We have conducted ANOVA test [48] on the results obtained (Table 2) from simulation.

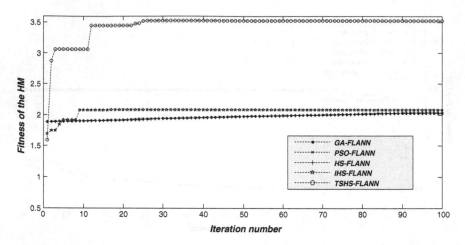

Fig. 5 Improvements in fitness of population in different iterations observed in MONK2 dataset

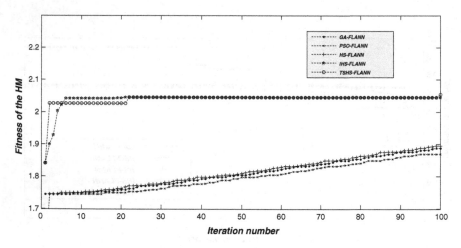

Fig. 6 Improvements in fitness of population in different iterations observed in Hayesroth dataset

5.1 ANOVA Test

In this paper, the statistics on all classifier's performance is compared by using ANOVA test (using SPSS (Version: 16.0) statistical tool) (Fig. 8) under null-hypothesis (H0). The test has been carried out with 95 % confidence interval, 0.05 significant level and linear polynomial contrast. To get the differences between the performances of classifiers, we have used Post hoc test by using mostly used Tukey test [49] and Dunnett test [50] (Fig. 9).

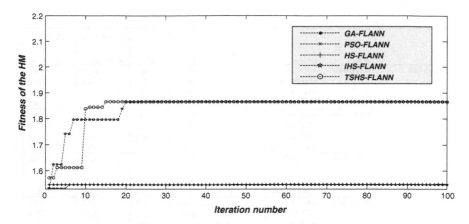

Fig. 7 Improvements in fitness of population in different iterations observed in Bupa dataset

Result

	N	Mean	Std. Deviation	Std. Error	95% Confidence Interval for Mean		Minimum	Maximum
					Lower Bound	Upper Bound		
GA-FLANN	120	2.0248	.34394	.03140	1.9626	2.0870	1.55	2.65
PSO-FLANN	120	2.0464	.34125	.03115	1.9847	2.1081	1.55	2.68
HS-FLANN	120	2.0972	.34568	.03156	2.0347	2.1597	1.55	2.68
IHS-FLANN	120	2.4384	.57896	.05285	2.3338	2.5431	1.87	3.46
TSHS-FLANN	120	2.6285	.66005	.06025	2.5092	2.7478	1.84	3.53
Total	600	2.2471	.53147	.02170	2.2044	2.2897	1.55	3.53

ANOVA

Result

			Sum of Squares	df	Mean Square	F	Sig.
Between Groups	(Combined)		35.306	4	8.827	39.225	.000
	Linear Term	Contrast	30.695	1	30.695	136.412	.000
		Deviation	4.611	3	1.537	6.830	.000
Within Groups			133.887	595	.225		
Total			169.193	599			

Fig. 8 One-way ANOVA test

H0: "all classifiers are same in performances and differences in performances are simply random".

As a conclusion of these tests, we noticed that, the mean differences (between-classifiers variability) among classifiers are larger than the standard errors (between-error variability) (except between TSHS-FLANN and IHS-FLANN) (Fig. 9). Also in Dunnett test (Fig. 9), we observed same as that of Tukey test. Out of five classifiers, the null-hypothesis is rejected for four no. of classifiers; hence the null-hypothesis can be clearly rejected in majority basis. All the models with their homogeneous group are presented in Fig. 10.

Multiple Comparisons

Dependent Variable: Result

(I) Algo	(J) Algo	Mean Difference (I-J)	Std. Error	Sig.	95% Confidence Interval Lower Bound	95% Confidence Interval Upper Bound
Tukey HSD						
GA-FLANN	PSO-FLANN	-.02161	.06124	.997	-.1892	.1460
	HS-FLANN	-.07239	.06124	.762	-.2400	.0952
	IHS-FLANN	-.41361*	.06124	.000	-.5812	-.2461
	TSHS-FLANN	-.60368*	.06124	.000	-.7712	-.4361
PSO-FLANN	GA-FLANN	.02161	.06124	.997	-.1460	.1892
	HS-FLANN	-.05078	.06124	.922	-.2183	.1168
	IHS-FLANN	-.39200*	.06124	.000	-.5596	-.2244
	TSHS-FLANN	-.58207*	.06124	.000	-.7496	-.4145
HS-FLANN	GA-FLANN	.07239	.06124	.762	-.0952	.2400
	PSO-FLANN	.05078	.06124	.922	-.1168	.2183
	IHS-FLANN	-.34122*	.06124	.000	-.5088	-.1737
	TSHS-FLANN	-.53129*	.06124	.000	-.6989	-.3637
IHS-FLANN	GA-FLANN	.41361*	.06124	.000	.2461	.5812
	PSO-FLANN	.39200*	.06124	.000	.2244	.5596
	HS-FLANN	.34122*	.06124	.000	.1737	.5088
	TSHS-FLANN	-.19006*	.06124	.017	-.3576	-.0225
TSHS-FLANN	GA-FLANN	.60368*	.06124	.000	.4361	.7712
	PSO-FLANN	.58207*	.06124	.000	.4145	.7496
	HS-FLANN	.53129*	.06124	.000	.3637	.6989
	IHS-FLANN	.19006*	.06124	.017	.0225	.3576
Dunnett t (2-sided)[a]						
GA-FLANN	TSHS-FLANN	-.60368*	.06124	.000	-.7536	-.4537
PSO-FLANN	TSHS-FLANN	-.58207*	.06124	.000	-.7320	-.4321
HS-FLANN	TSHS-FLANN	-.53129*	.06124	.000	-.6812	-.3814
IHS-FLANN	TSHS-FLANN	-.19006*	.06124	.007	-.3400	-.0401

*. The mean difference is significant at the 0.05 level.

a. Dunnett t-tests treat one group as a control, and compare all other groups against it.

Fig. 9 Tukey and Dunnett test

Result

	Algo	N	Subset for alpha = 0.05 - 1	Subset for alpha = 0.05 - 2	Subset for alpha = 0.05 - 3
Tukey HSD[a]	GA-FLANN	120	2.0248		
	PSO-FLANN	120	2.0464		
	HS-FLANN	120	2.0972		
	IHS-FLANN	120		2.4384	
	TSHS-FLANN	120			2.6285
	Sig.		.762	1.000	1.000
Duncan[a]	GA-FLANN	120	2.0248		
	PSO-FLANN	120	2.0464		
	HS-FLANN	120	2.0972		
	IHS-FLANN	120		2.4384	
	TSHS-FLANN	120			2.6285
	Sig.		.268	1.000	1.000

Means for groups in homogeneous subsets are displayed.

a. Uses Harmonic Mean Sample Size = 120.000.

Fig. 10 Models in homogeneous group

6 Conclusion

The proposed method has advantage over other related works as follows: (i) requires less mathematical computation and is free from complicated operators (like crossover in GA) and parameters (like c1, c2 in PSO), (ii) uses the concepts of dynamically changing bw and PAR and (iii) incorporates tournament selection strategy which allows better weight-sets to be included with high probability. Also performance of the TSHS-FLANN can be improved by fine-tuning the parameter tournament size 'ts' which can be used to effectively adjust selection pressure. We noticed the rejection of null-hypothesis after testing our method under ANOVA test. Hence, the proposed method is found to be better and outperforms other alternatives (GA-FLANN, PSO-FLANN, HS-FLANN and IHS-FLANN). Also the proposed method can be computed with a low cost due to less complex architecture of FLANN as compared to other HONNs. The future work is comprised of integration of similar improved variants of HS with FLANN in diverse applications of data mining.

Acknowledgments This work is supported by Department of Science and Technology (DST), Ministry of Science and Technology, New Delhi, Govt. of India, under grants No. DST/INSPIRE Fellowship/2013/585.

References

1. Akaike, H.: A new look at the statistical model identification. IEEE Trans. Autom. Control **19**, 716–723 (1974)
2. Söderström, T., Stoica, P.: System identification. Prentice-Hall, Upper Saddle River (1989)
3. Ljung, L.: System identification—theory for the user, 2nd edn. Prentice-Hall, Upper Saddle River (1999)
4. Pillonetto, G., De Nicolao, G.: A new kernel-based approach for linear system identification. Automatica **46**(1), 81–93 (2010)
5. Pillonetto, G., Chiuso, A., De Nicolao, G.: Prediction error identification of linear systems: a nonparametric Gaussian regression approach. Automatica **47**(2), 291–305 (2011)
6. Chen, T., Ohlsson, H., Ljung, L.: On the estimation of transfer functions, regularizations and Gaussian processes—revisited. Automatica **48**(8), 1525–1535 (2012)
7. Quinlan, J.R.: C4. 5: programs for machine learning. Morgan Kaufmann Publishers Inc., San Francisco, CA, USA (1993)
8. Yung, Y., Shaw, M.J.: Introduction to fuzzy decision tree. Fuzzy Net Syst. **69**(1), 125–139 (1995)
9. Hamamoto, Y., Uchimura, S., Tomita, S.: A bootstrap technique for nearest neighbour classifier design. IEEE Trans. Pattern Anal. Mach. Intell. **19**(1), 73–79 (1997)
10. Yager, R.R.: An extension of the naive Bayesian classifier. Inf. Sci. **176**(5), 577–588 (2006)
11. Zhang, G.P.: Neural networks for classification: a survey. IEEE Trans. Syst. Man Cybern. Part C Appl. Rev. **30**(4), 451–462 (2000)
12. Redding, N., Kowalczyk, A., Downs, T.: Constructive high-order network algorithm that is polynomial time. Neural Netw. **6**, 997–1010 (1993)
13. Goel, A., Saxena, S., Bhanot, S.: Modified functional link artificial neural network. Int. J. Electr. Comput. Eng. **1**(1), 22–30 (2006)

14. Mishra, B.B., Dehuri, S.: Functional link artificial neural network for classification task in data mining. J. Comput. Sci. **3**(12), 948–955 (Science Publications) (2007)
15. Dehuri, S., Cho, S.: A comprehensive survey on functional link neural networks and an adaptive PSO–BP learning for CFLNN. Neural Comput. Appl. **19**(2), 187–205 (2009)
16. Dehuri, S., Cho, S.-B.: Evolutionarily optimized features in functional link neural network for classification. Expert Syst. Appl. **37**, 4379–4391 (2010)
17. Dehuri, S., Roy, R., Cho, S., Ghosh, A.: An improved swarm optimized functional link artificial neural network (ISO-FLANN) for classification. J. Syst. Softw. 1333–1345, (2012)
18. Naik, B., Nayak, J., Behera, H.S.: A Novel FLANN with a Hybrid PSO and GA Based Gradient Descent Learning for Classification. Proc. 3rd Int. Conf. Front. Intell. Comput. (FICTA) Adv. Intell. Syst. Comput. **1**(327), 745–754 (2015)
19. Naik, B., Nayak, J., Behera, H.S.: A honey bee mating optimization based gradient descent learning—FLANN (HBMO-GDL-FLANN) for classification. In: Proceedings of the 49th Annual Convention of the Computer Society of India CSI—Emerging ICT for Bridging the Future, Advances in Intelligent Systems and Computing, vol. 338, pp. 211–220 (2015). doi:10.1007/978-3-319-13731-5_24
20. Naik, B., Nayak, J., Behera, H. S., Abraham, A.: A harmony search based gradient descent learning-FLANN (HS-GDL-FLANN) for classification. Comput. Intell. Data Min. **2**, 525–539 (2015) (Springer, India)
21. Naik, B., Nayak, J., Behera, H.S.: An efficient FLANN model with CRO based gradient descent learning for classification. International Journal of Business Information Systems (In Press)
22. Naik, B., Nayak, J., Behera, H.S.: An improved harmony search based functional link higher order ANN for non-linear data classification (In Press)
23. Pao, Y.H., Takefuji, Y.: Functional-link net computing: theory, system architecture, and functionalities. Computer **25**, 76–79 (1992)
24. Patra, J.C., Kot, A.C.: Nonlinear dynamic system identification using Chebyshev functional link artificial neural networks. IEEE Trans. Syst. Man Cybern. B Cybern. **32**, 505–511 (2002)
25. Klaseen, M., Pao, Y.H.: The functional link net in structural pattern recognition. In: TENCON 90. 1990 IEEE Region 10 Conference on Computer and Communication Systems, vol. 2, pp. 567–571 (1990)
26. Park, G.H., Pao, Y.H.: Unconstrained word-based approach for off-line script recognition using density-based random-vector functional-link net. Neurocomputing **31**, 45–65 (2000)
27. Liu, L.M., Manry, M.T., Amar, F., Dawson, M.S., Fung, A.K.: Image classification in remote sensing using functional link neural networks. In: Proceedings of the IEEE Southwest Symposium on Image Analysis and Interpretation, pp. 54–58 (1994)
28. Raghu, P.P., Poongodi, R., Yegnanarayana, B.: A combined neural network approach for texture classification. Neural Networks **8**(6), 975–987 (1995)
29. Abu-Mahfouz, I.-A.: A comparative study of three artificial neural networks for the detection and classification of gear faults. Int. J. Gen Syst **34**, 261–277 (2005)
30. Patra, J.C., Pal, R.N.: A functional link artificial neural network for adaptive channel equalization. Signal Process. vol. 43, pp. 181–195 (1995)
31. Teeter, J., Mo-Yuen, C.: Application of functional link neural network to HVAC thermal dynamic system identification. IEEE Trans. Ind. Electron. **45**, 170–176 (1998)
32. Abbas, H.M.: System identification using optimally designed functional link networks via a fast orthogonal search technique. J. Comput. **4**(2), 147–153 (2009)
33. Majhi, R., Panda, G., Sahoo, G.: Development and performance evaluation of FLANN based model for forecasting of stock markets. Expert Syst. Appl. **36**, 6800–6808 (2009)
34. Nayak, J., Naik, B., Behera, H.S., Abraham, A.: Particle swarm optimization based higher order neural network for classification. Comput. Intell. Data Min. **1**,401–414 (2015). Springer India
35. Nayak, J., Sahoo, N., Swain, J.R., Dash, T., Behera, H.S.: GA based polynomial neural network for data classification. In: International Conference on Information Technology (ICIT) 2014, pp. 234–239, IEEE (Dec, 2014)

36. Nayak, J., Naik, B., Behera, H.S.: A hybrid PSO-GA based Pi sigma neural network (PSNN) with standard back propagation gradient descent learning for classification. In: International Conference on Control, Instrumentation, Communication and Computational Technologies (ICCICCT) 2014, pp. 878–885, IEEE (July, 2014)

37. Nayak, J., Naik, B., Behera, H.S. (2015a). A novel chemical reaction optimization based higher order neural network (CRO-HONN) for nonlinear classification. Ain Shams Engineering Journal

38. Nayak, J., Kanungo, D.P., Naik, B., Behera, H.S.: A higher order evolutionary Jordan Pi-Sigma neural network with gradient descent learning for classification. In: International Conference on High Performance Computing and Applications (ICHPCA) 2014, pp. 1–6, IEEE (Dec, 2014a)

39. Nayak, J., Naik, B., Behera, H.S.: A novel nature inspired firefly algorithm with higher order neural network: Performance analysis. In: Engineering Science and Technology, an International Journal (2015b)

40. Nayak, J., Nanda, M., Nayak, K., Naik, B., Behera, H.S.: An improved firefly fuzzy C-Means (FAFCM) algorithm for clustering real world data sets. Smart Innovation, Syst. Technol. **27**, 339–348 (2014). doi:10.1007/978-3-319-07353-8_40

41. Rumelhart, D.E., Hinton, G.E., Williams, R.J.: Learning representations by back- propagating errors. Nature **323**(9), 533–536 (1986)

42. Geem, Z.W., Kim, J.-H., Loganathan, G.V.: A new heuristic optimization algorithm: harmony search. Simulation **76**(2), 60–68 (2001)

43. Karimi, M., Askarzadeh, A., Rezazadeh, A.: Using tournament selection approach to improve harmony search algorithm for modeling of proton exchange membrane fuel cell. Int. J. Electrochem. Sci. **7**, 6426–6435 (2012)

44. Mahdavi, M., Fesanghary, M., Damangir, E.: An improved harmony search algorithm for solving optimization problems. Appl. Math. Comput. **188**, 1567–1579 (2007)

45. Bache, K., Lichman, M.: UCI Machine learning repository (http://archive.ics.uci.edu/ml), Irvine, CA: University of California, School of Information and Computer Science (2013)

46. Alcalá-Fdez, J., Fernandez, A., Luengo, J., Derrac, J., García, S., Sánchez, L., Herrera, F.: KEEL data-mining software tool: data set repository, integration of algorithms and experimental analysis framework. J. Multiple-Valued Logic Soft Comput. **17**(2–3), 255–287 (2011)

47. Demsar, J.: Statistical comparisons of classifiers over multiple data sets. J. Mach. Learn. Res. **7**, 1–30 (2006)

48. Fisher, R.A.: Statistical methods and scientific inference, 2nd edn. Hafner Publishing Co., University of Michigan, New York (1959)

49. Tukey, J.W.: Comparing individual means in the analysis of variance. Biometrics **5**, 99–114 (1949)

50. Dunnett, C.W.: A multiple comparison procedure for comparing several treatments with a control. J. Am. Stat. Assoc. **50**, 1096–1121 (1980)

Heart Disease Prediction System Evaluation Using C4.5 Rules and Partial Tree

Purushottam Sharma, Kanak Saxena and Richa Sharma

Abstract Cardiovascular disease (CVD) is a big reason of morbidity and mortality in the current living style. Identification of Cardiovascular disease is an important but a complex task that needs to be performed very minutely and accurately and the correct automation would be very desirable. Every human being cannot be equally skilful and so as doctors. All doctors cannot be equally skilled in every sub specialty and at many places we don't have skilled and specialist doctors available easily. An automated system in medical diagnosis would enhance medical care and it can also reduce costs. In this study, we have designed a system that can efficiently discover the rules to predict the risk level of patients based on the given parameter about their health. Then we evaluate and compare this system using C45 rules and partial tree. The performance of the system is evaluated in terms of different parameter like rules generated, classification accuracy, classification error, global classification error and the experimental results shows that the system has great potential in predicting the heart disease risk level more efficiently.

Keywords C4.5 · Heart disease prediction system · CVD · CAD · PART

P. Sharma (✉)
R.G.T.U., Bhopal, M.P, India
e-mail: puru.mit2002@gmail.com

K. Saxena
Department of Computer Application, S.A.T.I., Vidisha, M.P, India
e-mail: kanak.saxena@gmail.com

P. Sharma · R. Sharma
Amity University, Noida, Uttar Pradesh, India
e-mail: s.richa.sharma@gmail.com

© Springer India 2016
H.S. Behera and D.P. Mohapatra (eds.), *Computational Intelligence in Data Mining—Volume 2*, Advances in Intelligent Systems and Computing 411, DOI 10.1007/978-81-322-2731-1_26

285

1 Introduction

In today's time at many places clinical test results are often produced based on doctors' intuition, skills and expertize rather than on the rich information available in many large databases. Many a times this process leads to error, unintentional biases and a huge medical cost. Sometimes it can affects the quality of service provided to patients drastically.

Today many hospitals installed some kind of patient's information collection systems to manage their healthcare or to collect patient data. These information systems usually generate large amounts of data which can be in different format like numbers, text, charts and images but unfortunately, this database that contains rich information is rarely used for clinical decision making. There is a lot of information stored in repositories that can be used effectively to support decision making in healthcare.

Here we focus on Heart Disease Prediction using data Mining techniques. The motivation for this study is the estimation given by WHO. As per the WHO estimation by year 2030, almost 23.6 million people will die due to Heart disease. So to minimize the risk, prediction of heart disease should be done. The most difficult and complex task in healthcare sector is diagnosis of correct disease. Heart disease prediction using different parameters of a patient diagnostic tests is a multi-layered issue which may lead to false presumptions and unpredictable effects. Now a day's Healthcare sector generating a huge amounts of raw data about patients, hospitals resources, disease diagnosis, electronic patient records, medical devices etc. This huge amount huge of raw data is the main resource that can be efficiently pre-processed and analysed for key information extraction that directly or indirectly motivates the medical society for cost-effectiveness and support decision making. Proper diagnosis of heart disease cannot be possible by using only human intelligence. There are lots of parameters that can affects the accurate diagnosis like less accurate results, less experience, time dependent performance, knowledge up gradation and so on. Lots of development and research happened in this field using multi-parametric attributes with nonlinear and linear features of Heart Rate Variability (HRV). A novel technique was proposed by Lee et al. [1]. To achieve this, many researchers have used many classifiers e.g. CMAR (Classification based on Multiple Association Rules), SVM (Support Vector Machine), Bayesian Classifiers and C4.5). Some of the latest techniques in this field described in [2]. In Healthcare, there is a very large scope and potential of Data mining applications usefulness but effectiveness of these application mostly reliable on accuracy of data and cleanliness. In this regard, it is very much desirable that the healthcare industry use such policies and methods so that data can be better prepared, stored, captured and mined. Some probable methods and methodology we suggested includes the clinical data standardization, analysis and the data sharing across the related industries to enhance the accuracy and effectiveness of data mining applications in healthcare [3]. It is also advisable to explore the use of text mining and image mining for expansion the nature and scope of data mining applications in healthcare sector. Data mining application can also be explored on digital diagnostic images for application effectiveness. Some progress has been made in these areas [4, 5].

The question can be arises out of this available data:

"How can we use this data to generate useful information that can be used by healthcare practitioners to make effective clinical decisions?" This is the main objective of this research.

2 Background

In recent time, many organizations in healthcare sector uses data mining applications intensively and extensively on large scale. Another reason is that the healthcare transactions generated by this sector are too voluminous and complex to be analysed and processed by traditional methods. Decision-making can be improved majorly by using data mining applications in discovering trends and patterns in large volumes of typical data [6]. In recent trends analysis on these large dataset has become necessary due to financial pressures on healthcare industries. This extracted information can be used for decisions making based on the regress analysis of medical and financial data. Knowledge extraction can influence industry operating efficiency, revenue and cost using knowledge discovery from database by maintaining a top level of care [7]. Research shows that if we uses data mining applications in healthcare organizations then these organizations would be in better position to meet their short term goals and long-term needs, Benko and Wilson argue [8]. We can get very useful results from healthcare raw data by transforming raw data into useful information. A great reason that enables researchers in this field is that this is very useful for all stake holder involved in the healthcare sector. Like, if we consider Insurance provider, they can detect abuse and fraud, practitioner in healthcare can gain assistance in decisions making, like in customer relationship management. Healthcare providers (hospitals, physician, test laboratories and patient etc.) can also use data mining applications in their respective expert zone for expert decision making for example, by finding best practices and correct and effective treatments.

3 UCI Heart Disease Dataset Description

Source Information:

(a) Creators of the used dataset: V.A. Medical Center, Long Beach and Cleveland
     Clinic Foundation: Robert Detrano, M.D., Ph.D.
(b) Donor: David W. Aha (aha@ics.uci.edu)

The "num" attributes indicate the presence and level or absence of heart disease in
the patient. The range of this attribute is from 0 (no presence) to 4 (severe).

Most of the experiments associated with Cleveland database are focused on
absence ("Num" value 0) and presence ("Num" values from 1 to 4) Due to personal
security patient's personal identification information replaced with dummy values.

Number of Instances: Cleveland: 303. The directory contains a dataset related
with heart disease diagnosis. The data was collected from the following locations:
Cleveland Clinic Foundation (cleveland.data).

The Cleveland database contains total 76 raw attributes, but in our experiments
only 14 of them is actually used because all published experiments till now using a
subset of 14 only and the data is also given only for these 14 attributes. The dataset
used in this experiment contains different important parameters like ECR, choles-
terol, chest pain, fasting sugar, MHR (maximum heart rate) and many more.

The detailed information about these attributes and their domain range are as
follows:

@relation Cleveland, @attribute age real [29.0, 77.0], @attribute sex real
[0.0, 1.0]
   @attribute cp real [1.0, 4.0], @attribute trestbps real [94.0, 200.0]
   @attribute chol real [126.0, 564.0], @attribute fbs real [0.0, 1.0]
   @attribute restecg real [0.0, 2.0], @attribute thalach real [71.0, 202.0]
   @attribute exang real [0.0, 1.0], @attribute oldpeak real [0.0, 6.2]
   @attribute slope real [1.0, 3.0], @attribute ca real [0.0, 3.0]
   @attribute thal real [3.0, 7.0], @attribute num {0, 1, 2, 3, 4}
   @inputs age, sex, cp, trestbps, chol, fbs, restecg, thalach, exang, oldpeak, slope,
ca, thal @outputs num

We have used the Classification model by covering rules (based on decision
trees) as C4.5 Rules [9–11] and partial tree on the above modified dataset and find
out the generated rule sets with different priority. We have also generated pruned
and classified rules. Further we have used WEKA tool [12] for dataset analysis and
KEEL [13, 14] to find out the classification decision rules and partial tree
generation.

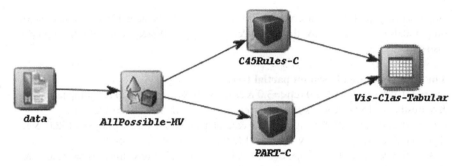

Fig. 1 Heart disease prediction model

4 Experiment Design with KEEL

We have used KEEL (Knowledge Extraction based on Evolutionary Learning) [14] tool. KEEL is an open source (GPLv3) Java software tool to assess evolutionary algorithms for Data Mining problems.

We have designed an Experiment using the Cleveland dataset as given in the Fig. 1. In the preprocessing phase we have used an AllPossible-MV [15] algorithm to fill the missing values in the dataset.

5 Classification Rule Generation

5.1 Pruned rule set generated by the experiments:
if(ca<=0.0 && thal>6.0 && oldpeak>0.3 && thalach<=129.0 && thalach>122.0) output=4 else if(cp>3.0 && ca>0.0 && slope<=2.0 && ca<=1.0 && age>62.0) output=2 else if(cp>3.0 && thal<=6.0 && restecg<=1.0 && exang>0.0 && age<=56.0) output=2 else if(cp>3.0 && slope<=2.0 && sex>0.0 && ca>1.0 && restecg<=1.0) output=3 else if(ca>0.0 && slope>2.0) output=3 else if(cp>3.0 && thal>6.0 && oldpeak<=0.3) output=1 else if(cp<=3.0 && oldpeak<=1.9 && fbs<=0.0 &&thalach>126.0) output=0 else if(cp<=3.0 && fbs>0.0 && oldpeak<=1.2) output=0 else if(ca<=0.0 && thal<=6.0 && restecg>1.0) output=0 else if(cp<=3.0 && oldpeak<=1.9 && age<=57.0) output=0 else if(cp<=3.0 && ca<=1.0 &&

slope<=1.0) output=0 else if(cp<=3.0 && fbs<=0.0 && thalach>126.0 && sex<=0.0)
output=0else if(ca<=0.0 && thal<=6.0 && restecg<=1.0 && exang<=0.0) output=0
else output=1

The rules generated based on partial tree:

(cp<=3.0 && sex<=0.0)-> 0 (cp<=3.0 && sex>0.0 && age<=63.0 && oldpeak<=2.0
&& trestbps<=152.0 && fbs<=0.0 && exang<=0.0 && thalach<=142.0)-> 1
(cp<=3.0 && sex>0.0 && age<=63.0 && oldpeak<=2.0 && trestbps<=152.0 &&
fbs<=0.0 && exang<=0.0 && thalach>142.0 && oldpeak<=0.5)-> 0
(cp<=3.0 && sex>0.0 && age<=63.0 && oldpeak<=2.0 && trestbps<=152.0 &&
fbs<=0.0 && exang<=0.0 && thalach>142.0 && oldpeak>0.5 && age<=57.0)-> 0
(cp<=3.0 && sex>0.0 && age<=63.0 && oldpeak<=2.0 && trestbps<=152.0 &&
fbs<=0.0 && exang<=0.0 && thalach>142.0 && oldpeak>0.5 && age>57.0)-> 1
(cp<=3.0 && sex>0.0 && age<=63.0 && oldpeak<=2.0 && trestbps<=152.0 &&
fbs<=0.0 && exang>0.0)-> 0 (cp<=3.0 && sex>0.0 && age<=63.0 && old-
peak<=2.0 && trestbps<=152.0 && fbs>0.0)-> 0

Partial tree generated by the experiments

```
if ( cp <= 3.000000 ) then
{
    if ( sex <= 0.000000 ) then
    {
        num = "0"
    }
    elseif ( sex > 0.000000 ) then
    {
        if ( age <= 63.000000 ) then
        {
            if ( oldpeak <= 2.000000 ) then
            {
            if ( trestbps <= 152.000000 ) then
                {
                if ( fbs <= 0.000000 ) then
                    {
                    if ( exang <= 0.000000 ) then
                        {
                        if ( thalach <= 142.000000 ) then
                            {
                            num = "1"
                            }
            elseif ( thalach > 142.000000 )  then
```

```
{
    if ( oldpeak <= 0.500000 )   then
    {
        num = "0" }elseif ( oldpeak >
0.500000           )            then
{ if ( age <= 57.000000 ) then
{ num = "0"

}
elseif ( age > 57.000000 )    then

{ num = "1"

} }     }   }
    elseif ( exang > 0.000000 ) then
        {
        num = "0"
        }
}
        elseif ( fbs > 0.000000 ) then
            {
            num = "0"
        }        }        }    }
}
```

6 Evaluation Results

We have used 5 folds for training and 5 folds for testing to evaluate the classification accuracy using different parameter.

6.1 Classification Results by Algorithm and by Fold

We have evaluated the classification accuracy of C4.5 Rules and Partial Tree classifier and the results using different classifier fold wise are as follows

Test Results using Partial Tree Classifier
Fold 0   CORRECT=0.540983606557377 N/C=0.0
Fold 1   CORRECT=0.5454545454545454 N/C=0.0
Fold 2   CORRECT=0.540983606557377 N/C=0.0
Fold 3   CORRECT=0.5238095238095238 N/C=0.0
Fold 4   CORRECT=0.5882352941176471 N/C=0.0

Global Classification Error + N/C: 0.45210668470070586
Stddev Global Classification Error + N/C: 0.02148925320086861
Correctly classified: 0.5478933152992942, Global N/C: 0.0

Train Results using Partial Tree Classifier
Fold 0   CORRECT=0.5503875968992248 N/C=0.0
Fold 1   CORRECT=0.5494071146245059 N/C=0.0
Fold 2   CORRECT=0.5503875968992248 N/C=0.0
Fold 3   CORRECT=0.5546875 N/C=0.0
Fold 4   CORRECT=0.5378486055776892 N/C=0.0

Global Classification Error + N/C: 0.451456317199871
Stddev Global Classification Error + N/C: 0.0056511365140908335
Correctly classified: 0.548543682800129, Global N/C: 0.0

Test Results using C4.5 Rules Classifier
Fold 0   CORRECT=0.5081967213114754 N/C=0.0
Fold 1   CORRECT=0.5 N/C=0.0
Fold 2   CORRECT=0.6065573770491803 N/C=0.0
Fold 3   CORRECT=0.47619047619047616 N/C=0.0
Fold 4   CORRECT=0.4558823529411765 N/C=0.0

Global Classification Error + N/C: 0.4906346145015383,
Stddev Global Classification Error + N/C: 0.05195453555220856
Correctly classified: 0.5093653854984617, Global N/C: 0.0

Train Results using C4.5 Rules Classifier
Fold 0   CORRECT=0.7093023255813953 N/C=0.0
Fold 1   CORRECT=0.6482213438735178 N/C=0.0
Fold 2   CORRECT=0.6550387596899225 N/C=0.0
Fold 3   CORRECT=0.62890625 N/C=0.0
Fold 4   CORRECT=0.6772908366533865 N/C=0.0

Global Classification Error + N/C: 0.33624809684035556,
stddev Global Classification Error + N/C: 0.02752991241516823
Correctly classified: 0.6637519031596444, Global N/C: 0.0

6.2   Global Average and Variance

The global average and variance measured using C4.5 Rules classifier and Partial
Tree classifier are given in Table 1.

Table 1 Global average and variance

C4.5 rules				Partial tree		
	Average correctly classified	Variance correctly classified	Not classified	Average correctly classified	Variance correctly classified	Not classified
Test	0.509365	0.002699	0.00	0.547893	0.000461	0.00
Train	0.663751	0.000757	0.00	0.548543	0.000031	0.00

Table 2 Classification rate by algorithm and by fold

		C4.5 rules		Partial tree	
		Correctly classified	Not classified	Correctly classified	Not classified
Test	Fold 0	0.5081967213	0.00000	0.5409836066	0.000000
	Fold 1	0.5000000000	0.00000	0.5454545455	0.000000
	Fold 2	0.6065573770	0.00000	0.5409836066	0.000000
	Fold 3	0.4761904762	0.00000	0.5238095238	0.000000
	Fold 4	0.4558823529	0.00000	0.5882352941	0.000000
Train	Fold 0	0.7093023256	0.00000	0.5503875969	0.000000
	Fold 1	0.6482213439	0.00000	0.5494071146	0.000000
	Fold 2	0.6550387597	0.00000	0.5503875969	0.000000
	Fold 3	0.6289062500	0.00000	0.5546875000	0.000000
	Fold 4	0.6772908367	0.00000	0.5378486056	0.000000

6.3 Classification Rate by Algorithm and by Fold

To evaluate the performance of C4.5 Rules classifier and Partial Tree classifier fold wise on test data set and training data set are given in the Table 2.

7 Conclusions

Heart Disease Prediction System evaluation analysis shows the evaluation of the two classifier on different parameter with different statistics measures. Results shows that C4.5 classifier can correctly classified the heart Disease up to 70.93 %. It has been also observed that C4.5 classifier supersedes the partial classifier on the given dataset.

References

1. Lee, H.G., Noh, K.Y., Ryu, K.H.: Mining Biosignal Data: Coronary Artery Disease Diagnosis using Linear and Nonlinear Features of HRV. LNAI 4819
2. Chhikara, S., Sharma, P.: Data Mining Techniques on Medical Data for Finding Locally Frequent Diseases. I JRASET, pp. 396–402. (2014)
3. Cody, W.F., Kreulen, J.T., Krishna, V., Spangler, W.S.: The integration of business intelligence and knowledge management. IBM Syst. J. **41**(4), 697–713 (2002)
4. Ceusters, W.: Medical natural language understanding as a supporting technology for data mining in healthcare. In: Cios, K.J. (ed.) Medical Data Mining and Knowledge Discovery, pp. 41–69. PhysicaVerlag Heidelberg, New York (2001)
5. Megalooikonomou, V., Herskovits, E.H.: Mining structure function associations in a brain image database. In: Cios, K.J. (ed.) Medical Data Mining and Knowledge Discovery, pp. 153–180. Physica-Verlag Heidelberg, New York (2001)
6. Biafore, S.: Predictive solutions bring more power to decision makers. Health Manag. Technol. **20**(10), 12–14 (1999)
7. Silver, M., Sakata, T., Su, H.C., Herman, C., Dolins, S.B., O'Shea, M.J.: Case study: how to apply data mining techniques in a healthcare data warehouse. J. Healthc. Inf. Manag. **15**(2), 155–164 (2001)
8. Benko, A., Wilson, B.: Online decision support gives plans an edge. Managed Healthc. Executive **13**(5), 20 (2003)
9. Quinlan, J.R.: C4.5: Programs for Machine Learning. Morgan Kauffman Publishers, San Mateo-California (1993)
10. Quinlan, J.R.: MDL and categorical theories (continued). In: Machine Learning: Proceedings of the Twelfth International Conference, pp. 464–470. Lake Tahoe, California. Morgan Kaufmann (1995)
11. Tang, T.-I., Zheng, G., Huang, Y., Shu, G., Wang, P.: A comparative study of medical data classification methods based on decision tree and system reconstruction analysis. IEMS **4**(1), 102–108 (2005)
12. Hall, M., Frank, E., Holmes, G., Pfahringer, B., Reutemann, P., Witten, I.H.: The WEKA data mining software: an update. SIGKDD Explor. **11**(1), 2009

13. Alcalá-Fdez, J., Sánchez, L., García, S., del Jesus, M.J., Ventura, S., Garrell, J.M., Otero, J., Romero, C., Bacardit, J., Rivas, V.M., Fernández, J.C., Herrera, F.: KEEL: a software tool to assess evolutionary algorithms to data mining problems. Soft Comput. 307–318 (2009)
14. Alcalá-Fdez, J., Fernandez, A., Luengo, J., Derrac, J., García, S., Sánchez, L., Herrera, F.: KEEL data-mining software tool: data set repository, integration of algorithms and experimental analysis framework. J. Multiple-Valued Logic Soft Comput. 17(2–3), 255–287 (2011)
15. Grzymala-Busse, J.W.: On the unknown attribute values in learning from examples. In: 6th International Symposium on Methodologies for Intelligent Systems (ISMIS'91). Lecture Notes in Computer Science, vol. 542, pp. 368–377. Springer, Charlotte (USA) (1991)

Prediction Strategy for Software Reliability Based on Recurrent Neural Network

Manmath Kumar Bhuyan, Durga Prasad Mohapatra
and Srinivas Sethi

Abstract Recurrent Neural Network (RNN) has been known to be very useful in predicting software reliability. In this paper, we propose a model that explores the applicability of Recurrent Neural Network with Back-propagation Through Time (RNNBPTT) learning to predict software reliability. The model has been applied on data sets collected across several standard software projects during system testing phase. Though the procedure is relatively complicated, the results depicted in this work suggest that RNN exhibits an accurate and consistent behavior in reliability prediction.

Keywords Software reliability prediction · Predictability measurement · Recurrent neural networks

1 Introduction

The standard definition of software reliability is the probability of the failure free operation of a computer system for a specified period in a specified environment [1–4]. Therefore, software reliability prediction becomes a crucial activity before releasing good quality software.

We propose a model *Recurrent Neural Network with Back-propagation Through Time (RNNBPTT)* that describes methodology to predict software reliability with the help of predictive parameters such as: Average Error (AE), Root Mean Square Error (RMSE), and Mean Absolute Error (MAE). Our scope of analysis in this

M.K. Bhuyan (✉) · D.P. Mohapatra · S. Sethi
CSEA Department, IGIT, Sarang, Utkal University, Bhabenswar, India
e-mail: manmathr@gmail.com

D.P. Mohapatra
e-mail: durga@nitrkl.ac.in

S. Sethi
e-mail: srinivas_sethi@igitsarang.ac.in

© Springer India 2016
H.S. Behera and D.P. Mohapatra (eds.), *Computational Intelligence in Data Mining—Volume 2*, Advances in Intelligent Systems and Computing 411, DOI 10.1007/978-81-322-2731-1_27

paper is on failure data at system testing time. Using sequence of failure data vectors (i.e. failure number) as input and a smaller feature sequence as a target (i.e. cumulative execution time to fail) the model compute the predictive parameters.

The rest of the paper is organized as follows: Sect. 2 describes the related work proposed so far in reliability prediction. Section 3 presents our proposed model RNNBPTT. The basic terminologies, application, architecture development, and step-by-step procedure for reliability prediction and training the proposed model is discussed in Sect. 4. Experimental result computation and observations are presented in Sect. 5. The conclusions and future work are presented Sect. 6.

2 Related Work

Artificial Neural Network (*ANN*) is a powerful technique for *Software Reliability Prediction* (*SRP*). The reliability of software can be measured at early stage of software development proposed by Singh et al. [5]. Yadav et al. [7] proposed a framework for product reliability estimation during development using fuzzy logic model. The authors [6–9] developed a connectionist model (i.e. feed forward network back-propagation) and taken failure data set as input to produce reliability as output. Raj Kiran et al. [11, 17] implemented the use of *wavelet neural networks* (*WNN*) and introduced three linear ensembles and one nonlinear ensemble to predict software reliability.

3 Back-Propagation Through Time Recurrent Neural Network Model Development

In this section, we discuss about our proposed model RNNBPTT architecture construction and its applicability in software reliability prediction. RNN is a dynamic network as it is a network with output feedback [12]. The RNN is a combination of temporal processing (i.e. using time-delay elements) of data and recurrent connections (i.e. feedback connections).

Figure 1 shows the RNN with BPTT learning that consists of cycles with its states and recursive relationship in nature. The network is unfolded in time, as shown in Fig. 1. It is nothing but the repetition of recurrent weights for an arbitrary number of times denoted as τ. Because of this, each node[1] sends activation along a recurrent connection, has at least τ number(s) of copies.

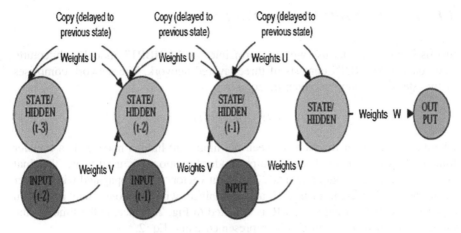

Fig. 1 An out-come of unfolding netword associated with recurrent back-propagation through time learning

4   Proposed Model RNNBPTT

As discussed in the earlier section, our proposed model RNNBPTT is a fully recurrent neural network, so the outputs of all nodes are recurrently connected to all nodes in the network. At each time, the input vector is propagated through a weight matrix V to the hidden layer and it is combined with the previous state activation through an additional recurrent weight layer as depicted in Fig. 2. A delay 'R' is introduced to simply delay the signal/activation until the next time step.

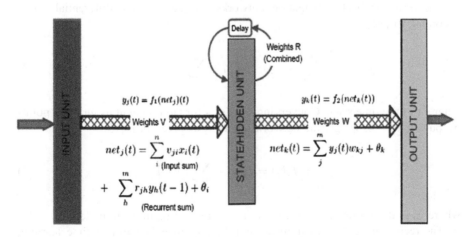

Fig. 2 A simple architecture of RNNBPTT

4.1 How RNNBPTT Model Work?

In this section, we discuss working principles of RNNBPTT model architecture. The conventional RNN consists of three-layered network. The network comprises of two steps activation as given in Eq. 1

$$y(t) = f_2(f_1(x(t))) \tag{1}$$

where f_1 is a transfer function between the input and hidden node, f_2 is a transfer function between the hidden and output node. Equation 1 is used to get the final output of the given input pattern 'x'. The input vector 'x' is propagated with a layer associated with weight matrix V and combined with previous state activation associated with recurrent weight R as depicted in Fig. 2. Hence, at the time 't', the input to the ith hidden unit can be represented using Eq. 2.

$$z_j(t) = f_1(net_j)(t) \tag{2}$$

The weight matrix W is associated with the hidden and output layers. Equation 3 is used to compute the output of the network.

$$y_k(t) = f_2(net_k(t)) \tag{3}$$

Here $net_k(t)$ is the net input to the model which receive external input (i.e. the failure number) and $y_k(t)$ as the output (predicted output). The desired state of unit i denoted as d_i corresponding to input x_i. The updated weight w_{ij} is calculated for each network copy that are summed up with inputs before individual weights are refined.

The recurrent back-propagation networks preside over the differential Eq. 4 according to [13].

$$\tau_i \frac{d\hat{y}i}{dx} = -y_i + f\left(x_i + \sum_{j=1}^{m} y_j w_{ij}\right) \tag{4}$$

where, τ_i is the time scale. Converging the Eq. 4 for target output using fixed points by putting $\hat{y}_i = 0$, we get the network output y_i as given in Eq. 5.

$$y_i = f(h_i) = f\left(x_i + \sum_{j=1}^{m} y_j w_{ij}\right) \tag{5}$$

where, x_i is the ith failure, $h_i = {}_j w_{ij} y_j + x_i$ is the net input to ith node.

The recurrent network is trained using back-propagation trough time learning with the training data and desired output patterns (as it is belongs to supervised learning rule). The output error (i.e. deference between the desired and predicted output) gradient is saved for each time step. Normally *Summed Square Error* (SSE)) 'L' in Eq. 6 use for training the neural network. It measures the deviation (i.e. difference between the desired and predicted values) of the network outputs $y_i(t)$ from the desired functions $d_i(t)$ from $t = t_0$ to $t = t_1$ for all copies of the output nodes.

$$L = \frac{1}{2} \sum_{k=1}^{n} (d_k(t) - y_k(t))^2 = \frac{1}{2} \sum_{k=1}^{n} L_k^2 \tag{6}$$

$$\text{where } L_k = \begin{cases} d_k - yk, & \text{for } k \text{ in } D \\ 0, & \text{otherwise} \end{cases}$$

Here, n is the total no of output nodes and is an index over the training sequence ($d_k(t)$ and $y_k(t)$ are the desired and predicted output functions of time respectively). In order to minimize the error weight updation is required. Using Gradient-descent method, this network gives a weight update rule w.r.t. the weights are presented in Eqs. 7 and 8.

$$\Delta W(h) = \gamma_m W(h-1) + \gamma_g L_k f_1'(h) y_k \tag{7}$$

Similarly,

$$\Delta V(h) = \gamma_m V(h-1) + \gamma_g L_k f_2'(h) x_k \tag{8}$$

where γ_m, $\gamma_g \varepsilon [0,1]$ are the constant parameters.

4.2 RNNBPTT Training

The training principle of RNNBPTT model is to change the free parameters of neural network according to the behavior of the network embedded in its environment. The aim of the training algorithm is to minimize the certain error using the cost function. The input layer outputs are feed-forward to the inputs of the next layer and the delayed layer outputs are fed back into the layer itself as shown in Fig. 2. So the hidden layer output at time (t − 1) for a certain layer acts as the state variable at time t for this layer.

We assume that there are multiple neurons present in the hidden layer. The input layer neurons are exempted from computation. The error is computed in the output layer and the difference is calculated between the predicted output and target output value. In our model, we used MSE as the stopping criterion. After the stopping criterion is satisfied, the cross-validation is carried out. The cross-validation process splits the entire representative data set into two sets: (a) a training data set, used to train the network, (b) a test data set used to predict the reliability of the system. We split the data set as follows: 2/3rd for training and 1/3rd for validation. The training data is in order sequence of failure sequence number) (CPU time) and desired output value (i.e. cumulative failure time). In this work, the logistic function binary sigmoidal $F(x) = 1/(1 + e^{-\lambda x})$ is used, where λ is the steepness parameter. The range of this transfer function varies from 0.0 to 1.0. The model is tested using MATLAB Version 7.10.0. The error precision here taken, is $E_{min} = 0.005$. The best weights are recorded for training data calculation and end-point-prediction of the reliability.

5 Experimental Results and Observations

In our reliability prediction experiment, we considered the failure data during system testing phase of various collected projects having defect severities 2 and 3 [14]. The data sets consist of (a) Failure Number (b) Time Between Failures (TBF) (c) Day of Failure, of this application. The list of some prediction parameters: Average Error (%): $AE_i = ((|F_i - A_i|/A_i|) * 100$, *Root Mean Square Error:* $RMSE = \left[\sqrt{\sum_1^n (Fi - Ai)^2} \right]$, *Normalized Root Mean Square Error:* $NRMSE = \left[\sqrt{\sum_1^n (F_i - A_i)^2} \right] / \sum_1^n F_i^2$ *and Mean Absolute Error:* $MAE == [\sum |F_i - A_i|]/n$, Here, F_i denotes the predicted output and A_i denotes the desire output of ith node.

The data set contains 191 numbers of failure histories in system testing phase. We experimented on the number of neurons present in the hidden layer varies from 25 to 50. Here, we analyze end-point prediction considering the data set. The values of various parameters such as AE, RMSE, and NRMSE for our experiment are listed in Table 1. Figure 3a–c summarize in terms of AE, RMSE and NRMSE.

In Table 1, the best results recorded for data set at 45 numbers of neurons in hidden layer for *long-term prediction (LTP)*. The LTP for measurement unit AE on training is shown in Fig. 3a (i.e. the desired output and predicted output) and deviation between actual and forecasted value. The performance and accuracy of the proposed network model against RMSE using data set is shown in Fig. 3c.

Table 1 Result as per hidden layer neurons	Neurons in each layer	AE	NRMSE	MAE	Epochs
	1,25,1	3.5123	0.0481	0.0348	735
	1,35,1	3.7634	0.0174	0.0313	101
	1,45,1	3.0029	0.0241	0.0321	1089
	1,50,1	5.6311	0.0498	0.0444	1202

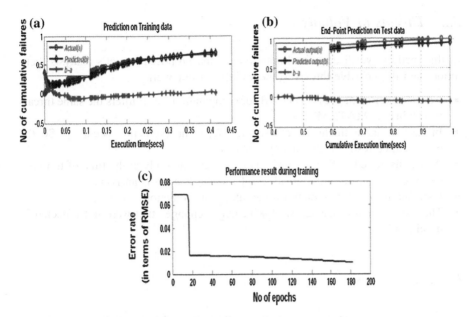

Fig. 3 End-point prediction using data. **a** Prediction on training data. **b** Prediction on training data. **c** Performance during training

Approach	NRMSE
Mohanty et al. [16]	0.076335
RNNBPTT	0.0241

Table 2 A comparison of NRMSE on data set

5.1 Observations

The accuracy and consistency of software are measured by the value of Normalized Root Mean Square Error (NRMSE). It is observed from Table 2 that for our proposed model NRMSE values is reasonably less (Table 2).

Model quality is observed if its predictions are close to the ideal line passing through the zero error [15]. Figure 3a–c show the prediction closeness between the desire value and predicted value.

Some advantages of using recurrent neural network with back-propagation learning rule are listed below:

- It supports the project manager to monitor testing, estimating the project schedule, and helping the researchers to evaluate the reliability model.
- If the data set is having more fluctuation then it will affect the prediction accuracy.

5.2 Threats to Validity

In this section, we discuss the limitations of the proposed model's analysis techniques and the possible threats to the validity of our work.

- As our experiment uses MATLAB for computation, so it suffers the same threats to validity as MATLAB does.
- In case of large fluctuate data set, it affects the prediction accuracy. So the performance result of the model is affected.
- As we discussed in Sect. 5, weights are taken randomly at the time of training, so for every run the same result on the same data set is an issue.
- Poor quality assurance methods can affect in prediction.
- The model cannot manage well with major changes that were not reflected in training phase.

6 Conclusions

In this model, we presented a novel technique for software reliability prediction using recurrent neural network with back-propagation through time technique. We presented experimental evidence showing that recurrent network using back propagation is giving the accurate result comparable to other methods. This model is easily compatible with deferent none fluctuate data set and projects. This work will be performed using hybrid techniques.

References

1. IEEE: Standard glossary of software engineering terminology. Standards Coordinating Committee of the IEEE Computer Society (1991)
2. Boland, P.J.: Challenges in software reliability and testing. Technical report, Department of Statistics National University of Ireland, Dublin Belfield—Dublin 4 Ireland (2002)
3. Khatatneh, K., Mustafa, T.: Software reliability modeling using soft computing technique. Eur. J. Sci. Res. **26,** 154–160 (2009)
4. Musa, J.D., Okumoto, K.: A logarithmic poisson execution time model for software reliability measurement. In: Proceedings of the 7th International Conference on Software Engineering, ICSE. IEEE Press, Piscataway, NJ, USA, pp. 230–238 (1984)
5. Singh, L.K., Tripathi, A.K., Vinod, G.: Software reliability early prediction in architectural design phase: overview and limitations. J. Softw. Eng. Appl. **4,** 181–186 (2011)
6. Yin, M.L., Hyde, C., James, L.E.: A petri-net approach for early-stage system-level software reliability estimation. In: Reliability and Maintainability Symposium (2000)
7. Yadav, O.P., Singh, N., Chinnam, R.B., Goel, P.S.: A fuzzy logic based approach to reliability improvement estimation during product development. Reliab. Eng. Syst. Saf. (Elsevier) **80,** 63–74 (2003)

8. Khoshgoftaar, T.M., Pandya, A.S., More, H.: A neural network approach for predicting software development faults. In: Proceedings of Third International Symposium on Software Reliability Engineering. Research Triangle Park, NC, pp. 83–89 (1992)
9. Singh, Y., Kumar, P.: Prediction of software reliability using feed forward neural networks. In: Conference, I. (ed.): Computational Intelligence and Software Engineering (CiSE), IEEE, pp. 1–5 (2010)
10. Thwin, M.M.T., Quah, T.S. (eds.): Application of neural network for predicting software development faults using object-oriented design metrics, vol. 5. In: Proceedings of the 9th International Conference on Neural Information Processing (ICONIP'02) (2002)
11. RajKiran, N., Ravi, V.: Software reliability prediction using wavelet neural networks, vol. 1. In: International Conference on Computational Intelligence and Multimedia Applications, pp. 195–199. Sivakasi, Tamil Nadu, IEEE (2007)
12. Hush, D.R., Herne, B.G.: Progress in supervised neural networks. IEEE Signal Process. Mag. **10,** 8–39 (1993)
13. Lin, C.T., Lee, C.G.: Neural Fuzzy Systems: A Neuro-Fuzzy Synergism to Intelligent Systems. Prentice Hall, Inc (1996)
14. Musa, J.D.: Software Reliability Data. Data & Analysis Center for Software (1980)
15. Karunanithi, N., Whitley, D.: Prediction of software reliability using feed-forward and recurrent neural nets. In: Baltimore, M.D. Neural Networks, IJCNN IEEE, vol. 1, pp. 800–805 (1992)
16. Mohanty, R., Ravi, V., Patra, M.R.: Hybrid Intelligent Systems for Predicting Software Reliability. Appl. Soft Comput. **13,** 189–200 (2013)
17. RajKiran, N., Ravi, V.: Software reliability prediction by soft computing techniques. J. Syst. Softw. **81,** 576–583 (2008)

A New Approach to Fuzzy Soft Set Theory and Its Application in Decision Making

B.K. Tripathy, T.R. Sooraj and R.K. Mohanty

Abstract Soft set theory is a new mathematical approach to vagueness introduced by Molodtsov. This is a parameterized family of subsets defined over a universal set associated with a set of parameters. In this paper, we define membership function for fuzzy soft sets. Like the soft sets, fuzzy soft set is a notion which allows fuzziness over a soft set model. So far, more than one attempt has been made to define this concept. Maji et al. defined fuzzy soft sets and several operations on them. In this paper we followed the definition of soft sets provided by Tripathy et al. through characteristic functions in 2015. Many related concepts like complement of a fuzzy soft set, null fuzzy soft set, absolute fuzzy soft set, intersection of fuzzy soft sets and union of fuzzy soft sets are redefined. We provide an application of fuzzy soft sets in decision making which substantially improve and is more realistic than the algorithm proposed earlier by Maji et al.

Keywords Soft sets · Fuzzy sets · Fuzzy soft sets · Decision making

1 Introduction

Uncertainty based models take care of handling modern day databases as uncertainty has become a part of these databases now. Several such models exist in literature [1]. There are several applications of such databases like data clustering [2–4], approximate reasoning [5, 6], cloud computing [7], medical diagnosis and optimization [8]. Several such applications of such models exist in the field of

B.K. Tripathy (✉) · T.R. Sooraj · R.K. Mohanty
School of Computing Science, VIT University, Vellore, Tamil Nadu, India
e-mail: tripathybk@vit.ac.in

T.R. Sooraj
e-mail: soorajtr19@gmail.com

R.K. Mohanty
e-mail: rknmohanty@gmail.com

© Springer India 2016   305
H.S. Behera and D.P. Mohapatra (eds.), *Computational Intelligence in Data Mining—Volume 2*, Advances in Intelligent Systems and Computing 411, DOI 10.1007/978-81-322-2731-1_28

Computational Intelligence [9]. Zadeh [10] initiated the concept of fuzzy sets in 1965 which is considered as generalization of classical or crisp sets. In the Zadehian definition, it has been accepted that the classical set theoretic axioms of exclusion and contradiction are not satisfied. In 1999, Molodtsov [11] introduced the concepts of soft sets, which is parameterized family of subsets defined over a universe and a set of parameters as a model to capture uncertainty and vagueness in data. Many new operations on soft sets were introduced by Maji et al. [12, 13] in a later paper. Maji et al. [14] put forward the concept of fuzzy soft sets, which is a hybrid model of fuzzy sets and soft sets. Membership function plays an important role in fuzzy set theory. The earlier approaches to define fuzzy soft set theory were not strictly following the membership functions approach in fuzzy set theory. We define membership functions for fuzzy soft sets in this paper. We also reframe the fundamental notions and operations on soft sets using the associated membership functions. Soft set theory has lot of applications. Some applications are discussed by Molodtsov in [11]. Maji et al. discussed an application of soft sets in decision making problems [4]. We identified some problems in decision making problem and made suitable changes. In this paper, we define union of fuzzy soft sets, intersection, complement and some other operations of fuzzy soft sets.

It has been observed that till date we don't have a definition of membership function of a fuzzy soft set. It is well known that sets are synonymous with their characteristic functions. However, no such definition was in existence for soft sets. Tripathy et al. [15] defined the notion of characteristic function of soft set. This new definition, besides being concise has systematized definitions of many existing operations on soft sets and using this new definition the proofs have become elegant also. In this paper, we define membership functions of fuzzy soft sets and related notions. In this paper we also introduce an application of fuzzy soft sets in decision making problems.

2 Definitions and Notions

Definition 2.1 (*Soft Set*) Let U be the universal set of elements and E be a set of parameters. The pair (U, E) is often regarded as a soft universe. Members of the universe and the parameter set are generally denoted by x and e respectively. Let A be the subset of E. A soft set over the soft universe (U, E) is denoted by (F, A), where

$$F : A \rightarrow P(U) \tag{1}$$

Definition 2.2 (*Soft Multiset*) A soft multiset over (U, E) is denoted by (M, A), where $A \subseteq E$ is such that $M : A \rightarrow P^*(U)$.

Every soft multiset (M, A) can be defined to be associated by a parametric family of count functions $\{C^a_{(M,A)}, a \in A\}$, where $C^a_{(M,A)} : U \rightarrow J$ is given by $C^a_{(M,A)}(x)$ denotes the number of times occurs in $M(a)$. Where $P^*(U)$ is the set of all sub multisets of U.

Definition 2.2 (*Fuzzy Soft Set*) Let U be an initial universal set and let E be a set of parameters. Let I^U denote the set of all fuzzy subsets of U. Let $A \subseteq E$. A pair (F, E) is called a fuzzy soft set over U, where F is a mapping given by

$$F : A \rightarrow I^U \tag{2}$$

3 Fuzzy Soft Sets

In this section, we introduce membership function for fuzzy soft sets. We also define all the operations of fuzzy soft sets based on the membership function. Let (F, A) be a fuzzy soft set over (U, E). Then we define the set of parametric membership functions $\mu_{(F,A)} = \left\{ \mu^a_{(F,A)} | a \in A \right\}$ of (F, A) as follows:

Definition 3.1 For any $\forall a \in A$, we define the membership function as follows.

$$\mu^a_{(F,A)} (x) = \alpha, \alpha \in [0, 1] \tag{3}$$

Definition 3.2 For any two fuzzy soft sets (F, A) and (G, B) over a common universe (U, E), the union of (F, A) and (G, B) is the fuzzy soft set (H, C) where $C = A \cup B$, and $\forall a \in C$ and $\forall x \in U$, we have

$$\mu^a_{(H,C)} (x) = \max \left\{ \mu^a_{(F,A)} (x), \mu^a_{(G,B)} (x) \right\} \tag{4}$$

Definition 3.3 For any two fuzzy soft sets (F, A) and (G, B) over a common universe (U, E), the intersection of (F, A) and (G, B) is the fuzzy soft set (H, C) where $C = A \cap B$, and $\forall a \in C$ and $\forall x \in U$, we have

$$\mu^a_{(H,C)} (x) = \min \left\{ \mu^a_{(F,A)} (x), \mu^a_{(G,B)} (x) \right\} \tag{5}$$

Definition 3.4 A fuzzy soft set (F, E) over the universe (U, E) is said to be absolute fuzzy soft set if $F(e) = U, \forall e \in E$. So, we have $\mu^e_U (x) = 1$.

Definition 3.5 A fuzzy soft set (F, E) over the universe (U, E) is said to be null fuzzy soft set if $F(e) = \phi, \forall e \in E$. So, we have $\mu^e_\phi (x) = 0$.

We note here that with the change of definition of absolute fuzzy soft set and null fuzzy soft set we have, for any soft set (F, A) defined over (U, E).

$$(F, A) \cup U = U. \tag{6}$$

$$(F, A) \cap \phi = (F, A) \tag{7}$$

Definition 3.6 Given two fuzzy soft sets (F, A) and (G, B) over a common soft universe (U, E), (F, A) is said to be fuzzy soft subset of (G, B), written as $(F,A) \subseteq (G,B)$, if $A \subset B$ and $\forall a \in A$, $\forall x \in U$,

$$\mu^a_{(F,A)}(x) \leq \mu^a_{(G,B)}(x) \tag{8}$$

Definition 3.7 For any two fuzzy soft sets (F, A) and (G, B) over a common soft universe (U, E), we say that (F, A) is equal to (G, B) written as $(F, A) = (G, B)$ if $A = B$ and $\forall x \in U$, $\forall a \in A$,

$$\mu^a_{(F,A)}(x) = \mu^a_{(G,B)}(x) \tag{9}$$

Definition 3.8 For any two fuzzy soft sets (F, A) and (G, B) over a common soft universe (U, E), we define the complement (H, C) of (G, B) in (F, A) as $\forall a \in A$ and $\forall x \in U$.

$$\mu^a_{(H,C)}(x) = \max\left\{0, \mu^a_{(F,A)}(x) - \mu^a_{(G,B)}(x)\right\} \tag{10}$$

Definition 3.9 The complement of a fuzzy soft set over a soft universe (U, E) can be derived from the Definition 3.8 by taking (F, A) as U and (G, B) as (F, A). We denote it by $(F,A)^c$ and clearly $\forall x \in U$ and $e \in E$,

$$\mu^a_{(F,A)^c}(x) = \max(0, \mu^a_U(x) - \mu^a_{(F,A)}(x)) \tag{11}$$

From definition (11) it can be easily derived that

$$\mu^a_{(F,A)^c}(x) = 1 - \mu^a_{(F,A)}(x)) \tag{12}$$

3.1 Lambda Cuts

Definition 3.1.1 A fuzzy soft set $(F,A)_\lambda$ is called the lambda cut of a fuzzy soft set (F, A), where

$$(F,A)_\lambda = \{x | \mu^a_{(F,A)}(x) \geq \lambda$$

3.1.1 Properties of Lambda Cuts

i. $((F,A) \cup (G,B))_\lambda = (F,A)_\lambda \cup (G,B)_\lambda \tag{13}$

ii. $((F,A) \cap (G,B))_\lambda = (F,A)_\lambda \cap (G,B)_\lambda \tag{14}$

Proof of (3.12)

$$\mu^a_{((F,A)\cup(G,B))}(x) = \max\left\{\mu^a_{(F,A)}(x), \mu^a_{(G,B)}(x)\right\}$$

$$\text{So, } x \in ((F,A)\cup(G,B))_\lambda \Leftrightarrow \max\left\{\mu^a_{(F,A)}(x), \mu^a_{(G,B)}(x)\right\} \geq \lambda$$

$$\Leftrightarrow \mu^a_{(F,A)}(x) \geq \lambda \text{ or } \mu^a_{(G,B)}(x) \geq \lambda$$

$$\Leftrightarrow x \in (F,A)_\lambda \text{ or } x \in (G,B)_\lambda$$

$$\Leftrightarrow x \in (F,A)_\lambda \cup (G,B)_\lambda$$

Proof of (3.13) is similar to the proof of (3.12).

4 Application of Fuzzy Soft Set in Decision Making

In [11] Molodtsov has given several applications of soft set theory in game theory, operations research, Riemann-integration, Perron integration, probability theory etc. In [12] Maji et al. has given an application of fuzzy soft sets in a decision making system. But the algorithm given in that paper has some issues. Some of those issues are:

(i) If there is some parameters like "Expensive" or "Distance", then the given decision making algorithm will give a wrong result. Because the values of these type of parameters affects the decision inversely. We call these parameters as negative parameter.

(ii) In comparison table, if one parameter of a house is greater than or equal to other, then the algorithm is taking the count as 1 and other as 0. This may affect the decision making result if the difference is very low.

(iii) If a beautiful house is slightly expensive then also someone may want to buy it; which is not taken into account in the decision making algorithm [12]. A customer's interest over any particular parameters is not entertained in [12].

All of the above issues are addressed in this paper below.
The parameters can be categorized as two types. (i) If the value of the parameter is directly proportional to the interest of a person than we say that, it is a positive parameter. (ii) If the value of the parameter is inversely proportional to the interest of a person than we say that, it is a negative parameter. For example 'Beautiful' is a positive parameter. If the value of parameter 'Beautiful' increases then the customer's interest will also increase. Whereas 'Expensive' is a negative parameter. Because, if the value of the parameter 'Expensive' increases then the interest of customer will decrease. So, we are restricting the user to give the priority for a −ve parameter in the interval $(-1,0)$. This will solve the first issue of algorithm for decision making in [12].

To tackle the second issue, we are taking the difference of two fuzzy values in comparison table instead of adding 0's and 1's.

To handle the third issue, we prioritize the parameters by multiplying with priority values given by the customer. The priority is a real number lying in the interval $[-1, 1]$. When a parameter value does not affect the customer's decision then the priority will be 0 (zero). If a parameter value affects positively to customer's decision then the priority will be $(0, 1]$ and if a parameter value affects negatively to customer's decision then the priority will be $[-1, 0)$. If priority value is not given for one or more parameters then the value of the priority is assumed to be 0 by default and that parameter can be eliminated from further computation. To get even more reduction in computation we can keep only one object if there is some objects with same values for all parameters. The customer's priority value will be multiplied by the parameter value. This will give the values for priority table.

The comparison table can be obtained by taking the difference of row sum of a house with others in priority table. The score of each house can be obtained by calculating row sum in comparison table. The house having more score will be more suitable to customer's requirement.

The example we consider is as follows:

Let U be a set of houses given by $U = \{h_1, h_2, h_3, h_4, h_5, h_6\}$ and E be the parameter set given by $E = \{$Beautiful, Wooden, Green Surrounded, Expensive, Distance$\}$.

Consider a fuzzy soft set (U, E) which describes the 'attractiveness of houses', given by $(U, E) = \{$Beautiful Houses $= \{h_1/0.1, h_2/1.0, h_3/0.3, h_4/0.7, h_5/0.3, h_6/0.9\}$, Wooden Houses $= \{h_2/0.6, h_3/0.1, h_4/0.7, h_5/0.4, h_6/0.5\}$, Green Surrounded Houses $= \{h_1/0.2, h_2/0.8, h_3/0.2, h_4/0.6, h_5/0.4, h_6/0.6\}$, Expensive Houses $= \{h_1/0.1, h_2/0.8, h_3/0.3, h_4/0.6, h_5/0.5, h_6/0.6\}$, Distance houses $= \{h_1/0.8, h_2/0.3, h_3/0.4, h_4/0.6, h_5/0.1, h_6/0.5\}\}$.

Suppose a customer Mr. X wants to buy a house out of given houses which suits his needs on the basis of choice parameters such as 'Beautiful', 'Wooden', 'Green Surrounded', 'Expensive', 'Distance'. That means out of all available houses, he needs to select a house according to his priorities which qualifies with maximum number of parameters of the parameter set.

Table 1 Tabular representation of the fuzzy soft set (U,E)

U	Beautiful	Wooden	Green surrounded	Expensive	Distance
h_1	0.1	0.0	0.2	0.1	0.8
h_2	0.9	0.6	0.8	0.8	0.3
h_3	0.3	0.1	0.2	0.3	0.4
h_4	0.7	0.7	0.6	0.6	0.6
h_5	0.3	0.4	0.4	0.5	0.1
h_6	0.9	0.5	0.6	0.6	0.5

4.1 Algorithm

1. Input the fuzzy soft set (U, E) and arrange that in tabular form (Table 1).
2. Input the priority given by the customer for every parameter. For positive parameters the priority must has to be in the interval (0,1) and for negative parameters the priority must has to be in the interval (−1,0). If priority of any parameter has not given, than take it as 0 (zero) by default and opt out from further computation.
3. Multiply the priority values with the corresponding parameter values to get the priority table (Table 2).
4. Compute the row sum of each row in the priority table.
5. Construct the comparison table by finding the entries as differences of each row sum with those of all other rows.
6. Compute the row sum for each row in the comparison table to get the score (Table 3).
7. Construct the decision table by taking the row sums in the comparison table (Table 4).
8. The object having highest value in the score column is to be selected. If more than one object is having the same score then the object having higher priority value is to be selected.

The priority given by the customer Mr. X for all parameters for Beautiful, wooden, Green Surrounded, Expensive, Distance respectively are 0.7, 0.0, 0.2, −0.5, −0.2. The priority value of the parameters 'Expensive' and 'Distance' is negative, which indicates that these parameters are negative parameters.

Decision Making: The Customer should go for the house which has highest score. If, there is same score for two houses, then the greater value in highest priority column will decide the best suitable house and so on. If, for any reason, the customer don't want that then he/she can choose next highest and so on.

Table 2 Priority table

U	Beautiful	Green surrounded	Expensive	Distance
h_1	0.07	0.04	−0.05	−0.16
h_2	0.63	0.16	−0.40	−0.06
h_3	0.21	0.04	−0.15	−0.08
h_4	0.49	0.12	−0.30	−0.12
h_5	0.21	0.08	−0.25	−0.02
h_6	0.63	0.12	−0.30	−0.10

Note that there is no column for the parameter 'wooden', because its priority is zero.

Table 3 Comparison table

U	h_1	h_2	h_3	h_4	h_5	h_6
h_1	0.00	−0.43	−0.12	−0.29	−0.12	−0.45
h_2	0.43	0.00	0.31	0.14	0.31	−0.02
h_3	0.12	−0.31	0.00	−0.17	0.00	−0.33
h_4	0.29	−0.14	0.17	0.00	0.17	−0.16
h_5	0.12	−0.31	0.00	−0.17	0.00	−0.33
h_6	0.45	0.02	0.33	0.16	0.33	0.00

Table 4 Decision table

Houses	Score
h_1	−1.41
h_2	1.17
h_3	−0.69
h_4	0.33
h_5	−0.69
h_6	1.29

5 Conclusions

The definition of soft set using the characteristic function approach was provided by Tripathy et al. recently, which besides being able to take care of several definitions of operations on soft sets and could make the proofs of properties very elegant. For fuzzy soft sets no such approach was in existence. In this paper we introduced the membership function for fuzzy soft sets which extends the notion of characteristic function introduced by Tripathy et al. in 2015. With this new definition we redefined many concepts associated with fuzzy soft sets and established some of their properties. Earlier fuzzy soft sets were used for decision making by Maji et al. Their approach had many flaws. In this paper we pointed out these flaws and provided solutions to rectify them, so that the decision making becomes more efficient and realistic.

References

1. Tripathy, B.K.: Rough sets on fuzzy approximation spaces and intuitionistic fuzzy approximation spaces. In: Abraham, A., Falcon, R., Bello, R. (eds.): Rough Set Theory: A True landmark in Data Analysis, pp. 3–44. Springer International Studies in Computational Intelligence, vol. 174 (2009)
2. Tripathy, B.K., Tripathy, A. and Govindarajulu, K.: Possibilistic rough fuzzy C-means algorithm in data clustering and image segmentation. In: Proceedings of the IEEE ICCIC2014, pp. 981–986 (2014)

3. Tripathy, B.K., Tripathy, A., Govindarajulu, K., Bhargav, R.: On kernel based rough intuitionistic fuzzy C-means algorithm and a comparative analysis. In: ICACNI 2014, Advanced Computing Networking and Informatics, vol. 1. Smart Innovation Systems and Technologies, vol. 27, pp. 349–359 (2014)
4. Tripathy, B.K., Tripathy, A. Govindarajulu, K.: On PRIFCM algorithm for data clustering, image segmentation and comparative analysis. In Accepted for Presentation at the IACC Conference. Bangalore (2015)
5. Tripathy, B.K., Jawahar, A. and Vats, E.: An analysis of generalised approximate equalities based on rough fuzzy sets. In: Proceedings of the International Conference on SocPros 2011, Dec 22–24. IIT-Roorkee, AISC 130, Springer India, vol. 1, pp. 333–346 (2012)
6. Tripathy, B.K., Vats, E., Jhawar, A., Chan, C.S.: Generalised approximate equalities based on rough fuzzy sets & rough measures of fuzzy sets. In: IEEE FUZZ, pp. 1–6 (2013)
7. Tripathy, B.K., Shivalkar, P.: Rough set based green cloud computing in emerging markets. In: Encyclopedia of Information Science and technology. IGI publications, Chapter 103, pp. 286–295 (2014)
8. Tripathy, B.K., Arun, K. R.: Soft set theory and applications. In: Accepted for Publication in IGI Volume by Sunil Jacob John (2015)
9. Tripathy, B.K.: Rough sets and approximate reasoning. In: Tripathy, B.K., Acharjya, D. P. (eds): Global Trends in Intelligent Computing Research and Development-Advances in Computational Intelligence and Robotics Book Series, pp. 180–228. IGI Publications (2014)
10. Zadeh, L.A.: Fuzzy sets. Inf. Control **8**, 338–353 (1965)
11. Molodtsov, D.: Soft set theory—first results. Comput. Math Appl. **37**, 19–31 (1999)
12. Maji, P.K., Biswas, R., Roy, A.R.: An application of soft sets in a decision making problem. Comput. Math Appl. **44**, 1007–1083 (2002)
13. Maji, P.K., Biswas, R., Roy, A.R.: Soft set theory. Comput. Math Appl. **45**, 555–562 (2003)
14. Maji, P.K., Biswas, R., Roy, A.R.: Fuzzy soft sets. J. Fuzzy Math. **9**(3), 589–602 (2001)
15. Tripathy B.K., Arun K.R.: A new approach to soft sets, soft multisets and their properties. Int. J. Reasoning-Based Intell. Syst. **7**(3/4), 244–253 (2015)

Design of Optimized Multiply Accumulate Unit Using EMBR Techniques for Low Power Applications

K.N. Narendra Swamy and J. Venkata Suman

Abstract Composite operations of arithmetic are extensively used in the applications of Digital Signal Processing (DSP). An optimized Multiply Accumulator Unit using fused Add-Multiply (FAM) operator by exploring structured and proficient recoding methods utilizing them. This paper deals with the study of performance comparisons of 16-bit and 32-bit MAC design based on EMBR techniques in terms of look up tables and power utilization with 8-bit and 16-bit recoding form of Modified Booth (MB) multiplier.

Keywords Efficient modified booth recoding · Fused add multiply · Signed multiplier · Multiply accumulate

1 Introduction

In previous years, multiplication was typically executed via a series of addition and shift operations. Presently, electronic applications of making sufficient use of Digital Signal Processing, based on whole host of arithmetic operations. Multiplication is one of them, recurring additions. System's presentation primarily depends on multiplier performance. Three different schemes incorporated in Fused Add-Multiply (FAM) [1, 2] design for the reduction related to power. The actual process of multiplication involves two integers (multiplicand and multiplier). In this process, first one that comes is multiplicand and the second is multiplier. All the partial products produced in each step are summed. Reiterated mode of addition, which recommends in arithmetic model, is deliberate which substitutes the algorithm. Multipliers are decomposed into two parts. The formation of partial products

K.N. Narendra Swamy (✉) · J. Venkata Suman
Department of ECE, GMR Institute of Technology, Rajam, India
e-mail: nariswami27@gmail.com

© Springer India 2016
H.S. Behera and D.P. Mohapatra (eds.), *Computational Intelligence in Data Mining—Volume 2*, Advances in Intelligent Systems and Computing 411, DOI 10.1007/978-81-322-2731-1_29

315

is the first part, collecting and adding them is the second one. MAC [3, 4] results of 16-bit and 32-bit for signed numbers to be generated using efficient techniques of modified booth recoding (EMBR) in three various schemes of FAM design. The final Carry Look Ahead (CLA) [5] adder and the Carry Save Adder (CSA) tree used for speed operation.

Actually, in case of real-time signal processing necessitates fast and efficient multiplier-accumulator (MAC) unit which put away less power, to attain a high performance. Low power MAC unit with efficient modified booth recoding technique to save power is to be designed. This design of MAC is done with relevant geometries which provide improved delay and power. The delay based on controller that overcomes the flow of data between the MAC [4] blocks in case of low power.

2 Conceptual Briefing

2.1 Motivation

The propagation of carry is overwhelming, and is repeated for each and every partial product for summation. An analogous method was initially proposed by Booth [3]. The unique Booth's algorithm uses adjoining series of 1's with the possession that: $2 + 2(n - 1) + 2(n - 2) + \cdots + 2n) = 2(n + 1) - 2(n - m)$. Even though set of rules for Booth yields the majority of N/2 programmed partial products using N bit operand, thus formed differs from others. Because of this only, modified forms of Booth's Algorithm are designed and simulated using Xilinx tool which can be used in applications of low power.

2.2 The Unit of Multiplier Accumulator (MAC)

Processor units involve MAC, which greatly influences the speed of it. MAC is collection of multiplier, adder, and accumulator. The MAC inputs were repossessed from memory and given to multiplier unit of MAC which performs multiplication and results to adder, accumulator and restore to location of memory. Complete procedure is performed by one clock period. The design of two MAC units contains 8-bit Modified Booth Multiplier, 16-bit accumulator and 16-bit multiplier, 32-bit accumulator. The values of U and V are multiplied by using Modified Booth multiplier [6] as a substitute of conventional multiplier because that boost the MAC devise speed and decrease multiplication complication. The product $U \times V$ is fed to the 64-bit accumulator and it is summed subsequently to product $U_i \times V_i$. This unit

Fig. 1 Block diagram of
multiplier and accumulator

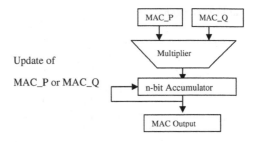

of MAC is proficient of multiply and add the preceding product successively several times. The common structural design of this MAC is given in Fig. 1. The act of multiplication is executed by P with multiplicand Q, summation of earlier multiplication to accumulator as output in gathering at the final stage.

2.3 Modified Form of Booth

The new form of Booth, as Modified Booth (MB) [6, 7] is utilized in multiplication. Here generation of $n/2 + 1$ partial product is done. This reduces half of the partial products in multiplication related to other representation radix.

2.3.1 Algorithm

1. LSB is padded with one zero.
2. MSB is padded with 2 0's if n is even or 1 '0' if n is odd.
3. Overlapping groups into 3-bits by dividing the multiplier.
4. Partial product scale factor is determined from modified booth into encoding table.
5. The Multiplicand Multiples are computed.
6. Partial Products are then summed.

The encoding process can be done by considering three bits at first and then to add multiplicand times −2, −1, 0, 1 and 2. Meanwhile Booth recoding [2] must get cleared of 3's; creating of partial products is very much easy. When grouping of partial products completed, are added, weighted appropriately, using a Carry-Save Adder (CSA) tree. Formation of a carry signal in case of two circumstances using carry look ahead adder: (1) when both bits e and f are 1, or (2) one of them is 1 and the carry-in is 1. Using Carry Look Ahead adder (CLA) the problem of carry delay is resolved by computing the carry in advance, focusing on the input signals. This process of addition lessens all partial-products to carry-save numerals by summing up them in an adder tree (Table 1).

Table 1 Grouping table

y_{2j+1}	y_{2j}	y_{2j-1}	y_j^{MB}	s_j	one_j	two_j	c_{inj}
0	0	0	0	0	0	0	0
0	0	1	1	0	1	0	0
0	1	0	1	0	1	0	0
0	1	1	2	0	0	1	0
1	0	0	-2	1	0	1	1
1	0	1	-1	1	1	0	1
1	1	0	-1	1	1	0	1
1	1	1	0	1	0	0	0

$one_j = y_{2j-1} \oplus y_{2j}$ $\quad two_j = (y_{2j+1} \oplus y_{2j}) \cdot \overline{one}_j$ $\quad s_j = y_{2j+1}$

3 Recoding Techniques—Sum to Modified Booth (S-MB)

3.1 Signed Arithmetic Structure

For the reduction of quantity of partial products and their summation, both signed and conventional HAs and FAs are used [8], each of the schemes can be certainly applied in both signed and un-signed numbers that comprise of even (2 k) bits. In this S-MB recoding performance technique, the sum of two successive bits of the input $E\left(e_{2j}, e_{2j+1}\right)$ along with two bits of the input $F\left(f_{2j}, f_{2j+1}\right)$ into one MB digit y_j^{MB} more precisely recoded. In the process of recoding, two versions of signed HAs used which can be denoted as HA* and HA**. Truth tables of corresponding HA*, HA** and their Boolean equations are considered [1].

$$V = E + F = y_k \cdot 2^{2k} + \sum_{j=0}^{k-1} y_j^{MB} \cdot 2^{2j} \tag{1}$$

where $y_j^{MB} = -2s_{2j+1} + s_{2j} + c_{2j}$.

3.2 S-MB Recoding Techniques

3.2.1 Input Numbers in Signed Form

In case of signed input, the most significant bits numbers E and F are filled with 1 (negative number). Figures 2, 3 and 4 representation of S-MB schemes of even (two most significant digits change) bit-width of E and F, regarding the signs of the most significant bits of E and F is done here. These each schemes are implemented in FAM model [1, 2].

(1) Recoding Technique of S-MB1:

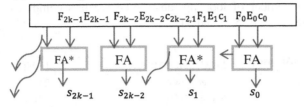

Fig. 2 Even bit signed S-MB1

(2) Recoding Technique of S-MB2:

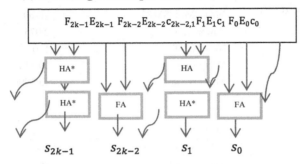

Fig. 3 Even bit signed S-MB2

(3) Recoding Technique of S-MB3:

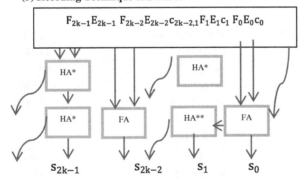

Fig. 4 Even bit signed S-MB3

4 Results

Performance comparison between 16-bit and 32-bit MAC design in FAM [2] model EMBR based is designed simulated using Xilinx simulation tool. In case of 16-bit MAC with 8-bit multiplier, power and delay of third scheme is recommended as the number of look up tables used are less compared to other two schemes. In case of

32-bit MAC with 16-bit multiplier, power and delay of second scheme is recommended as the number of look up tables used are less compared to other two schemes (Tables 2, 3, 4, 5, 6 and 7).

Table 2 16-bit MAC using SMB1

MAC_Mul_SMB1	16-bit
Power (mw)	2.99
No of LUT's	171
Delay (ns)	19.298
Memory (kb)	139,996

Table 3 16-bit MAC using SMB2

MAC_Mul_SMB2	16-bit
Power (mw)	2.975
No of LUT's	170
Delay (ns)	19.353
Memory (kb)	139,996

Table 4 16-bit MAC using SMB3

MAC_Mul_SMB3	16-bit
Power (mw)	2.88
No of LUT's	165
Delay (ns)	19.109
Memory (kb)	139,996

Table 5 32-bit MAC using SMB1

MAC_Mul_SMB1	32-bit
Power (mw)	11.44
No of LUT's	654
Delay (ns)	28.616
Memory (kb)	146,140

Table 6 32-bit MAC using SMB2

MAC_Mul_SMB2	32-bit
Power (mw)	11.04
No of LUT's	631
Delay (ns)	28.463
Memory (kb)	147,164

Table 7 32-bit MAC using SMB3

MAC_Mul_SMB3	32-bit
Power (mw)	11.27
No of LUT's	644
Delay (ns)	29.164
Memory (kb)	147,164

4.1 RTL Schematic and Simulation of 16-Bit MAC Using SMB3

See Figs. 5 and 6.

4.2 RTL Schematic and Simulation of 32-bit MAC using SMB2

See Figs. 7 and 8.

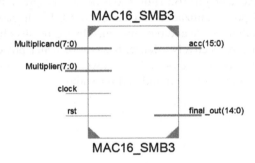

Fig. 5 RTL schematic diagram of 16-bit MAC using SMB3

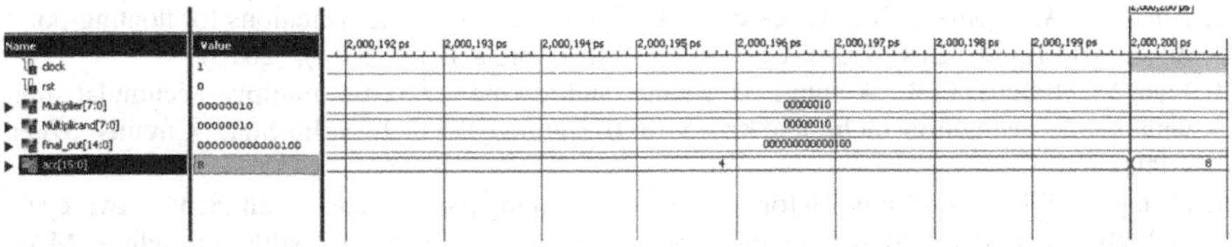

Fig. 6 RTL schematic diagram of 16-bit MAC using SMB3

Fig. 7 RTL schematic
diagram of 32-bit MAC using
SMB2

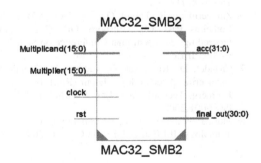

Fig. 8 RTL schematic diagram of 32-bit MAC using SMB2

5 Conclusion

In this paper, the study of performance comparisons of multiply accumulate unit of 16-bit and 32-bit in fused add-multiply using EMBR Techniques is tabulated. The design of 16-bit MAC using SMB3 shows considerable progress then the other two schemes in terms of power utilization and number of LUT's. In case of 32-bit MAC using SMB2 is the efficient one and recommended by the results obtained compared to other two schemes respectively. Hence, because of better power efficiency, the design is used in the applications of Low Power. In future, this model can be extended for more number of even and odd bit widths.

References

1. Taoumanis, K., Xydis, S., Efstathiou, C., Moschopoulos, N., Pekmestzi, K.: An optimized modified booth recoder for efficient design of the add-multiply operator. IEEE Trans. Comput. Syst. I Regular Papers **61**(4), (2014)
2. Amaricai, A., Vladutiu, M., Boncalo, O.: Design issues and implementations for floating-point divide–add fused. IEEE Trans. Circuits Syst. Ii: Express Briefs **57**(4), (2010)
3. Liao, Y., Roberts, D.B.: A high-performance and low-power 32-bit multiply–accumulate unit with single-instruction-multiple-data (SIMD) feature. IEEE J. Solid-State Circuits **37**(7), (2002)
4. Hoang, T.T., Larsson-Edefors, P.: A high-speed, energy-efficient two-cycle multiply-accumulate (MAC) architecture and its application to a double-throughput MAC unit. IEEE Trans. Circuits Syst. I: Regular Papers **57**(12), (2010)
5. Chaudhary, M., Narula, M.S.: High speed modified booth's multiplier for signed and unsigned numbers. ITM University, Gurgaon, Haryana, (Dept. of EECE), ITM University, GurIn Thisgaon, Haryana
6. Zimmermann, R., David, Q.: Tran Design Ware, Solutions Group, Synopsys, Inc. 2025 NW Cornelius Pass Rd., Hillsboro, OR 97124: optimized synthesis of sum-of-products. In: Proceedings 37th Asilomar Conference on Signals, Systems, and Computers, November 2003 © 2003 IEEE
7. Chandel, D., Kumawat, G., Lahoty, P., Chandrodaya, V.V., Sharma, S.: Arya institute of engineering & technology. Jaipur (Raj.), India: Booth Multiplier: Ease of Multiplication. Int. J. Emerg. Technol. Adv. Eng. Certified J. **3**(3), (2013). www.ijetae.com. ISSN:2250-2459, ISO 9001:2008
8. Mori, J., et al.: A 10 ns 54 54b parallel structured full array multiplier with 0.5 m units cmos technology. IEEE J. Solid-State Circuits **26**(4), 600–605 (1991)

9. Sukhmeet Kaur, S., Manna, M.S.: Implementation of modified booth algorithm (radix 4) and its comparison with booth algorithm (radix-2). Adv. Electron. Electr. Eng. **3**(6), 683–690 (2013). ISSN:2231-1297

10. Uma, R., Vijayan, V., Mohanapriya, M., Paul, S.: Area, delay and power comparison of adder topologies. Int. J. VLSI Design Commun. Syst. (VLSICS) **3**(1), (2012). doi:10.5121/vlsic. 2012.3113 153 Research Scholar, Department of Computer Science Pondicherry University, Department of Electronics and Communication Engineering

11. Narendra Swamy, K.N., Venkata Suman, J.: Design of efficient and fast multiplier using MB recoding techniques. Int. J. Emerg. Res. Manag. Technol. **4**(6), (2015). ISSN:2278-9359

Decrease in False Assumption for Detection Using Digital Mammography

Chiranji Lal Chowdhary, Gudavalli Vijaya Krishna Sai and D.P. Acharjya

Abstract Our research work elaborated in the design and construction of a method that bring sustenance for a reduction in false assumptions during the detection of breast cancer. Our key drive of this research work was to elude the false assumptions in the detection practice in a cost effective manner. We proposed a unique method to decrease false assumption in breast cancer detection cases and split this method in three different modules as preprocessing, formation of homogeneous blocks and color quantization. The preprocessing convoluted in eradicating the extraneous slices. The formation homogeneous blocks sub-method was to do segmentation of the image. The task of the third sub-method (i.e. color quantization) was to break the colors amid different regions.

Keywords Color quantization · False assumption · Digital mammography · Computer aided system

1 Introduction

Many researchers produced a quality research work in the area, medical sciences, but cancer is still being the perplexing disease. Atypical growth of cells became the main reason of cancer. Due to cancer, there may be the materialization of lumps and called as a tumor. The growth of tumors recompenses the body cells. One's a tumor formation start in the human body, it can grow repeatedly. Such tumors disturb new tissues too. According to American National Institute, there is a presence of over

C.L. Chowdhary (✉) · G.V.K. Sai
School of Information Technology and Engineering, VIT University, Vellore, India
e-mail: c.l.chowdhary@gmail.com

G.V.K. Sai
e-mail: krishnasai.6931@gmail.com

D.P. Acharjya
School of Computing Science and Engineering, VIT University, Vellore, India
e-mail: dpacharjya@gmail.com

© Springer India 2016
H.S. Behera and D.P. Mohapatra (eds.), *Computational Intelligence in Data Mining—Volume 2*, Advances in Intelligent Systems and Computing 411, DOI 10.1007/978-81-322-2731-1_30

100 categories of cancer and amid all the type of breast cancer is a common disease found in women [1]. Breast cancer progresses in lining of the milk ducts in the breast and called as lobules. According to current studies, the ratio breast cancer affecting women is about 1/8 [2].

Several screening methods as mammography, clinical and self-breast exams, genetic screening, ultrasound, and magnetic resonance imaging, are available for breast cancer detection. Over these mammography is most common and easiest method presented for early detection of cancer. In this technique, X-ray images are used to scrutinize the internal assembly of the breast. Such X-ray images are mammograms and used for advance screening progression.

With this approach we can go for early detection of cancer avoiding any indicators. Direct mammogram images are not visible for the screening of breast cancer due to diverse dynamics of each woman. In some cases when radiologists detect any doubtful spot during his observation, he will acclaim for the surgery or for next investigative route. Such decision may lead towards false-positive consequences in case when a radiologist came to a decision that patient is having cancer but in reality it was not true.

Many approaches are available for early detection in literature and been used in medical applications so if a radiologist make false redemptions by considering as a tumor that should be avoided. So in our study we will try to decrease the cases of false detection at early stages. On other issues like false negative results normal result to the infected patient are also possible during screening process when mammography is used. They may also go for bad decision of surgeries or other health's impediment for the patient is not having cancer.

Our study provides an opportunity to construct an organism to reduce the false redemptions. Segmentation part was used to implement the formation of homogeneous blocks by using this technique for identifying the abnormalities. The last step was color quantization that differentiates the masses by using different colors.

Excluding the introduction Section, our paper is having 4 fold. Section 2 is about the literature survey, whereas proposed system and implementation are in Sect. 3. Experimental evaluations are discussed in Sect. 4. Finally Sect. 5 is covering conclusions.

2 Literature Survey

A number of researchers, medical organizations and international agency, including world health organization (WHO), are doing research work on breast cancer detection. This is a conjoint disease perceived in the world. Through confirmed sources of this area, it is observed a growth of about 30 % of different cancers is found only in urban area of India [3]. Mammograms are a useful source for radiologists for detection of breast cancer. Over available techniques, digital mammography is best screening algorithm [4]. Fear and others [5] worked on microwave imaging for cancer detection and offered there method. In conditions when the patient breast had absorbed in a liquid, the images were taken. Such liquid had comparable belongings to breast tissue and a chain of antennas is placed in a liquid and sited in a slice around the breast. Three-dimensional approach was an alternative tactic for cancer detection [6]. Every mammogram images had irritating areas and vital for the screening of cancer. It was compulsory to identify the process enough before detachment to such areas. Two important methods for the extracting Region of Interest (ROI) was initial extraction and gradient adjustment for GVF-Snake [7].

Sharma et al. [3] suggested an idea in which binary homogeneity enhancement algorithm and edge detection algorithm for detection of breast boundary. This was found in literature that pectoral muscle for the screening of cancer and in this we have to remove the unwanted area enough before advance process. The pectoral muscle is situated in the corner of the image and having extraordinary intensity. The unsolicited area can be distant with conventional list estimation based on the Region of Interest (ROI) and was completed with extracting the border of the image [4]. Gaussian derivative, ridge detector, oriented bins and line operator scheme support at each pixel of orientation and line-strength [8]. This research work will help in detection of dark or bright linear structure of mammograms. Sumari, Raman and Patrick [9] worked on the automated seed selection, region partitioning, feature extraction and mass detection phases recognize masses growth in the breast. One of them region partitioning used for different threshold values for individual district, extracted breast expanse is recognized and in the end the seed selection in that is one point allow growth in the Region of Interest (ROI) [9].

3   Proposed System and Implementation

See Fig. 1.

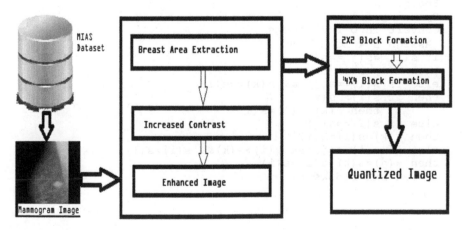

Fig. 1 Architecture of proposed system

3.1 Pre-processing of Mammogram Image

In our proposed work we have implemented two types of preprocessing techniques. The first step involved in removal of labels in the image and the second step used to identify the pectoral muscle and to remove that portion from the mammogram image.

3.2 Formation of Homogeneous Blocks

In this part the mammogram image was segmented into small blocks. Therefore, 1024 × 1024 sized mammogram images were disintegrated into 8 × 8 pixel blocks. Initially the formation to 8 × 8 pixel homogeneous blocks was difficult, so we have started with 2 × 2 pixel block and later converted into 4 × 4 pixel block. Further, this was degenerated into 8 × 8 pixel block. There are certain conditions to be fulfilled before conversion to a homogeneous 2 × 2 pixel block. In a 2 × 2 pixel block, 3 pixel values are same. Convert the 4th pixel into the remaining three pixel value. Therefore, in a 2 × 2 pixel values was having maximum intensity, then convert the remaining values to that value. If remaining two values are different then converting the remaining pixel blocks with the maximum occurred pixel value. If all the values in the block are different then finding the maximum pixel value in the block and convert the remaining values to that value.

Pseudo code

```
Input: Preprocessed mammogram Image (img)
s = [0 0 0 0] comparison for values
i, j, k, l values are in between 1,2,3,4 i ≠ j ≠ k ≠ l
Output: Segmented image
Begin:
Open img file
Convert into 2x2 blocks
 Assign s ←2x2 block
if s(i)==s(j)==s(k)
then s(l)=s(i)
else if s(i)==s(j) && s(k)==s(l)
then if s(i)> s(k)
        then s(k)=s(l)=s(i)
else if s(i)==s(j)
then s(k)=s(l)=s(i)
else if s(i)>s(j) && s(i)>s(k)&& s(i)>s(l)
then s(j)=s(k)=s(l)=s(i)
Assign 2x2 block←s
```

3.3   Color Quantization

The color quantization technique was helpful to break the color space of the mammogram image into eight equal size regions. Therefore, we have used mat2-gray function that helps to convert the gray image values in between the range 0–1. That value is multiplied with the required number of gray levels and later it was converted into an integer. Further, the values are varying from 0 to the required number of gray levels. Later gray image was quantized and then it should fill with different colors. We have used a color map with 64 colors.

Pseudo code

```
Input: Preprocessed mammogram Image (img)
Map= hot
r = zeros(1024,1024);
g = zeros(1024,1024);
b = zeros(1024,1024);
Begin:
Open img file
Img ←mat2gray(img)
Convert Img to integer
Get r,g,b values using map function
Res ←Concat(r,g,b)
```

4   Experimental Evaluation

In our proposed study we have used the Mammogram Image Association Society (MIAS) Database [10]. This database had many images of two types as cancer affected or non-cancer affected images.

4.1   Evaluation Criteria

In evaluation criteria we had a random image from the database and proceed with the pre-processing techniques to eliminate the annoying areas. Our next step was towards implementation by formation of homogeneous blocks to reduce the complexity of the texture. The last part was the color quantization technique to help in breaking the colors in regions and maps.

We opted for a normal image (mdb025) to perform pre-processing. The preprocessing step mainly detects the labels, pectoral muscle and reduces noise in the image. If there was an absence of any appearance of any label, the figure will remain same as the previous image.

After removal of label from the image (Fig. 2), pectoral muscle was recognized which was shown in the right corner of the Fig. 3a. Figure 3b helped us to find the image after removal of pectoral muscles. In the next step (Fig. 4a–c), the image was

(a)   (b)

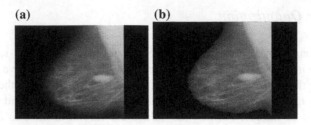

Fig. 2 **a** Mammogram without label removal. **b** Mammogram with label removal

(a)   (b)

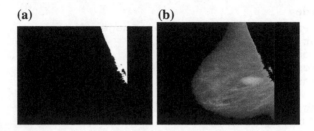

Fig. 3 **a** Pectoral muscle. **b** Mammogram without pectoral muscle

(a)   (b)   (c)

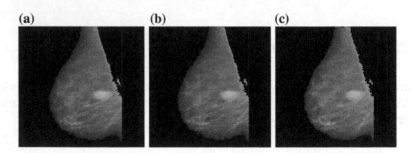

Fig. 4 **a** 2 × 2 block. **b** 4 × 4 block, **c** 8 × 8 block

divided into homogenous blocks as 2 × 2, 4 × 4 and 8 × 8. This step helped us to expand the abnormality in the adjacent region to assist in identification of the tumor affected area. In the below mentioned figures we can observe the differences easily:

Figure 5a, b had the image partitioned into some regions, depending upon the behavior of the region different color will be allocated. Figure 5a helped us to identify the different regions as different gray levels. Figure 5b helped us to identify the different regions with different colors. The white region has the highest intensity, so it was detected as the tumor.

(a)   (b)

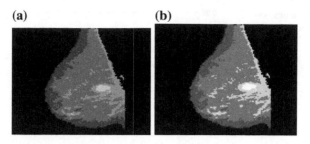

Fig. 5 **a** Color quantized image. **b** Color quantized image with different colors

4.2 Result Analysis

Result analysis was done on the final image produced by the application. Each image has shaped dissimilar textures and texture will help to classify the type of the image. Texture features comprise both normal and abnormal regions.

We have used gray level co-occurrence matrix (GLCM) for feature extraction of the enhanced image. The matrices of GLCM were fabricated using a distance d (in case when d is equal to 1 for analyzing the texture) and 4 different directions (i.e. 0, 45, 90, and 135). Now from all the directions, the texture information was investigated for both abnormal and normal regions. We have analyzed textures with diverse features in the GLCM matrices. We have selected gray-level vales as contrast, energy, homogeneity and correlation.

Based on the final outcome of the images, Table 1 has represented the different values of the contrast in benign, malignant and normal images (Fig. 6). In Table 1, values are benign, malignant and normal images. At 00 benign image values were ranging from 0.030471 to 0.038093, malignant image values were ranging from 0.028118 to 0.043377 and the normal image 0.029279 to 0.055687. Therefore, from these values we can justify between those ranges we can identify the type of the image.

Tables 2, 3, and 4 represent the different corresponding values of the contrast, homogeneity, and correlation for benign malignant and normal images.

Table 1 Contrast values for benign, malignant and normal images

Image	0°	45°	90°	135°
mdb025	0.038093	0.062731	0.026943	0.061917
mdb080	0.030471	0.044167	0.015793	0.045294
mdb099	0.032732	0.048429	0.016809	0.047326

Fig. 6 Contrast values for benign, malignant and normal image graph

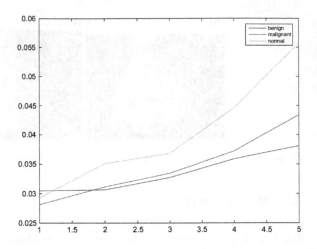

Table 2 Energy values for benign, malignant and normal images

Image	0°	45°	90°	135°
mdb025	0.453105	0.450368	0.453513	0.450356
mdb080	0.636233	0.634901	0.637047	0.634869
mdb099	0.562417	0.560738	0.5634	0.560765

Table 3 Homogeneity values for benign, malignant and normal images

Image	0°	45°	90°	135°
mdb025	0.994972	0.990586	0.995152	0.990551
mdb080	0.996799	0.994583	0.997544	0.994581
mdb099	0.995633	0.992666	0.996706	0.992674

Table 4 Correlation values for benign, malignant and normal images

Image	0°	45°	90°	135°
mdb025	0.994717	0.991304	0.996263	0.991417
mdb080	0.995715	0.993793	0.997779	0.993634
mdb099	0.995900	0.993938	0.9978950	0.994076

5   Conclusions

It was very difficult to decide about a cancerous or non-cancerous tumor. Mammography found as the best way of diagnosing the breast cancer to do early detection. Development in technology directed to the improvement in computer aided systems for detecting breast cancer. In this study, we focused mainly on the diagnosis part. Our work has decreased the false assumptions in breast cancer detection system by removing the irrelevant parts of mammogram images, by the formation of homogeneous blocks and color quantization to break the colors among different regions. Our system reduces the false assumptions during the detection of breast cancer. In future we may extend this work for implementing the classification technique.

References

1. National Cancer Institute. www.cancer.gov
2. U.S. Breast Cancer Statistics. http://www.breastcancer.org/symptoms/understand_bc/statistics (2015)
3. Sharma, J., Rai, J.K., Tewari, R.P.: Identification of pre-processing technique for enhancement of mammogram images. In: International Conference on Medical Imaging, m-Health and Emerging Communication Systems, pp. 115–119. (2014)
4. Kwok, S.M., Chandrasekhar, R., Attikiouzel, Y., Rickard, M.T.: Automatic pectoral muscle segmentation on mediolateral oblique view mammograms. IEEE Trans. Med. Imaging **23**, 1129–1140 (2004)
5. Fear, E.C., Stuchly, M.A.: Microwave detection of breast cancer. IEEE Trans. Microwave Theory Tech. **48**, 1854–1863 (2000)
6. Vladimir, A., Cherepenin, V.A., Karpov, A.Y., Korjenevsky, A.V., Kornienko, V.N., Kultiasov, Y.S., Ochapkin, M.B., Trochanova, O.V., Meister, J.D.: Three-dimensional EIT imaging of breast tissues: system design and clinical testing. IEEE Trans. Med. Imaging **2**, 662–667 (2002)
7. Yu, S.S., Tsai, C.Y., Liu, C.C.: A breast region extraction scheme for digital mammograms using gradient vector flow snake, pp. 715–720. Department of Electronic Engineering National Chin-Vi University of Technology Taiping, Taichung (2010)
8. Zwiggelaar, R., Astley, S.M., Boggis, C.R.M., Taylor, C.J.: Linear structures in mammographic images: detection and classification. IEEE Trans. Med. Imaging **23**, 1077–1086 (2004)
9. Sumari, P., Raman, V.: Patrick: digital mammogram tumor preprocessing segmentation feature extraction and classification. In: World Congress on Computer Science and Information Engineering, pp. 507–511. (2009)
10. MIAS Database. http://peipa.essex.ac.uk

different regions. Our system reduces the false accounting, during the detection of breast cancer. In future we may extend this work to characterizing the classification technique.

References

1. [illegible]

Test Case Prioritization Using Association Rule Mining and Business Criticality Test Value

Prateeva Mahali, Arup Abhinna Acharya
and Durga Prasad Mohapatra

Abstract Regression Testing plays a vital role for the improvement in quality of product during software maintenance phase. This phase ensures that modification made to the system under test doesn't adversely affect the performance of the existing features. Hence regression testing incurs more cost and time. Test case prioritization, which is one of the techniques of regression testing, is an efficient technique to minimize the cost and time of testing. In this paper, the author has proposed an approach for test case prioritization by maintaining information of previous and current release of the project in a repository. To represent the behavioral aspect of the system, it is modeled using both UML activity and sequence diagram. Then frequent pattern is generated by applying Association Rule Mining on the information stored in the repository. Finally the prioritization is carried out using the generated frequent patterns and Business Criticality Value of the different features.

Keywords Regressiontesting · Test case prioritization · Association rule mining · Business criticality value · Activity diagram · Sequence diagram

1 Introduction

Today's software industry aims at developing qualitative product rather than quantitative product. In previous days, users are satisfied with a new product with less feature due to unawareness of technical knowledge about the features.

P. Mahali (✉) · D.P. Mohapatra
Department of Computer Science & Engineering, National Institute
of Technology, Rourkela 769008, India
e-mail: prateevamahali@gmail.com

D.P. Mohapatra
e-mail: durga@nitrkl.ac.in

A.A. Acharya
School of Computer Engineering, KIIT University, Bhubaneswar 751024, India
e-mail: aacharyafcs@kiit.ac.in

© Springer India 2016
H.S. Behera and D.P. Mohapatra (eds.), *Computational Intelligence
in Data Mining—Volume 2*, Advances in Intelligent Systems
and Computing 411, DOI 10.1007/978-81-322-2731-1_31

Nowadays users are more involved in the development process like agile development process etc. Therefore developing companies are conscious about the quality of service and quality of product. The quality can only be achieved if all the defects in the product can be identified and the software testers are able to fix all the defects in due time. In reality, it is not feasible to fix all the bugs due to shortage of time and resources. Another aspect of regression testing is retesting, which also leads to increase in time and cost for testing. The retesting of complete product with the modification to ensure that it should not affect the existing product is called as *Regression Testing* [1]. There are different techniques available to reduce testing time and cost such as test case selection, test suite minimization and test case prioritization [2]. In test case selection method, a subset of test cases are selected for the retesting purpose an in test suite minimization, the size of test suite is reduced by eliminating the redundant test cases or by eliminating some less priority test cases. Test case prioritization is the process of reordering or sorting of test cases base on some criteria, insuch a way that highest priority test case will be executed first [3]. The advantages of this technique are it improves fault detection process and debugging process with minimum testing time and cost.

Nowadays software failure happens due to less fault detection capability and presence of undiscovered faults. To resolve this problem we have proposed an approach for test case prioritization process. Here system models are taken as input to the proposed approach. The system is modelled using Unified Modelling Language (UML) i.e. Activity Diagram (AD) and Sequence Diagram (SD). Then an intermediate graph i.e. Activity Sequence Graph (ASG) is generated from the combination of activity and sequence diagram. The graph is traversed to generate test scenarios. Then all the modified and corresponding affected nodes are collected from the graph using forward slicing algorithm proposed by Acharya et al. [4]. Frequent pattern of affected nodes are generated by applying Association Rule Mining (ARM) on the information collected from the graph. In our previous approach [4], we have considered each features having the same contribution towards the business value of the system under test. But in this paper we defined the Business Criticality Value (BCV) of each node and accordingly prioritization has been carried out. Business Criticality Value (BCV) of feature in a system under test is defined as the extend of contribution feature towards the successful execution of the business process of the application. Business Criticality Value (BCV) is calculated for each node present in the frequent pattern and Business Criticality Test Value (BCTV) for all test cases is calculated using the formula proposed by Khandai et al. [5]. Then test case prioritization is performed based on business criticality test value.

To implement this proposed approach we have to know some basic knowledge about ARM and BCTV. Association Rule Mining (ARM) is a popular methodology of data mining to discover regularities between products in large scale database [4, 6]. In ARM, calculation of two parameters are required for generation of frequent pattern i.e. *support* and *confidence* value [6]. Business Criticality Test Value (BCTV) is the summation of business criticality value of all the functions encountered during the execution of the test case.

The rest of the paper is organized as follows: Sect. 2 discusses the related work and its analysis and the proposed approach is discussed in Sect. 3. Section 4 presents the working of the proposed approach using a case study ofLibrary Management System (LMS) and Sect. 5 discusses the conclusion and future work.

2　Related Work

In this section, different related research work proposed by researcher is discussed. All the mechanisms are based on test case generation and test case prioritization using different approaches in model based testing [7–10]. Khandai et al. [5] proposed an approach for test case prioritization using Business Criticality Value (BCV) in model based testing. In that paper, the author has maintain a repository which contain the detail information of project. Then they matched new version of the project with the previous version from the repository and assigns business criticality value to each affected function. Finally they have calculated Business Criticality Test value (BCTV) for all test cases and performs prioritization as per BCTV. But the author has not considered non-functional requirements in their approach. They have considered only functional requirements of project in their approach.

Acharya et al. [4] described an approach for test case prioritization by generating frequent pattern of affected nodes. The pattern was generated using Association Rule Mining (ARM). First they had also maintained a historical database containing both graph data and observation data. Then graph data is found out by applying forward slicing algorithm and the modified nodes and the corresponding affected nodes are traced out. That affected nodes are used for generation of frequent pattern using ARM. The resultant frequent pattern was followed for test case prioritization by assigning priority value to the nodes present in the pattern. When we analyse this approach we found some limitation such as only single type of modification is taken for consideration. There might have multiple modification done with all version of project.

Muthusamy et al. [11] proposed an approach for test case prioritization process by identifying the severe faults. Here test case prioritization is performed based on four practical weight factor such as customer allotted priority, developer observed code execution complexity, change in requirements, fault impact, completeness and traceability.

3　Proposed Approach

In this paper, we discuss our proposed approach for test case prioritization by Frequent Pattern (FP) generation using ARM and BCTV calculation. Frequent pattern generation and test case prioritization is done based on the model based testing concept. Here we have taken Sequence Diagram (SD) and Activity

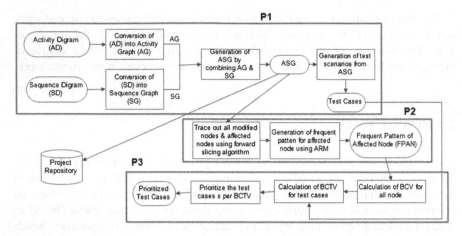

Fig. 1 Proposed framework for test case prioritization

Diagram (AD) as system model to represent the system behaviour. The proposed framework for test case prioritization is shown in Fig. 1 and working of the approach is discussed in Sect. 4 with a case study of Library Management System (LMS).

The proposed framework is divided into three sections

P1 → In this section test cases are generated from Activity Sequence Graph (ASG).
P2 → In this section Frequent Pattern for Affected Nodes (FPAN) is generated using Association Rule Mining (ARM).
P3 → In this section Frequent Pattern (FP) of affected nodes and Business Criticality Value (BCV) of test cases are used for test case prioritization.

P1: Test case generation from Activity Sequence Graph (ASG)

This section describes generation of test cases from Activity Sequence Graph (ASG). First Activity Diagram (AD) is converted into an intermediate dependency graph called as Activity Graph (AG). Similarly Sequence Diagram (SD) is converted into an intermediate graph called as Sequence Graph (SG). Then feature of both the graph (i.e. AG and SG) combined to generate a dependency graph called Activity Sequence Graph (ASG). From the resultant ASG, test scenarios are generated by traversing the graph from source to destination. Here each test scenarios are considered as test cases.

Here all the information of ASG is stored in project repository. The project repository contains Project ID, Total number of Node, Modified Nodes and Affected Nodes for each modified node, Test cases, BCV of all node, BCTV of test cases in tabular format.

P2: Frequent Pattern Generation

This section discusses about the generation of frequent pattern from the affected nodes using Association Rule Mining (ARM). From ASG, all the modified and corresponding affected nodes are collected using forward slicing algorithm for the current version of the project. The forward slicing algorithm was proposed in our previous paper [4]. Then all this information are stored in project repository. To generate frequent pattern, all the modified and affected nodes of current version as well as previous version of project are used as input. Frequent pattern (FP) is generated by applying Association Rule Mining (ARM). First, the itemsets are collected from the repository. Then support value of individual item and combination of items are calculated. This support value is compared with user-specified *minimum support* value and resultant itemset is called as frequent itemset. Then the confidence value of frequent itemset is calculated and compared with user-specified *minimum confidence* value. The resultant itemset or pattern is called as Frequent Pattern of Affected Node (FPAN).

P3: Prioritization of test cases using Business Criticality Test Value (BCTV)
This section defines test case prioritization process using Business Criticality Test Value (BCTV). To calculate BCTV of test cases, first we have to calculate Business Criticality Value (BCV) of each node by using the proposed Khandai et al. [5] and given in Eq. 1. Then BCTV of test cases are calculated by adding BCV of all node present in that test case.

Then prioritization of test cases is done as per BCTV. Test cases with higher BCTV is considered as higher priority and lower BCTV is considered as lower priority. In this way, prioritized test cases are generated and stored in the repository for future reference.

4 Working of Proposed Approach with a Case Study

To discuss the working of proposed approach, we have considered the case study of Library Management System (LMS) for book issue. The application includes different features like checking the availability of book, validating the membership, issue book and update the status of book. The working of proposed approach with the case study of library management system is discussed in the following sections.

4.1 Construction of Activity Sequence Graph (ASG) and Test Scenario Generation

In this section, the test case generation from activity sequence graph is discussed. Here Activity Diagram (AD) and Sequence Diagram (SD) is used for system modelling. We have taken these two diagrams to represent message sharing between objects and activity flow between different activities [12]. The AD and SD of Library Management System (LMS) is shown in Figs. 2 and 3 respectively.

Fig. 2 Activity diagram of library management system (LMS) for book issue

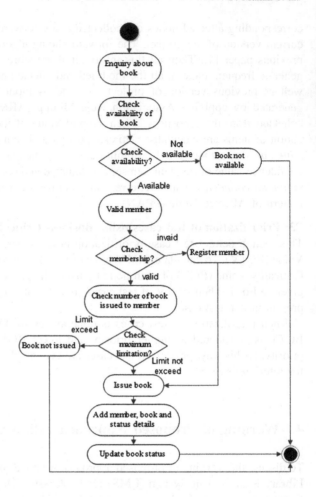

The node number sequence from A1 to A15 are mapped with associated activities present in activity diagram of LMS i.e. *Start, Enquiry about book, Check availability of book, Check availability, Book not available, Validate member, Check membership, Register member, Check number of book issued to member, Check maximum limitation, Book not issued, Issue book, Add member, book and status details, Update book status* and *End* respectively. Similarly the node number sequence from S1 to S9 are mapped with associated massages present in the sequence diagram of LMS i.e. *check availability of book(), Book available(), Valid member(), Check number of book issued (), book can be issued (), create (), add member* and *book details (), update book status (), update member record ()* respectively. Then the information's from both the UML diagrams are combined to a generate dependency graph called as Activity Sequence Graph (ASG). Here mapping is done as per the behaviour of activities and functionalities. The ASG of LMS is shown in Fig. 4. After that, test scenarios are generated from ASG by

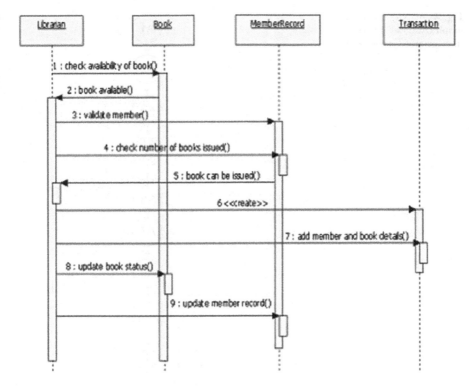

Fig. 3 Sequence diagram of library management system (LMS) for book issue

traversing the graph from source to destination. For this case study, four test scenarios are possible and these test scenarios are considered as test cases. The test scenarios are shown in Table 1. All the information of ASG is stored in project repository.

4.2 Frequent Pattern Generation

This section represents the generation of frequent pattern of affected nodes using ARM. For this we need to identify the modified nodes and affected nodes. These nodes are calculated using forward slicing algorithm [4]. In the new project we have done two changes i.e. C1 and C2 in node 7 and node 13 respectively. The identified modified nodes with corresponding affected nodes are

Fig. 4 Activity sequence
graph (ASG) of library
management system (LMS)

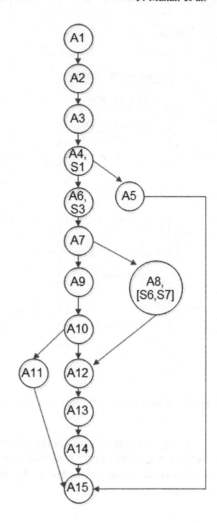

Table 1 Test scenarios of LMS

S. no. (path)	Test case id	Independent path
1	T1	A1→A2→A3→A4, S1→A5→A15
2	T2	A1→A2→A3→A4, S1→A6, S3→A7→A8, [S6,S7]→A12→A13→A14→A15
3	T3	A1→A2→A3→A4, S1→A6, S3→A7→A9→A10→A12→A13→A14→A15
4	T4	A1→A2→A3→A4, S1→A6, S3→A7→A9→A10→A11→A15

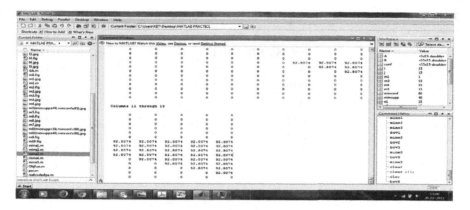

Fig. 5 Confidence values of all nodes present in FPAN

$$C1 : A7, (A8, [S6, S7]), A9, A10, A11, A12, A13, A14, A15$$
$$C2 : A13, A14, A15$$

By using MATLAB, we have generated Frequent Patter of Affected Nodes (FPAN) by applying ARM technique. Here we have taken 40 % as minimum support value and 80 % as minimum confidence value. The confidence value of each nodes are given in Fig. 5.

Now the FPAN is {A7, (A8, [S6, S7]), A9, A10, A11, A12, A13, A14, A15}.

4.3 Prioritization of Test Cases Using Business Criticality Test Value (BCTV)

The itemsets is responsible for test case prioritization using BCTV. For this we have calculate BCV of all nodes present in FPAN.

$$BCV(N) = \frac{No. \, of \, times \, N \, encounter}{Total \, no. \, of \, factor \, being \, affected} \qquad (1)$$

Now BCV of each node has been calculated and represented in Table 2.

As per our proposed approach BCTV of test cases are calculated by adding BCV of all node present in test case. If the node is not present then its value is zero otherwise the value can be extracted from Table 2. BCTV of test cases is shown in Table 3. Now the prioritized test cases are T3, T2, T4, T1.

Table 2 Business criticality value (BCV) table

Node no.	No. of time node encounter	BCV
A7	1	0.111
A8, [S6, S7]	1	0.111
A9	1	0.111
A10	1	0.111
A11	1	0.111
A12	1	0.111
A13	2	0.222
A14	2	0.222
A15	2	0.222

Table 3 Business criticality test value (BCTV) table

S. no.	Test case id	Business criticality test value (BCTV)
1	T1	0.222
2	T2	0.999
3	T3	1.11
4	T4	0.666

5 Conclusion and Future Work

The authors have divided the complete proposed approach into three different phases. In the first phase, the author has generated test cases from combination of two UML diagrams (i.e. activity diagram and sequence diagram). In the second phase, all modified and corresponding affected nodes are identified by using forward slicing algorithm and frequent pattern is generated using ARM. In last phase, test case prioritization is performed using BCTV of all test cases. But in this approach, the author has considered only single type of modification and both functional and non-functional requirements. In future research, we will consider all kind of modifications with functional and non-functional requirements.

References

1. Mall, R.: Fundamental of Software Engineering, 3rd edn. PHI Learning Private Limited, New Delhi (2009)
2. Chauhan, N.: Software Testing Principles: Practices, 3rd edn. Oxford University Press, New Delhi (2010)
3. Haidry, S.-e-Z., Miller, T.: Using dependency structures for prioritization of fundamental test suites. IEEE Trans. Softw. Eng. **39**(2), (2013)
4. Acharya, A.A., Mahali, P., Mohapatra, D.P.: Model based test case prioritization using association rule mining. Comput. Intell. Data Mining **3**, 429–440 (2015)

5. Khandai, S., Acharya, A.A., Mohapatra, D.P.: Prioritizing test cases using business test criticality value. Int. J. Adv. Comput. Sci. Appl. **3**(5), 103–110 (2011)
6. Han, J., Kamber, M.: Data Mining: Concepts and Techniques, 2nd edn. Morgan Kaufmann Publishers, 500 Sansome Street, Suite 400, San Francisco, CA 94111 (2010)
7. Kundu, D., Samanta, D.: A novel approach to generate test cases from uml activity diagram. J. Object Technol. **8**(3), 65–83 (2009)
8. Tyagi, Manika, Malhotra, S.: An approach for test case prioritization based on three factors. Int. J. Inf. Technol. Comput. Sci. **4**, 79–86 (2015)
9. Sarma, M., Mall, R.: Automatic test case generation from uml models. In: 10th International Conference on Information Technology, pp. 196–201. (2007)
10. Swain, S.K., Mohapatra, D.P.: Test case generation from behavioural uml models. Int. J. Comput. Appl. **6**(8), 5–11. (2010)
11. Muthusamy, T., Seetharaman, K.: A new effective test case prioritization for regression testing based on prioritization algorithm. Int. J. Appl. Inf. Syst. (IJAIS) **6**(7), 21–26 (2014)
12. UML 2.4 Diagrams Overview. http://www.uml-diagrams.org/uml-24diagrams.html

Nearest Neighbour with Priority Based Recommendation Approach to Group Recommender System

Abinash Pujahari, Vineet Padmanabhan and Soma Patel

Abstract Group Recommender System is one of the categories of recommender system, where the recommendation of things is for a group of users rather than for any individual. These system combines the preferences of each user present in the group and then predicts things which are suitable for the users of the group. Various grouping strategies are available, which are used to generate to group recommendations, but most of them are suitable when used for specific purpose only. In this paper we have proposed a novel approach to group recommender system using collaborative filtering technique, which can be applicable to all the real world scenarios where the data set uses rating system to distinguish among users' preferences. We have made use of nearest neighbor algorithm to create a group of users with similar likeness. We have also applied the priority among users of the group as there are some members whose preferences might affect the whole group. We have validated our results with the movie lens data set which is the standard data set for recommender system testing.

Keywords Recommender system · Collaborative filtering · Nearest neighbor

1 Introduction

Recommender Systems [1–6] have created much attention in the field of research, because its use and popularity has been increased significantly these days. Almost every web applications, i.e. search engines to shopping sites, business portal to

A. Pujahari (✉) · S. Patel
Sambalpur University Institute of Information Technology,
Sambalpur University, Jyoti Vihar 768019, India
e-mail: abinash.pujahari@gmail.com

S. Patel
e-mail: somapatel5@gmail.com

V. Padmanabhan
School of Computer and Information Sciences, University of Hyderabad,
Hyderabad 500046, India
e-mail: vcpnair73@gmail.com

© Springer India 2016   347
H.S. Behera and D.P. Mohapatra (eds.), *Computational Intelligence in Data Mining—Volume 2*, Advances in Intelligent Systems and Computing 411, DOI 10.1007/978-81-322-2731-1_32

government etc., are using recommender system these days to help their end users. Generally, recommender systems uses either of the two techniques, i.e. collaborative filtering and content-based filtering to produce predicted things for users. The collaborative filtering is based on analyzing previous data to predict things for a user that might be suitable for him based on his similarity of likeness to other users. While, the content-based recommender system uses any machine learning algorithm to learn rules using previous preferences on things and based on that rules it predicts the unseen items which might be suitable for the users.

Recommender Systems are broadly classified into two categories. They are Personal Recommender System (PRS) and Group Recommender System (GRS) [5, 7]. PRS are the kind of recommender system which generate recommendation of things for a single user only while, GRS generates recommendation of things for a group of users. GRS aggregates the preferences of all the users of the group and then predicts things or items for them. But aggregating preferences of users is not so easy, because all the users in the group may have different kind of preferences. So a group recommender system is effective only when we analyze the preference of all users correctly and the members of the group has similar kind of preferences. So in this paper we have sued the nearest neighbor algorithm to make homogeneous groups i.e. members having similar likeness and for generating recommendations, we have used the priority to different users in the group. In the next section we have discussed some of the previous approaches to GRS and in later sections we have described the algorithms used in our approach to GRS and the obtained experimental results in details.

2 Recommendation Procedure

As we mentioned earlier, in this paper we have followed the collaborative filtering [2, 6, 8] approach for recommendation which analyzes large volume of information for generating recommendation. In fact, this technique basically finds the similarity among users or things to predict similar kind of items for users. There are various algorithms available to find the degree of similarity among two variables like Pearson Correlation, cosine similarity etc. We are considering the ratings of different users to different items, to find the degree of similarity among users. Because we can only know any user's preference from the ratings given to items by that particular user.

2.1 Finding Similarity

There are various methods for finding similarity between two objects. The Pearson Correlation [5, 6] technique finds the degree of similarity among two variables. The two variables may be two different persons or things. The range of similarity coefficient is in between −1 to +1. The positive coefficient value indicates positive similarity between two variables whereas the negative coefficient value indicates the negative similarity or the dissimilarity among two variables. The formula for finding the similarity coefficient is given in Eq. 1. We have taken the ratings given by users to different things in the past as the attribute for finding similarity.

$$sim(u,v) = \frac{\sum_{i \in I}(r_{u,i} - \overline{r_u})(r_{v,i} - \overline{r_v})}{\sqrt{\sum_{i \in I}(r_{u,i} - \overline{r_u})^2}\sqrt{\sum_{i \in I}(r_{v,i} - \overline{r_v})^2}} \qquad (1)$$

where u and v are two different users or variables for whom we are going to find the degree of similarity, $r_{u,i}$ is the rating given by user u to instance i and $\overline{r_u}$ is the mean of all the ratings given by user u. If we want to calculate the similarity between two different items i and j, then the similarity coefficient can be calculated by using Eq. 2.

$$sim(i,j) = \frac{\sum_{u \in U}(r_{u,i} - \overline{r_i})(r_{u,j} - \overline{r_j})}{\sqrt{\sum_{u \in U}(r_{u,i} - \overline{r_i})^2}\sqrt{\sum_{i \in I}(r_{u,j} - \overline{r_j})^2}} \qquad (2)$$

2.2 Prediction of Things

After finding the similarity coefficient value among users or things, we then need to generate recommendation. The standard equation used for prediction [5, 6] of things for a single user is given in Eq. 3.

$$p_{u,i} = \frac{\sum_{j \in S} sim(i,j) r_{u,j}}{\sum_{j \in S}|sim(i,j)|} \qquad (3)$$

Equation 3 gives prediction values to items (j), which are similar to item I and those items having highest prediction values are recommended. In this equation we need to find the items for S i.e. the similar items, so that it will produce good recommendations. In out recommender system we have taken the similar items set to be consist of 10 items for testing.

3 Proposed Method for GRS

In the previous section we have discussed regarding recommendation for an individual. But, our proposed problem is a group recommender system. So we need to aggregate all those things to produce group recommendation. In order to produce effective group recommendation we need to first generate homogeneous groups. So in this paper we have followed the k-nearest neighbor algorithms to make homogeneous groups and then generate recommendations for the group.

3.1 Proposed Method for Group Generation

The k-nearest neighbor algorithm uses the 'distance' measure to find the closeness of two classes. Here in our problem we have taken the rating information of users to calculate the distance and closeness information among different users. Algorithm 1 written below makes a homogeneous group by considering the ratings given by different users. Let 'U' be the vector of users and 'I' be the vector of instances, examples from which the recommendation will be done. Let $r[][]$ be the two dimensional vector that contains the rating information of all the users where the first subscript represents the user no. and the second subscript represents the movie no. The rating information is collected from previous ratings given by the users in the past. Hence, the collaborative filtering technique requires collecting large amount of information in order to produce better recommendation.

Algorithm 1: Algorithm for generating homogeneous group

1. G be the set of groups, $G \leftarrow \emptyset$
2. **for** users $u \in U$ **do**
3.   **for** all user v other than u **do**

$$d[u][v] = \sum_{i \in I} \sqrt{\left(r_{u,i} - r_{v,i}\right)^2}$$

4.   **end for**
5. **end for**
6. Find the group of users g with common lowest distance in 'd'
7. $U \leftarrow U - g$
8. $G \leftarrow G \cup g$
9. **if** $|U| \neq 0$
10.   **Go to** step 6;
11. **end if**

The distance or closeness between two users' u and v is calculated in Algorithm 1 by summation of squares of the difference between ratings given by both to each items or things. Then we remove the users from the set I who have common least distance between then using distance vector and make them a group. We then add this group to the set G as a homogeneous group. After generating groups we then go for recommendation for them.

3.2 Proposed Method for Group Recommendation

After generating homogeneous groups using Algorithm 1, written above we need to generate recommendation for the group. The proposed Algorithm 2, written below will generate recommendation for a selected group. As stated earlier in our approach to group recommender system we have followed a weighted user approach, where after selecting a group of users we need to give their priorities among the group members as weights. The priorities of the members of the group is also considered while generating recommendations. Algorithm 2, starts with an empty recommendation list i.e. R and also it asks for the value of k, i.e. the number of items or instances to be recommended or we can say it generates top-k recommendation. This algorithm uses Eq. 3 to find the prediction value of all the instances and stores that instance that has the maximum prediction value for all the users in a vector called Q. Now the most common instances in Q will be added to the recommended list and those are removed from the original set of instances for further recommendation. If the number of instances in R is already reaches k then it stops adding more and the instances that are present in R will be recommended.

Algorithm 2: For group recommendation

1. Select a group g from G.
2. Recommended list $R \leftarrow \emptyset$, initially empty
3. Input the value of k, i.e. the no of things to be recommended
4. **for** all users $u \in g$ **do**
$\qquad w[u] =$ Input weight of the user
5. **end for**
6. **for** each user $u \in g$ **do**

$$Q[u] = \underset{i}{\mathrm{argmax}} \left(\frac{\sum_{j \in s} sim(i,j)(w[u]) * r_{u,j}}{\sum_{j \in s} |sim(i,j)|} \right)$$

7. **end for**
8. $i[\ ] \leftarrow$ Most common instances in Q.
9. $R \leftarrow R \cup i$
10. Remove the instances (i) from the total instances I.
11. **If** $|R| < k$
$\qquad$ **Go to** step 4.
12. **else**
$\qquad$ **List** the instances in R as recommendation.
13. **end if**

After generating recommendation for a group of users then we need to test the efficiency of our recommender system by using some real world data and some evaluation criteria. The detailed experimental results along with the evaluation criteria and the efficiency of our proposed method is discussed in the next section.

4   Experimental Results

In order to test the proposed group recommender system we have taken the movie
lens data set which contains movies information. So our tested group recommender
system is a group movie recommender system. The data set we used consists of two
files, one contains the movie information and the other contains the ratings of users
to those movies. The movie file consists of 1682 movies and their details like which
type of movie they are (genres), release date etc. The ratings file contains the ratings
of different users to movies and each user have rated at least 20 movies. The users'
ratings are between 1 (worst) and 5 (excellent). Based on these ratings we have
generated the recommendation of new movies to a group of users. To evaluate our
built recommender system we have calculated the precision [5, 6] of the recommended results by varying parameters like the no. of members in the group or the
no of items being recommended. The criteria used for evaluation of our group
recommender system is given in Eq. 4, where R represents the recommended
movies by the recommender system and 'U' is the set of movies used by the current
user and $R \cap U$ is the common movies between R and U. This is the precision value
for a single user only. For calculating the precision of the recommender system we
need to find the precision of all the users and

$$Precision = \frac{R \cap U}{R} \tag{4}$$

then find the average precision of them. The obtained precision values for different
groups is shown in Fig. 1. Here we have varied the no. of members in the group to
check the variation of precision value with the change in no. of members in the
group. The figure tells that the precision gradually decreases with the increase in the
no. of members in the group.

Next variation is the precision obtained for the no. of instances being recommended i.e. the k-value. The precision of the recommendation with respect to no of

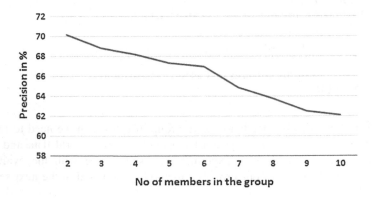

Fig. 1 Precision among various groups

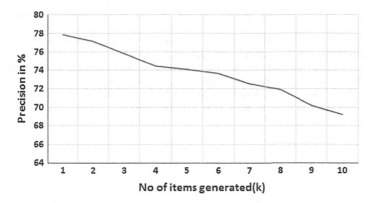

Fig. 2 Precision with respect to no. of items

items being recommended by keeping the group no. fixed is shown in Fig. 2. The figure indicates that, the precision of the group recommendation also decreases with increase in the no. of instance being recommended.

5   Conclusion

The experimental results indicates that, the generated group recommendation is good when the users of the group has similar kind of preferences and the number of generated items is of reasonable amount. If we increase the number of members in the group or increase the number of items being generated then the precision of the group recommendation decreases gradually. This is because when the number of members increased in a group the degree of similarity among them also decreases and also there is less common items between them. In our future work we are concentrating on building a hybrid recommender system which will use both the techniques of collaborative and content based filtering to produce even better recommendation for a group of users.

References

1. Cho, Y.H., Kim, J.K., Kim, S.H.: A personalized recommender system based on web usage mining and decision tree induction. Expert Syst. Appl. **23**(3), 329–342 (2002)
2. Cosley, D., Lam, S.K., Albert, I., Konstan, J.A., Riedl, J.: Is seeing believing? How recommender system interfaces affect users' opinions. In: Proceedings of the SIGCHI Conference on Human Factors in Computing Systems, pp. 585–592. ACM (2003)
3. Noh, G., Oh, H., Lee, K.H., Kim, C.K.: Toward trustworthy social network services: a robust design of recommender systems. J. Commun. Netw. **17**(2), 145–156 (2015)
4. Resnick, P., Varian, H.R.: Recommender systems. Commun. ACM **40**(3), 56–58 (1997)

5. Ricci, F., Rokach, L., Shapira, B., Kantor, P.B.: Recommender Systems Handbook, vol. 1. Springer (2011)
6. Schafer, J.B., Frankowski, D., Herlockeer, J., Sen, S.: Collaborative filtering recommender systems. In: The Adaptive Web, pp. 291–324. Springer (2007)
7. Garcia, I., Sebastia, L., Onaindia, E.: On the design of individual and group recommender systems for tourism. Expert Syst. Appl. 38(6), 7683–7692 (2011)
8. Ekstrand, M.D., Riedl, J.T., Konstan, J.A.: Collaborative filtering recommender systems. Found. Trends Human Comput. Inter. 4(2), 81–173 (2010)

Image Super Resolution Using Wavelet Transformation Based Genetic Algorithm

Sudam Sekhar Panda and Gunamani Jena

Abstract Super resolution became one of the best techniques to obtain high resolution images as of a number of low-resolution images because of its simplicity and wide range of application in many fields of science and technology. There are several methods exist for super resolution but, wavelet transformation is chosen because of its minimalism and the constraints used to get better image restoration result. In this paper first Wavelet Transformation is considered to restore better image. Further Genetic algorithm is used to smooth the noise and better frequency addition into the image to get an optimum super resolution image.

Keywords PSNR: peak signal to noise ratio · Regularization · LR: HR low/high resolution · GA: genetic algorithm · Wavelet transforms

1 Introduction

In 1960s Harris and Good man developed a method using single image recovery concept called Super-resolution image reconstruction [1]. Many researchers subsequently conducted a study, and have proposed a variety of recovery methods. From the results of current research and application point of view, super-resolution reconstruction algorithm can be classified as frequency domain methods and spatial methods [2]. Early research focused on the frequency domain, but with the image

S.S. Panda (✉)
AMET University, Chennai, India
e-mail: sudamshekhar@gmail.com

S.S. Panda
Sir CR Reddy College of Engineering, Eluru, India

G. Jena
Roland Institute of Technology, Berhampur, India
e-mail: g_jena@reduffmail.com

© Springer India 2016
H.S. Behera and D.P. Mohapatra (eds.), *Computational Intelligence in Data Mining—Volume 2*, Advances in Intelligent Systems and Computing 411, DOI 10.1007/978-81-322-2731-1_33

degradation model for more general considerations, the latter part of the majority of research has focused on the space domain.

Schultz and Stevenson [3] In early 1992, will greatly posterior probability estimation methods used in the Huber-Markov random field image as a priori knowledge of the interpolation to improve image clarity. Regularization term first appeared in the field of mathematics. Near the beginning regularization methods like Tikhonov regularization are used to solve this type of problem where they depend on some regularization parameter that controls how much filtering is introduced by the regularization [4, 5].

Later, Geman D. et al. used the "half-quadratic regularization" method, to get the optimal solution [6], while, Lagendijk et al. and MGKang et al. used Regularized iterative image restoration algorithm [7, 8], to solve a nonlinear optimization problem to decrease the blur and look after the edges of the images.

Recently, wavelet-based iterative regularization image restoration algorithm is widely used because it takes care of noise control and edge preserving image textures. In this work, the Wavelet transformation is used to reconstruct an HR image by choosing an iterative regularization parameter to stabilize the problem and to look after the effect of the noise. Further genetic algorithm is used to get an optimized super resolution image.

2 Problem Definition

The mathematical model for High Resolution (HR) and low-resolution images (LR) frame may be written as

$$y_k = DBM_kX + n_k = H_kx + n_k, \quad 1 \le k \le p \tag{1}$$

where y_k stands for the k frame image degradation, X is denoted as the high-resolution images which is to be restored. n_k is the line for the additive noise row vector, and can be measured an additional white Gaussian noise (AWGN). Further D is considered as under-sampled imaging system operator, B is the fuzzy operators, M_k is the frame displacement operator, and H_k stands for the Degradation of the k-frame matrix. Also p is the total number of low-resolution images.

Suppose X is an image of size $L_1N_1 \times L_2N_2$ may be expressed as a vector $X = [x_1, x_2, \ldots, x_N]$, with $N = L_1N_1 \times L_2N_2$. If L_1 and L_2 are horizontal and vertical sampling factor, then the low-resolution images of each test y_k is of size $N_1 \times N_2$. Then the expression $y_k = [y_{k1}, y_{k2}, \ldots, y_{kM}]$, k = 1, 2,...p and $M = N_1 \times N_2$ is the kth frame of the low resolution image.

The model stated above calculate the value of X approximately, which determine that in what form one can use the reconstruction algorithm. The well known super-resolution image reconstruction uses all possible technically feasible information from the given degraded images y and reconstruct a high resolution image.

3 Wavelet Domain Regularization

Described in previous Section, relative to the noise distribution is based on the method in the spatial domain of the noise distribution estimate of the noise in the wavelet domain where the distribution of estimated effect will be even better. This is because: (1) the original image wavelet coefficients and the existence of spatial correspondence between the original image and transform coefficients are highly correlated; (2) different sub-band wavelet transform information can complement each other, for accurate judgments edge and noise distribution is very useful. (3) the use of wavelet decomposition levels and in different sub-band selecting multiple regularization parameters is usually a good feature.

In decomposition level, the degraded image can be divided into low-frequency sub-bands LL and three high frequency sub-bands that is perpendicular to the direction of the horizontal high-frequency low-frequency sub-bands HL, vertical high-frequency low-frequency sub-bands LH horizontal and diagonal direction of the high frequency sub-bands HH. As for the low-frequency sub-bands LL, and noise are merely located in the HL, LH and HH sub-bands, so when estimating the local noise variance, LL sub-band cannot be considered. In the high frequency sub-bands the estimated noise variance and the noise-directional is a very useful information, so the high frequency sub-bands in all the variance should be relatively close in; the edge of the side with the more obvious direction, and its various high frequency sub-bands should be large differences in variance, select the smallest sub-band variance as its variance can be estimated accurately reflect the true value of the edge by the noise level.

In the algorithm implementation, first select the horizontal, vertical and two diagonal directions of the simple uniform four different operators, respectively, with the HL, LH, HH and HH sub-bands for convolution. Then select the appropriate size of the window, calculate the fuzzy sub-band before and after the wavelet coefficients of the local variance of the difference, whichever is the minimum estimate of the noise as the local image variance, resulting in an estimate of the local noise variance matrix. Note that the level of wavelet decomposition, the size of $M \times N$ of the degraded image of its sub-band size $M/2 \times N/2$, so the local noise variance matrix size is $M/2 \times N/2$. The local noise variance matrix interpolation algorithm used to get the size of $M \times N$ of the new local noise variance matrix, the matrix element values is the degraded image at each pixel location is estimated noise variance. Finally, calculated in accordance with regularization weights W (i, j), complete the iterative image reconstruction.

4 Genetic Algorithm

Genetic Algorithm (GA) is derived from evolution and natural selection which is useful for finding the optimum solution. It helps to solves many complex problems and gives a perfect solution [9]. It always produces original population randomly and generates next generation by the process of crossover and mutation. Forming a new population of N individual (Fig. 1) is also an important task of it by satisfying the conditions determined by the fitness function.

In this work genetic algorithm is used to get an optimum solution from a set of solution, and the algorithm is designed as:

Step 1. Generate number of Chromosomes; decide mutation rate and crossover rate
Step 2. Create number of population, and initialize the genes randomly
Step 3. Decide the generation limit (L) and Execute the steps 4–7 till L reaches
Step 4. Calculate the objective function to find the fitness value of chromosomes
Step 5. Selection of Chromosomes
Step 5. Crossover and Mutation
Step 6. New Chromosomes (Offspring)
Step 7. Optimum Solution (Best Chromosomes)

Fig. 1 Block diagram for the proposed algorithm

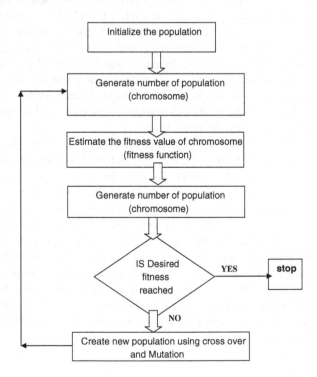

Table 1 Iterative_adaptive_R

L1	L2	Dx	PSNR
200	204	1.3346e+07	50.3682
200	204	1.2478e+07	53.1927
204	204	1.5601e+07	51.3615

Where L1: Horizontal sampling, L2: Vertical sampling factor

5 Experimental Results

The regularization parameter alpha and beta can be obtained by the experiment. L1 (horizontal sampling factor) and L2 (Vertical sampling factor) with the difference Dx with corresponding PSNR value is given in the Table 1 (Fig. 2).

To optimize the result and to get a better restoration of the image genetic algorithm is implemented and it is given bellow. By iterative adaptive and genetic algorithm the following result of alpha and beta are obtained

It shows that (Table 2) the values of alpha Beta and the time cost between them is significantly more in Genetic algorithm as compare to Iterative Adaptive algorithm. This is the reason we got a good optimize super resolution image.

Optimal image by genetic algorithm

This result is compared with the literature [10, 11], it is observed that sharpness is improved by adjusting the parameter in the algorithm, further here implementation of genetic algorithm given an added advantage to increase the resolution with the expectation.

Fig. 2 Initialization of the image by converting the image into low, high and its combination of the image

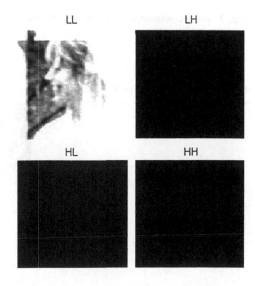

Table 2 Genetic algorithm

Algorithm	Alpha	Beta	Alpha-Beta time cost
Wavelet transforms	4.062500e+00	7.539063e+00	110.6429
Genetic algorithm	8.125000e+00	9.570313e+00	105.5145

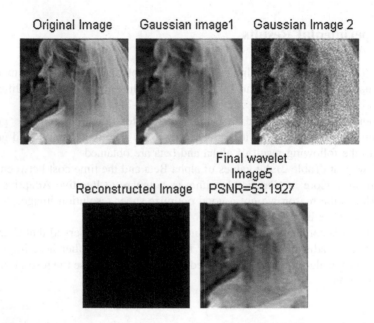

Fig. 3 The reconstructed image by Wavelet transformation method by decomposing the noise

Fig. 4 The optimize image of super resolution image

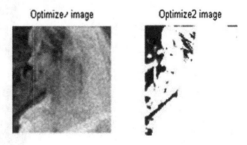

6 Conclusion

For better estimation of the registration of image, wavelet transforms method is used to decompose the high frequency and to extract the distribution of information, while the genetic algorithm is used to have a good image restoration result, Since the Alpha- Beta time cost is close to each other in both Wavelet transforms and genetic algorithm (Table 2) we can get a better compatible environment with a good result but not better than the result found by using adaptive regularization and genetic algorithm (Figs. 3 and 4) [11]. Also it is observed that combination of these two algorithms does not have much better result comparatively [11]. However one can observe that a significant improvement is there very good result in enhancing the image resolution but not taken care of high frequency removal successfully. This part can be taken as a future work.

References

1. Bing, T., Qing, X., Xun, G., Shuai, S.: Super-resolution image reconstruction technology development status of the Information Engineering University. **12**(4), 4 (2003)
2. Huang, L.-L., Xiao, L., Wei, Z.-H.: Efficient and effective total variation image super-resolution: a preconditioned operator splitting approach. Math. Prob. Eng. (2011)
3. Borman, S., Stevenson, R.: Spatial resolution enhance-ment of low-resolution image sequences a comprehensive reviewwith directions for future research [J/OL]. http://citeseer.nj.nec.com
4. Hansen, P.C, O'Leary, D.P.: The use of the L-curve in the regularization of discrete ill-posed problems. SIAM J. Sci. Comput. **14**(6), 1487–1503 (1993)
5. Wu gold.: Adaptive regularized image restoration (Ph.D. Thesis). National Defense University, (2006)
6. Geman, D., Yang, C.: Nonlinear image recovery with half-quadratic regularization [J]. IEEE Trans. Image Process. **4**(7), 932–946 (1995)
7. Lagendijk, Reginald L., Biemond, Jan, Boekee, Dick E.: Regularized iterative image restoration with ringing reduction. IEEE Trans. Acoust. Speech Signal Process. **36**(12), 1874–1888 (1988)
8. Kang, M.G., Katsaggelos, A.K., Schafer, R.W.: A regularized iterative image restoration algorithm. IEEE Trans. Signal Process. **39**(4), 4 (1991)
9. Xi, Y., Chai, T., Yun, W.: Summarization of genetic algorithm. Control Theory Appl. (6), 697–708 (1996)
10. Efrat, N., et al. Accurate blur models vs. image priors in single image super-resolution. 2013 IEEE International Conference on Computer Vision (ICCV) IEEE, (2013)
11. Panda, S.S., Jena, G., Sahu, S.K.: Image super resolution reconstruction using iterative adaptive regularization method and genetic algorithm. In: Computational Intelligence in Data Mining—Volume 2, pp. 675–681. Springer India (2015)

Implementation of Operating System Selection Using AHP-Entropy Model

Neha Yadav, S.K. Divyaa and Sanjay Kumar Dubey

Abstract Operating system selection problem is to choose the optimum OS based on user preferences on different factors. Functionality is one such attribute selected from software quality model which comprises several other sub factors preferred to analyse the quality of system. Analysis is done using AHP approach for weight calculations and results are validated using Entropy method on Functionality factors of Software quality model ISO 9126. This analysis helps in decision making for users which concerns about how well outcomes of any execution are achieved. A model AHP-Entropy is proposed which can be used to rank alternatives on both subjective and quantitative factors.

Keywords AHP · Entropy · Functionality · Operating system · Rank analysis

1 Introduction

In today's world, technology is rapidly increasing which is unfolding the limits of development, so more people are adapting these technologies because of their usability and dependability. A communication channel should be present for interaction between humans and technology. Operating System (OS) is used as an interacting medium between computer user and hardware of the computer system. OS provide an executing environment to the user where programs can be executed to derive results efficiently and in convenient manner [1]. To select an appropriate operating system is a complex task. Some attributes such as memory management,

N. Yadav (✉) · S.K. Divyaa · S.K. Dubey
Amity University, Sec-125, Noida, U.P., India
e-mail: nehayadav0403@gmail.com

S.K. Divyaa
e-mail: diyaas14@gmail.com

S.K. Dubey
e-mail: skdubey1@amity.edu

© Springer India 2016   363
H.S. Behera and D.P. Mohapatra (eds.), *Computational Intelligence in Data Mining—Volume 2*, Advances in Intelligent Systems and Computing 411, DOI 10.1007/978-81-322-2731-1_34

storage management, process management, security and requirements for selecting operating system [2, 3].

As OS is software so it should meet software quality standards. Functionality is one such factor laid in ISO 9126 software quality model which can be analysed to select a better functional OS which will provide an ease to users those are result driven [4]. Software quality model outlines the different parameters and metrics on which quality of any software can be analysed and functionality is one of the important factors in this model. Functionality refers to the capability of preparing different operations.

Selection is a complex and cognitive decision making process in which multiple alternatives are analysed based on different attributes and decision makers preferences to select the appropriate alternative. Various mathematical approaches are present such as AHP (Analytical Hierarchy Process) [5], Electra [6], TOPSIS [7], FuzzyAHP [8], ANP (Analytical Network Process) [9], etc. which helps decision makers for selecting many products, services, and locations in real world.

In this paper, AHP method is used for selecting a better functional OS among the variety of alternatives available in market on the basis of functionality factors of ISO 9126 software quality model. AHP is used for weight calculation based on the data obtained from experts after conducting a survey and Entropy method is used to validate the results obtained from AHP for making a decision for selecting alternatives presented in the hierarchy [10].

2 Literature Review

Wu and Chang applied AHP to select web services based on QOS by which users can select the service using web user interfaces they evaluated the results using scalability [11]. Wang presented an innovative AHP model to solve the problem of supplier selection by combining the influential factors based on historical data to reduce the complexity of calculation [12]. Jun and Wei, applied AHP for selecting routes for Hazardous Material transportation [13]. Zhanjiang and Guanghua applied AHP with Stenier tree problem to select the location for constructing distribution centre [14]. Yang et al. applied AHP and SVM (Support vector machines) to select site for electricity transmission of 500 kV substation [15]. Hsu et al. applied AHP for selecting the best online shopping site to meet the customer needs [16]. Jiang et al. solved the inconsistency in the weights for multiple alternative evaluation problems for decision making and proposed a new model AHP-TFN [17]. Wan et al. applied AHP for GIS oil and gas pipeline route selection [18]. Lee et al. applied Entropy decision model for selecting Enterprise Resource Planning System.

They proposed an entropy method to evaluate the weights of attributes and ERP alternatives [19]. Ke et al. applied AHP to select country wind farm location. They analysed the impact of factors that affects the wind power site selection [20]. Taguchi et al., applied entropy to select online motion which is used for registration of 3D objects by robots [21]. Gupta and Chaturvedi, applied entropy method to select multiuser MIMO (Multiple input multiple output) system. The results of this algorithm are closed to the optimal solution and have less complexity than brute force search and higher sum rate [22]. Zhang et al. proposed a new selection algorithm for selecting network in wireless environment and the performance of network is evaluated in real time by calculating subjective and objective weights using entropy theories [23]. Pandey and Dukkipati, developed a maximum entropy method for gene selection. This method can be applied for finding the number of moments and feature selection [24]. Wengjie and Guang, applied Entropy-AHP weights to select nuclear grade equipment supplier [25]. Mishra and Dubey used AHP method for reliability evaluation purpose of object-oriented software system [26]. Zhang et al. applied Entropy method to calculate the unique content and information that is required for selecting the scene matching area. This model improves the reliability of the system and has minimum error rate with high probability of matching [27].

3   Steps for Analytical Hierarchy Process

3.1   Construct a Hierarchical Tree for Problem

In the topmost layer the problem is to be defined then find the factors (C) based upon which evaluation of decision is to be made then sub factors and so on. These factors can be chosen from some standard model or from users preferences gained from perception and experiences. The last layer in the hierarchy will comprise of the alternatives (A) among which the best is selected as a decision.

3.2   Construct the Pair Wise Comparison Matrix

A pair wise comparison matrix is constructed for each factor comparing it with another factor in the same level hierarchy, then importance is evaluated by rating the factor between 1/9 to 9. If number of factors are n then an n × n matrix is formed in which each element a_{ij} in the (i,j)th position represents the importance of factor x_i when compared with x_j. The comparison matrix satisfies:

$$a_{ij} > 0, \ a_{ii} = 1 \ \text{and} \ a_{ji} = 1/a_{ij} \ \text{for} \ i \neq j; \ \text{where} \ i, \ j = 1 \ \text{to} \ n.$$

3.3 Calculate the Eigen Vector

3.3.1 Find the nth root for each row in the comparison matrix by multiplying each row elements and finding nth root of the product.

$$u_i = \prod a_{ij}a_{ik} \ \text{for all} \ k = 1 \ \text{to} \ n.$$

3.3.2 Find the weighted Sum

$$u_n = \sum u_i \ \text{for} \ i = 1 \ \text{to} \ n.$$

3.3.3 Normalize the nth root by column.

$$W_n = u_i / \sum u_i \ \text{for} \ i = 1 \ \text{to} \ n. \ \text{Here} \ w_n \ \text{is eigen vector.}$$

3.3.4 Calculate $\lambda_{avg(max)}$ using eigen vector

$$Aw_i = a_{ik} \times w_i \ \text{for all} \ j = 1 \text{ton}$$
$$Aw_i = \lambda_i w_n$$
$$\lambda_{avg(max)} = \sum \lambda_i / n$$

4 Steps for Entropy Method

The Entropy method is a MADM (Multiple Attribute Decision Making) based technique which helps decision makers to select best among the available alternatives by analyzing different factors. In this alternative Performance Matrix m × n is created for factors n and alternatives m. The entropy is an objective method used to measure uncertainty which evaluates weights only by using mathematical model without considering the preferences of decision makers.

4.1 Measure the Uncertainty of Attributes by Calculating Entropy

4.1.1 $E_i = -c \sum P_{ij} * \ln(P_{ij})$ for all i = 1 ton, j = 1 to m, where c = 1/m (m is number of alternatives). This entropy vector E_i must satisfies $0 \leq E_i \leq 1$.

4.1.2 Calculate degree of diversification by $D_i = 1 - E_i$ for all i = 1 to n

4.1.3 Calculate Weight Matrix for entropy by $w_i = D_i / \sum D_i$ for i = 1 to n.

5 Logical Analysis

Functionality (C) is a factor of ISO Software quality model [4] which defines a set of sub factors considered for evaluating alternative (A). The sub factors Suitability (C1) represents how much the system is appropriate for providing specific functions of the software and how much the software meets the needs of users, Accurateness (C2) represents appropriateness of a function whenever a operation in software is executed then its results is evaluated for its accuracy which deals with real time operations, Interoperability (C3) represents extent of connectivity or dependability of one component of a system with other. To what extent a component in a system can be accessed by another without any restricted access, Security (C4) factor maintains the authorized access of software stopping all the malicious activity by intruders and Functionality Compliance (C5) allows in making decision about how much software compiles with application standards, protocols and regulations of laws for some software. By analyzing these factors, OS selection problem is formulated. For this purpose hierarchical structure is made which comprises 3 levels as shown in Fig. 1.

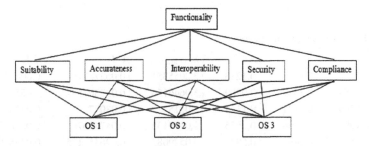

Fig. 1 The hierarchal model of operating system selection

5.1   Alternatives of OS

According to market analysis there are three popular OS which users are adapting for usage so considering these OS as alternatives for analysis. These are Windows (A1), Unix (A2), Mac (A3). These operating systems are selected by analysing the current market trends and because of popularity among users.

5.2   Pair Wise Comparison Matrices for AHP

Each factor is compared with other factors based on the data collected in survey which is filled and rated by experts, from which consistent data is analysed by AHP (shown in Table 1). Then matrices for each sub factor (C) are analysed with their alternatives (A) using AHP and if their consistency ratio is less than 0.1 then considered consistent and analysed (shown in Table 2).

The pair wise comparison matrix is analysed by AHP to rank the alternatives based on the larger priority number result of analysis (shown in Table 3). The Table indicates that the Alternative 2 has rank 1.

Table 1 Matrix for criteria

	C1	C2	C3	C4	C5	Eigen vector
C1	1	3	1	1/5	4	0.185
C2	1/3	1	½	1/3	5	0.12
C3	1	2	1	½	7	0.229
C4	5	3	2	1	5	0.424
C5	¼	1/5	1/7	1/5	1	0.042

$\lambda_{avg(max)}$ = 5.417324729, C.I = 0.1043311, C.R = 0.0931552841

Table 2 Eigen vectors for sub factors

	A1	A2	A3	$\lambda_{AVG(MAX)}$	C.I	C.R
C1	0.2764	0.5953	0.1283	3.0055351	0.00276775	0.0047717
C2	0.1884	0.7307	0.0809	3.0646150	0.032307532	0.055702643
C3	0.1769	0.7121	0.1114	3.06154	0.03077	0.053051724
C4	0.1991	0.7334	0.0675	3.0949135	0.04745676	0.081822001
C5	0.157	0.249	0.594	3.0529083	0.026454175	0.045610646

Table 3 AHP ranking order

	C1	C2	C3	C4	C5	Priority vector	Rank
W_i	0.185	0.12	0.229	0.424	0.042	–	–
A1	0.2764	0.1884	0.1769	0.1991	0.157	0.2052	2nd
A2	0.5953	0.7307	0.7121	0.7334	0.249	0.6823	1st
A3	0.1283	0.0809	0.1114	0.0675	0.594	0.1125	3rd

Table 4 Entropy calculations table

	C1	C2	C3	C4	C5
E_i	0.84437	0.68008	0.72152	0.66509	0.86132
D_i	0.15563	0.31992	0.27848	0.33491	0.13868
W_i	0.12677	0.26060	0.22684	0.27281	0.11296

Table 5 Ranking of alternatives using entropy

Alternatives	A1	A2	A3
RI	0.19631545	0.65617888	0.14813002
Rank	2nd	1st	3rd

6 Result Analysis

The result of experimental work is validated by using Entropy rank analysis method in following subsection. The procedure is already described in Sect. 4. Following procedure (Table 4) is validating weight Matrix of Functionality by using Entropy method of the results obtained by AHP to check the feasibility of Rank (Table 5).

7 Conclusions

Selection of operating system according to its functionality is an important concern for a result driven user. In this paper selection problem is analyzed using AHP approach and the results of analysis shows that Unix OS is more functional as compared to Windows and Mac. To make these results certain for decision making ranks are validated from Entropy method. A new model is proposed AHP-Entropy with which rank of alternatives can be achieved for subjective attributes. This methodology will help to select best selections alternatives of operating systems. In future more parameters will be selected for experimental purpose and by using different methodologies.

References

1. Silberschatz, A., Galvin, P.B., Gagne, G.: Operating System Concepts, 7th edn. Wiley (2004)
2. Tanenbaum, A.S., Woodhull, A.S.: Operating Systems: Design and Implementation, 3rd edn. Prentice Hall (2006)
3. McKusick, M.K., Neville-Neil, G.V.: The Design and Implementation of the FreeBSD Operating System. Addison Wesley (2004)
4. ISO/IEC 9126-1, Institute of Electrical and Electronics Engineers, Part 1, 2, 3: Software Quality Model
5. Al-Azab, M.F.G., Ayu, M.: Web based multi criteria decision making using AHP method. In: 2010 International Conference on Information and Communication Technology for the Muslim World (ICT4 M), pp. A6–A12. IEEE (2010)
6. Bírgün, S., Cíhan, E.: Supplier selection process using ELECTRE method. In: 2010 International Conference on Intelligent Systems and Knowledge Engineering (ISKE), pp. 634–639. IEEE (2010)
7. Xing, L., Tu, K., Ma, L.: The performance evaluation of IT project risk based on TOPSIS and vague set. In: 2009 International Conference on Computational Intelligence and Software Engineering, pp. 1–5. (2009)
8. Kaboli, A., Aryanezhad, M.-B., Shahanaghi, K., Niroomand, I.: A new method for plant location selection problem: a fuzzy-ahp approach. In: IEEE International Conference on Systems, Man and Cybernetics, 2007. ISIC., pp. 582–586. IEEE (2007)
9. Pandey, A.K., Agrawal, C.P.: Analytical network process based model to estimate the quality of software components. In: 2014 International Conference on Issues and Challenges in Intelligent Computing Techniques (ICICT), pp. 678–682. IEEE (2014)
10. Saaty, T.L.: The analytic hierarchy process. McGraw-Hill, New York (1980)
11. Wu, C., Chang, E.: Intelligent web services selection based on ahp and wiki. In: Proceedings of the IEEE/WIC/ACM International Conference on Web Intelligence. IEEE Computer Society. (2007)
12. Wang, Y.: An application of the ahp in supplier selection of maintenance and repair parts. In: 2009 1st International Conference on Information Science and Engineering (ICISE), pp. 4176–4179. IEEE (2009)
13. Jun, Y., Wei, C.: AHP application in the selection of routes for hazardous materials transportation. In: 2010 International Conference on Optoelectronics and Image Processing (ICOIP), vol. 1, pp. 189–191. IEEE (2010)
14. Guanghua, W, Zhanjiang, S.: Application of AHP and steiner tree problem in the location-selection of logistics' distribution center. In: 2010 2nd International Conference on Networking and Digital Society (ICNDS), vol. 1, pp. 290–293. IEEE (2010)
15. Yang, Y., Du, Q., Zhao, J.: The application of sites selection based on AHP-SVM in 500KV substation. In: 2010 International Conference on Logistics Systems and Intelligent Management, vol. 2, pp. 1225–1229. IEEE (2010)
16. Hsu, C.-H. Yang, C.-M., Chen, T.-C., Chen, C.-Y.: Applying AHP method select online shopping platform. In: 2010 7th International Conference on Service Systems and Service Management (ICSSSM), pp. 1–5. IEEE (2010)
17. Jiang, Z., Feng, X., Feng, X., Shi, J.: An AHP-TFN model based approach to evaluating the partner selection for aviation subcontract production. In: 2010 2nd IEEE International Conference on Information and Financial Engineering (ICIFE), pp. 311–315. IEEE (2010)
18. Wan, J., Qi, G., Zeng, Z., Sun, S.: The application of AHP in oil and gas pipeline route selection. In: 2011 19th International Conference on Geoinformatics, pp. 1–4. IEEE (2011)
19. Lee, M.-C., Chang, J.-F., Chen, J.-F.: An entropy decision model for selection of enterprise resource planning system. Int. J. Comput. Trends and Technol. 1–9 (2011)
20. Ke, C., Fuling, D., Yong, J.: Application of AHP in the Selection of Our Country Wind Farm's Location. In: 2011 International Conference on Management and Service Science (MASS), pp. 1–4. IEEE (2011)

21. Taguchi, Y., Marks, T.K., Hershey, J.R.: Entropy-based motion selection for touch-based registration using rao-blackwellized particle filtering. In: 2011 IEEE/RSJ International Conference on Intelligent Robots and Systems (IROS), pp. 4690–4697. IEEE (2011)
22. Gupta, G., Chaturvedi, A.K.: Conditional Entropy based User Selection for Multiuser MIMO Systems. IEEE Commun. Lett. **17**(8), 1628–1631 (2013)
23. Zhang, D., Zhang, Y., Lv, N., He, Y.: An access selection algorithm based on GRA integrated with FAHP and entropy weight in hybrid wireless environment. In: 2013 7th International Conference on Application of Information and Communication Technologies (AICT), pp. 1–5. IEEE (2013)
24. Pandey, G.K., Dukkipati, A.: Minimum description length principle for maximum entropy model selection. In: 2013 IEEE International Symposium on Information Theory Proceedings (ISIT), pp. 1521–1525. IEEE (2013)
25. Guang, Y., Wenjie, H.: Application of the TOPSIS based on entropy-AHP weight in nuclear power plant nuclear-grade equipment supplier selection. In: International Conference on Environmental Science and Information Application Technology, 2009. ESIAT 2009, vol. 3, pp. 633–636. IEEE, (2009)
26. Mishra, A., Dubey, S.K.: Evaluation of reliability of object oriented software system using fuzzy approach. In: Confluence The Next Generation Information Technology Summit (Confluence), 2014 5th International Conference-, pp. 806–809. IEEE (2014)
27. Zhang, X., He, Z., Liang, Y., Zeng, P.: Selection method for scene matching area based on information entropy. In: 2012 Fifth International Symposium on Computational Intelligence and Design (ISCID), vol. 1, pp. 364–368. IEEE (2012)

Modified Firefly Algorithm (MFA) Based Vector Quantization for Image Compression

Karri Chiranjeevi, Uma Ranjan Jena, B. Murali Krishna and Jeevan Kumar

Abstract Firefly algorithm optimization is based on the attractiveness/brightness of the firefly. In firefly algorithm, a lighter (lesser fitness function) firefly move towards the brighter firefly (higher fitness function) with amplitude proportional to Euclidean distance between the lighter and brighter firefly. If no such brighter firefly is found then it moves randomly is search space. This random move causes chance of decrement in brightness of the brighter firefly depending on the direction in which it is move. We proposed a modified firefly algorithm in which movement of brighter fireflies is towards the direction of brightness instead of random move. If this direction of brightness is not in the process then firefly is in same position. We call this novel algorithm as MFA-LBG. Experimental results shows that modified firefly algorithm reconstructed image quality and fitness function value is better than the standard firefly algorithm (FA-LBG) and LBG algorithms. It is observed that that modified firefly algorithm convergence time is less than the standard firefly algorithm.

Keywords Vector quantization · LBG · Firefly algorithm · Modified firefly algorithm · Attractiveness · Fitness function

K. Chiranjeevi · U.R. Jena (✉)
Department of Electronics and Tele-Communication Engineering,
Veer Surendra Sai University of Technology (VSSUT), Burla 768018
Odisha, India
e-mail: urjena@rediffmail.com

K. Chiranjeevi
e-mail: chiru404@gmail.com

B. Murali Krishna · J. Kumar
Department of Electronics and Communication Engineering,
Sri Sivani Institute of Technology, Srikakulam, Andhra Pradesh, India
e-mail: muralikrishna.bonthu@gmail.com

J. Kumar
e-mail: Jeevan.vajja@gmail.com

© Springer India 2016
H.S. Behera and D.P. Mohapatra (eds.), *Computational Intelligence in Data Mining—Volume 2*, Advances in Intelligent Systems and Computing 411, DOI 10.1007/978-81-322-2731-1_35

1 Introduction

Image compression plays a major role in our day by day life. Coding of image to be compressed with less number of bits is called image compression. In literature there are number of image compression methods such as transform method (Discrete Cosine Transform, Discrete Wavelet Transform, K-L Transform etc.....), fractal method and optimization methods. Recently it is observed that vector quantization (VQ) outperforms the scalar quantization. Linde-Buzo-Gary (LBG) algorithm is the first VQ technique that guarantees the best codebook for each iteration but it suffer with local minimum i.e. it could not generate global codebook that minimize the error between original image and reconstructed image [1]. Menez et al. proposed an Optimum Quantizer Algorithm for Real-Time Block Quantizing [2]. Ra and Kim proposed a fast mean-distance-ordered partial codebook search algorithm in 1993 [3] they filter the false codewords for an input vector based on the squared mean distance (SMD). Principal Component Analysis (PCA) is another method to design a codebook for vector quantization. PCA is based on correlated variables are transformed to uncorrelated variable. Huang and Harris proposed a directed search binary-splitting (DSBS) method in 1993 [4]. In their scheme, PCA is used to select the initial codebook to reduce the dimension of the training vectors. The most important thing in Vector Quantization is to design a better codebook that minimizes the distortion between original image and reconstructed image. Last few years it is observed that with optimization techniques for codebook design present a better reconstructed image quality. Franti et al. proposed [5] a Genetic Algorithm (GA) based vector quantization for codebook design by considering the codebook as chromosome. Genetic Algorithm designs an efficient codebook by the process of crossover and mutation operation. Patane and Russo proposed enhanced LBG (ELBG) (in 2001) [6] based on the utility of codebook so algorithm overcomes the problem of local minimum. Utility of jth codebook is defined as the ratio of total distortion of jth cluster to mean value of the cluster. Rajpoot et al. have designed a codebook by using ant colony system (ACS) algorithm [7]. Chun-Wei Tsai et al. proposed a fast ant colony optimization for codebook generation [8] by observing the redundant operations performed during convergence of ant colony optimization algorithm. In addition, Practical swarm optimization (PSO) vector quantization [9] outperforms LBG algorithm which is based on updating the global best (gbest) and local best (lbest) solution. The Hsuan-Ming Feng, Ching-Yi Chen and Fun Ye showed that Evolutionary fuzzy particle swarm optimization algorithm gives better codebook than LBG [10]. Quantum particle swarm algorithm (QPSO) was proposed by Wang (2007) to solve the 0–1 knapsack problem [11]. Ming-Huwi Horng proposed a firefly algorithm for vector quantization in 2012 [12].

In Modified firefly algorithm for vector quantization we assumed the current iteration best solution as a brightest firefly among all. The major problem in normal firefly algorithm is that these fireflies are moving randomly so there is chance of decrement in the brightness of fireflies. This kind of move results decrement in fitness function and also effects on performance of the algorithm. So to overcome this problem we are moving this type of fireflies towards the direction where chance of increment in brightness of fireflies is there. So that this movement will not decrease the fitness function hence generate a global codebook for efficient vector quantization leading to improved image compression.

2   Vector Quantization Methods

2.1   LBG Vector Quantization

The most commonly used method in VQ is the Generalized Lloyd Algorithm (GLA) which is also called Linde-Buzo-Gary (LBG) algorithm. The algorithm is as follows:

Step 1:   Begin with initial codebook C1 of size N. Let the iteration counter be $m = 1$ and the initial distortion $D1 = \infty$

Step 2:   Using codebook $Cm = \{Yi\}$, partition the training set into cluster sets Ri using the nearest neighbor condition

Step 3:   Once the mapping of all the input vectors to the initial code vectors is made, compute the centroids of the partition region found in step 2. This gives an improved codebook $Cm + 1$

Step 4:   Calculate the average distortion $Dm + 1$. If $Dm - Dm + 1 < T$ then stop, otherwise $m = m + 1$ and repeat step 2–4

The distortion becomes smaller after recursively executing the LBG algorithm. Actually, the LBG algorithm can guarantee that the distortion will not increase from iteration to the next iteration. However, it cannot guarantee the resulting codebook will become the optimum one and the initial condition will significantly influence the results. Therefore, in the LBG algorithm we should pay more attention to the choice of the initial codebook.

2.2 Firefly Algorithm Vector Quantization

Firefly algorithm works with two assumptions: (1) Irrespective of sex of fireflies all the fireflies are attracted towards the other fireflies (2) second assumption is that lighter fireflies are always attracted by the brighter fireflies so lighter one always moves towards the brighter. If no brighter in the search space then brightest one moves randomly in search space.

Here our assumption is fitness function is equal to brightness of the fireflies i.e. lower fitness value fireflies move towards the higher fitness value. According to theory of physic the brightness/fitness function is monotonically decreasing function. Suppose a distance $r_{i,j} = d(X_j, X_i)$ is the Euclidean distance between the chosen jth firefly to ith firefly which is given by Eq. (1).

$$\text{Euclidean distance } r_{ij} = \|X_I - X_J\| = \sqrt{\sum_{k=1}^{N_c} \sum_{h=1}^{L} (X_{i,k}^h - X_{j,k}^h)^2} \tag{1}$$

$$\beta = \beta_0 e^{-\gamma r_{i,j}} \tag{2}$$

where β_0 is the brightness at $r_{i,j} = 0$ and β is the light absorption coefficient at the source. With this β and Euclidean distance, the lighter firefly i move towards the brighter firefly j as per the following Eq. (3)

$$X_{i,k} = (1 - \beta)X_{i,k} + \beta X_{j,k} + u_{i,k} \tag{3}$$

$$u_{i,k} = \alpha(\text{rand } 1 - \frac{1}{2}) \tag{4}$$

where α is a random number.

If no brighter firefly in the search space than firefly X_i then it moves randomly with Eq. (5)

$$X_{i^{\max},k} = X_{i^{\max},k}^+ u_{i^{\max},k} \tag{5}$$

where rand1 value is lies between 0 to 1. The objective function in firefly algorithm is equal to the brightness of the firefly.

The details of FF-LBG algorithm is as follows:

Step 1: Run the LBG algorithm once and assign it as brighter codebook

Step 2: Initialize α, β and γ parameters, Initialize rest codebooks with random numbers

Step 3: Find out fitness values by Eq. (6) of each codebook

$$\text{Fitness}(C) = \frac{1}{D(C)} = \frac{N_b}{\sum\limits_{j=1}^{N_c} \sum\limits_{i=1}^{N_b} u_{ij} \bullet \left\| X_i - C_j \right\|^2} \tag{6}$$

Step 4: randomly select a codebook and record its fitness value. If there is a codebook with higher fitness value, then it moves towards the brighter codebook (highest fitness value) based on the Eqs. (7)–(9).

$$\text{Euclidean distance } r_{ij} = \left\| X_I - X_J \right\| = \sqrt{\sum\limits_{k=1}^{N_c} \sum\limits_{h=1}^{L} \left(X_{i,k}^h - X_{j,k}^h \right)^2} \tag{7}$$

X_i is randomly selected firefly, X_j is brighter firefly

$$\beta = \beta_0 e^{-\gamma_{ij}} \tag{8}$$

$$X_{j,k}^h = (1 - \beta) X_{i,k}^h + \beta X_{j,k}^h + u_{j,k}^h \tag{9}$$

where u is random number between 0 and 1, k = 1, 2,...,Nc, h = 1, 2,....L.

Step 5: if no firefly fitness value is better than the selected firefly then it moves randomly is search space with following equation

$$X_{i,k}^h = X_{i,k}^h + u_{j,k}^h \quad k = 1, 2 \ldots N_c, h = 1, 2, \ldots L \tag{10}$$

3 Modified Firefly Algorithm

Surafel et al. [13] proposed a modified firefly algorithm in the year 2012 [13]. In ordinary firefly algorithm, if no brighter firefly in the search space then the firefly under consideration moves randomly where as in modified firefly algorithm the random movement is towards the firefly whose brightness is going to be increased [14]. To know the brightness directions generate randomly Y unit vectors i.e. Y_1, Y_2, Y_3.... Y_m and then choose a firefly randomly whose brightness is increased in next iteration. The movement of the brightest firefly follows the following Eq. (11)

$$Y = Y + \alpha R \tag{11}$$

where α is a random step length. If firefly does not find such kind of firefly then it stays in the current position.

The detailed algorithm is as follows:

Step 1:   Run the LBG algorithm once and assign it as brighter codebook

Step 2:   Initialize α, β and γ parameters, Initialize rest codebooks with random numbers

Step 3:   Find out fitness values by Eq. (6) of each codebook

Step 4:   randomly select a codebook and record its fitness value. If there is a codebook with higher fitness value, then it moves towards the brighter codebook (highest fitness value) based on the Eqs. (7)–(9)

Step 5:   if no firefly fitness value is better than the selected firefly then it moves with Eq. (11) else it stays in the same position

4   Results and Discussions

The modified firefly algorithm, current firefly algorithm and LBG have been implemented using Matlab 7.9.0 (R2009b), and then these three algorithms are tested and compared with each other using the two selected images. Images were used for comparison are peppers (.png format) and baboon (.jpg format). Figure 1 shows the images used for comparison of three methods. The image statistics namely, fitness function and PSNR (Peak Signal to Noise Ratio) afford the performance measure of an image compression technique. Figure 2a–c shows the reconstructed baboon image or decompressed baboon images of three algorithms and it is observed that reconstructed image quality of modified firefly algorithm is

(a)   (b)

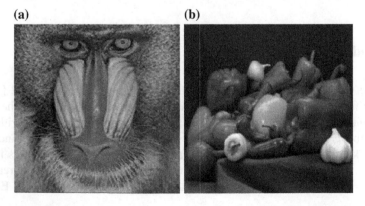

Fig. 1 Standard image used for comparison for three methods. **a** Baboon. **b** Peppers

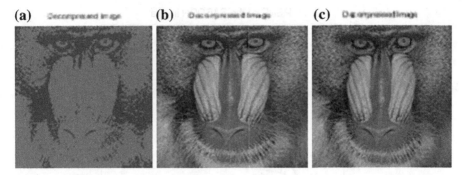

Fig. 2 Decompressed baboon images of three methods. **a** LBG. **b** FA–LBG. **c** MFA–LBG

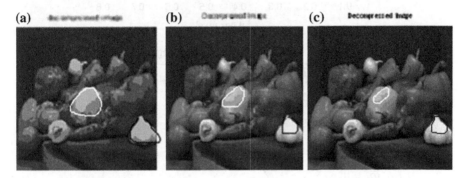

Fig. 3 Decompressed pepper images of three methods. **a** LBG. **b** FA–LBG. **c** MFA–LBG

better than firefly algorithm and LBG algorithms. Figure 3a–c shows the reconstructed peppers image or decompressed peppers image of three algorithms. The indicated yellow and red marks on peppers image shows that reconstructed image quality of modified firefly algorithm is better than firefly algorithm and LBG algorithms.

Figure 4a, b shows the peak signal to noise ratio verses bit rate for three methods. From the figures it is observed that peak signal to noise ratio of modified firefly algorithm is better than firefly algorithm and LBG algorithms. From practical observation the modified firefly algorithm convergence speed is better than ordinary firefly algorithm. From Fig. 5 represents the Cost of fitness function. From Figures fitness function of modified firefly algorithm is better than firefly algorithm and LBG algorithms. Figure 5a, b shows the reconstructed image or decompressed images of three algorithms and it is observed that reconstructed image quality of modified firefly algorithm is better than firefly algorithm and LBG algorithms.

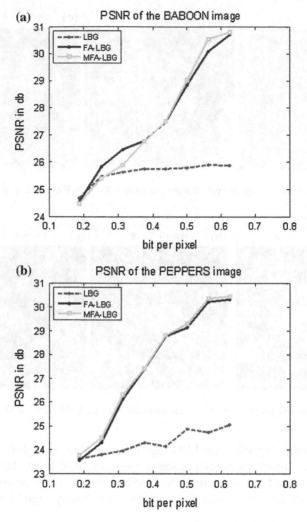

Fig. 4 Average peak signal to noise ratio of three methods. **a** Baboon image. **b** Peppers image

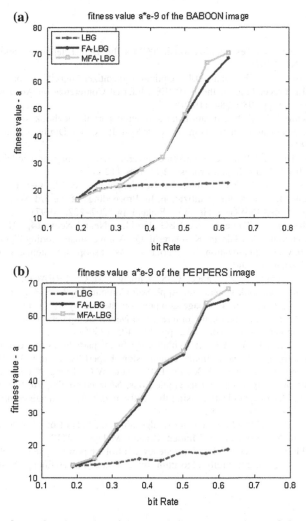

Fig. 5 Average fitness function value of three methods. **a** Baboon image. **b** Peppers image

5 Conclusions

In this paper, a Modified Firefly algorithm (MFA) optimized vector quantization for image compression is proposed. From the simulation results it is observed that modified firefly algorithm peak signal to noise ratio and fitness function value is better than the firefly algorithm (FA-LBG) and LBG algorithms. From the simulation results it is observed that that modified firefly algorithm convergence time is less than the standard firefly algorithm.

References

1. Lloyd, S.P.: Least squares quantization in PCM's. In: Bell Telephone Laboratories Paper, Murray Hill, NJ (1957)
2. Menez, J., Bceri, F., Esteban, D.J.: Optimum quantizer algorithm for real-time block quantizing. In: Proceedings of the I979 IEEE Internal Conference on Acoustics, Speech, & Signal Processing, pp. 980–984. (1979)
3. Ra, S.W., Kim, J.K.: A fast mean-distance-ordered partial codebook search algorithm for image vector quantization. IEEE Trans. Circuits Syst. II: Analog Digital Signal Process. **40**(9), 576–579 (1993)
4. Huang, C.M., Harris, R.W.: A comparison of several vector quantization codebook generation approaches. IEEE Trans. Image Process. **2**(1), 108–112 (1993)
5. Franti, P., Kivijarvi, J., Kaukoranta, T., Nevalainen, O., et al.: Genetic algorithms for codebook generation in vector quantization. In: Proceedings of the 3rd Nordic Workshop on Genetic Algorithms (3NWGA), Helsinki, Finland, pp. 207–222. (1997)
6. Patane, G., Russo, M.: The enhanced LBG algorithm. Neural Netw. **14**, 1219–1237 (2002)
7. Rajpoot, A., Hussain, A., Saleem, K., Qureshi, Q.: A novel image coding algorithm using ant colony system vector quantization. In: International Workshop on Systems, Signals and Image Processing, Poznan, Poland
8. Tsaia, C.-W., Tsengb, S.-P., Yangc, C.-S., Chiangb, M.-C.: PREACO: a fast ant colony optimization for codebook generation. Appl. Soft Comput. **13**, 3008–3020 (2013)
9. Chen, Q., Yang, J.G., Gou, J.: Image compression method by using improved PSO vector quantization. In: First International Conference on Neural Computation (ICNC 2005), Lecture Notes on Computer Science, vol. 3612, pp. 490–495. (2005)
10. Feng, H.-M., Chen, C.-Y., Ye, F.: Evolutionary fuzzy particle swarm optimization vector quantization learning scheme in image compression. Expert Syst. Appl. **32**, 213–222 (2007)
11. Wang, Y., Feng, X.Y., Huang, Y.X., Pu, D.B., Zhou, W.G., Liang, Y.C.: A novel quantum swarm evolutionary algorithm and its applications. Neurocomputing **70**, 633–640 (2007)
12. Horng, M.-H.: Vector quantization using the firefly algorithm for image compression. Expert Syst. Appl. **39**, 1078–1091 (2012)
13. Tilahun, S.L., Ong, H.C.: Modified firefly algorithm. In: School of Mathematical Sciences, Universiti Sains Malaysia, 11800 Minden, Penang, Malaysia (2012)
14. Chandra Sekhar, G.T., et al.: Load frequency control of power system under deregulated environment using optimal firefly algorithm. Int. J. Electr. Power Energy Syst. **74**, 195–211 (2016)

Performance Analysis of Selected Data Mining Algorithms on Social Network Data and Discovery of User Latent Behavior

Santosh Phulari, Parag Bhalchandra, Santosh Khamitkar,
Nilesh Deshmukh, Sakharam Lokhande, Satish Mekewad
and Pawan Wasnik

Abstract This paper is summery for experimental research work carried out on internet usage activities of students on social network sites. The summarized research work has proposed a novel and efficient method for discovery of user latent behavior from the student's datasets. A comparison with existing standard method is also presented which show that the proposed approach is better on the basis of accuracy, error rates, time sharing etc. features.

Keywords Data mining · Social network analysis · ARM · LDA · Usage patterns

P. Bhalchandra · S. Khamitkar · N. Deshmukh · S. Lokhande · S. Mekewad · P. Wasnik
School of Computational Sciecnes, S.R.T.M. University, Nanded 431606, MS, India
e-mail: srtmun.parag@gmail.com

S. Khamitkar
e-mail: s.khamitkar@gmail.com

N. Deshmukh
e-mail: nileshkd@yahoo.com

S. Lokhande
e-mail: sana_lokhande@rediff.com

S. Mekewad
e-mail: satishmekewad@gmail.com

P. Wasnik
e-mail: pawan_wasnik@yahoo.com

S. Phulari (✉)
Department of Computer Science, New Model Degree College, Hingoli, MS, India
e-mail: santoshphulari@gmail.com

© Springer India 2016
H.S. Behera and D.P. Mohapatra (eds.), *Computational Intelligence in Data Mining—Volume 2*, Advances in Intelligent Systems and Computing 411, DOI 10.1007/978-81-322-2731-1_36

1 Introduction

Now days, the Internet, is used to bridge the digital divide between geographically scattered people. It has become common platform for formal and informal communications via email, text and video chats. Internet is also used for participative, processes, feedbacks, community platforms and social network sites [1, 2]. Therefore it has become an extension of our self [1] in a way that allows us to construct and extend our social ties and form social networks within this new medium. There is tremendous trend in people to go online and have online conversations with other peers in terms of tweets, updates, feedbacks, likes, share, etc. [3, 4]. The online interactions are always in terms of groups. The Social networks have become virtual community and have been fairly studied in the general context of analyzing interactions between people and determining the important structural patterns in such interactions [1]. Now days we have online social networks like *Face book, LinkedIn, Tweeter, Instagram* and *MySpace* where social networks are internet embedded applications. Such social networks have rapidly grown in popularity [5]. These are geographical limitations free, unconventional, interactive social networks. They have provisions for friendships, information/emotion shearing, face-to-face meetings, or personal friendships [3]. The interactions on them are related to social activities, hobbies, feelings, education levels. For instance, students come from various Universities are always found connected to each other by the school names these people from online, a special group [4, 6]. Analysis of social behavior patterns in Social networks is very hard. People usually deploy machine learning methods to extract hidden relationships from these social networks and predict people into different groups related to their preference, hobbies, and education levels. Such machine learning practices are known as Social Network Analysis (SNA) [7, 8]. In past years, social network analysis (SNA) has become a new discipline. New groups are created very day. SNA has been more and more used as to analyze informal relations between groups [9–11]. SNA primarily work for discovery of invisible patterns of social collaboration [12, 13]. Such discovery is called as discovery of user latent behavior [14, 15].

2 Scope and Outline of the Research Work

Discovery of user latent behavior for computer social networks communities from social network database is major research issue for more than decade. By using regular statistical packages discovery of user latent behavior requires large time and bull work. It leads to high processing budget to extract hidden inter-relationship amongst various complex attributes of user's behavior pattern. With common purpose statistical packages like *Lotus, MS-Excel, SPSS,* etc. discovery of user latent behavior is unattainable accurately. In the underlined research work, focus is on discovery of user latent behavior for computer social networks communities i.e.

finding common clusters of user behavior. However, the efforts needed to discover user latent behavior becomes complicated and exponential with study area. The considered study is automated one and aim for detection of clusters of user's behavior for computer social networks communities by reducing efforts with good reliability of results. The general purpose of automated discovery of hidden patterns is to give high accuracy results without human help. Discovery of user latent behavior for computer social networks communities can be used in many applications such as education policy enforcement, social issues, Human resource planning and development, etc. Despite extensive research in computer social networks communities for hidden pattern recognition and reconstruction, no accurate solutions are yet discovered. This is because user have a wide variety of attributes and their complex inter-relationship which increase huge numbers of permutation and combinations to define and detect hidden pattern. Our study offers a solution to the problem of discovery of user latent behavior for computer social networks communities; primarily for the recognition and reconstruction of user's latent behavior patterns. It uses Association Rule Mining (ARM) algorithms. We use Apriori algorithm on dataset generated on questionnaire as shown in Fig. 2, is prepared on the behavior of students who use the *Face book* at *School of Computational Sciences, Swami Ramanand Teerth Marathwada University, and Nanded*. The hidden patterns of these users are extracted. This activity is counter checked with *the Face book* dataset of students of *University of Illinois*. Keeping above steps in mind, a prototype is proposed which is based on Nystrom method. Implementation and comparison of well known data mining algorithms like Association Rule Mining (ARM), Latent Dirichlet Allocation (LDA) and Clustering Classification (CC) algorithm on the dataset is carried out. The outcome of these three algorithms is compared with our newly proposed prototype implemented and compared on computer social networks community user's online database. The Association Rule Mining (ARM) searches hidden relationships among items [16, 17]. The Latent Dirichlet Allocation (LDA), work for extraction and recognisation user behavior from social network database. However, it has been applied to social network online database user behavior extraction in the context of topic mining.

3 Data Collection and Pre Processing

The methodology for recognition and extraction of user latent behavior from social network communities' i.e. University of Illinois student database involves following steps. This dataset is obtained from www.facebookproject.com which is free to all users for research purpose. This is created from a survey of social network browsing students of University of Illinois at Urbana Champaign with 77 attributes. There are 74 different records. During experimentations, we followed standard Data mining procedures including data cleaning, attribute grading, etc. The original dataset contained some missing values for various attributes. To proceed with the work, those missing values were replaced as if the people answered don't know or blanked. Further analysis including redundancy reduction and co relational filtering are also carried out. We compared the ARM, LDA, CC and proposed prototype on the basis of error rates, access time, accuracy and data packet receive rate. The details are narrated in the Experimentation section. Extraction of hidden patterns of user from social network communities' database our modified prototype is pre-eminent. The results are very interesting and leads to promising future works.

4   Experimental Set Up

This section focuses on implementing the ARM, LDA and CC algorithm on University of Illinois database and also on defining the prototype which is capable to add all advantageous features of ARM, LDA and CC algorithm and trying to eradicate complexity and append features which increase the percentage of accuracy and decreases the access time. At the time of implementation of prototype time-accuracy trade-off problem is raised. To solve this problem some control parameter sets are used by setting appropriate threshold values which results in best-fitted solution for the discovery of user latent behavior for computer social networks communities' database. The database containing student data of University of Illinois was used to implement ARM, LDA, CC and prototype data mining algorithms using *MATLAB 2009b*. These dataset works as the input for these algorithms. The resulting information can be employed to increase accuracy in evaluating attributes from database. The prototype based on Nystrom method evaluates attributes effectively and efficiently than the others such as—ARM, LDA and CC. Using our prototype the accuracy level increased and error rates, access time was decreased; hence user latent behavior is clearly extracted and recognized using the same. We used *MATLAB 2009b* on database of University of Illinois students. The *MATLAB* provides wide variety of library functions for performing different operations and also provides programming environment to define user defined functions by which we can define prototype. It supports different types of algorithms. We choose ARM, LDA and CC algorithm for finding similar attributes from which we extract latent behavior of user/student. Following are the step wise implementation of experiment as per methodology defined in previous section.

1. **Database collection**: We have used database from University of Illinois students using questionnaire (Dataset 2). This database is available on www.thefaceproject.com for the researchers who working on the *Face book*. For our research FBP2k7.xls database is used which contains 77 columns and 74 rows.
2. **Database conversion into MAT format**: We use FBP2k7.xls database to implement our methodology. In this research *MATLAB 2009b* is used to implement the different types of algorithm on it. We convert database file format from.xlsx to.mat by default option in MS-Excel and the input file 'FBP2k7.xlsx' is converted into 'FBP2k7.mat'.

3. **Implementation of ARM, LDA and CC algorithms**: The ARM algorithm implemented on database using *MATLAB 2009(b)* [18]. To avoid redundancy, the output of this stage is used for comparison with LDA, CC and proposed prototype which is in the last stage of this methodology. Likewise LDA and cluster classification algorithms were also implemented.

5   Results and Observations—Part A (Comparison of Data Mining Algorithms)

The Table 1 shows that our proposed prototype is producing the less error rates in comparison with the other algorithms such as ARM, LDA and CC. ARM algorithm produces 0.65 error rates that is highest error rates in this comparison table. But our prototype produces 0.11 that is lowest error rates than the others. LDA produces 0.62 error rates and CC produces 0.54 error rates which are less in comparison with ARM but more than prototype. The graphical representation of the comparison table on the basis of error rates is shown below Fig. 1. The Table 2 shows Access Time of different algorithms and Fig. 2 shows graphical representation comparison between existing algorithms with proposed algorithm keeping in mind the access time.

Table 1 Comparison on the basis of error rates

Algorithm	ARM	LDA	CC	Proposed prototype
Error rates	0.65	0.62	0.54	0.11

Fig. 1 Comparison on the basis of error rates

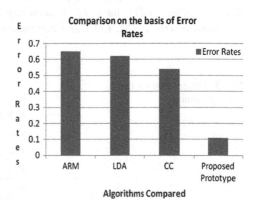

Table 2 Comparison on the basis of access time

Algorithm	ARM	LDA	CC	Proposed prototype
Access time (in s)	98	89	75	41

Fig. 2 Comparison on the
basis of access time

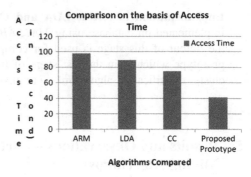

Table 3 Comparison on the
basis of accuracy

Algorithm	ARM	LDA	CC	Proposed prototype
Accuracy (in %)	93.50	92	88	96

The Table 2 shows that our proposed prototype takes the less access time in comparison with the other predefined algorithms such as ARM, LDA and CC. ARM algorithm takes 98 s access time that is highest access time in this comparison table. But our prototype takes 41 s that is lowest access time than the others. LDA takes 89 s access time whereas CC takes 75 s access time which is less in comparison with ARM but more than prototype. The Table 3 shows Accuracy of different algorithms and Fig. 3 shows graphical representation comparison between existing algorithms with proposed algorithm on the basis of accuracy. The table shows that our proposed prototype is more accurate in comparison with the other predefined algorithms such as ARM, LDA and CC. ARM algorithms' accuracy level is 93.50 %. But our prototype has accuracy 96 % that is highest accuracy level than the others. LDA have 92 % accuracy and CC have 88 % which is less in comparison with ARM, LDA and our prototype.

Fig. 3 Comparison on the
basis of accuracy

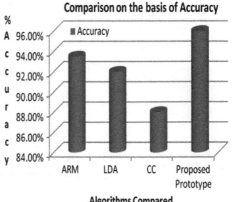

Fig. 4 Comparison on the basis of data packet received

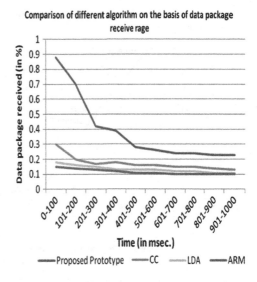

Comparison of different algorithm on the basis of data package receive rage

Time (in msec.)

━━ Proposed Prototype ━━ CC ━━ LDA ━━ ARM

The Fig. 4 shows that our proposed prototype is faster than the other predefined algorithms such as ARM, LDA and CC. ARM algorithm received 1.16 % of data in first second (1000 ms). After 1 s it shows steady performance i.e. 0.1 % data received in every 100 ms interval which required 98 s to receive complete data. LDA algorithm received 1.34 % data in first second. After 1 s it shows steady performance i.e. 0.11 % data is received in every 100 ms interval which required 89 s to receive the complete data. CC algorithm received 1.74 % data in first second. After 1 s it shows steady performance i.e. 0.13 % data is received in every 100 ms interval which required 75 s to receive the complete data. As compared with other predefined algorithms, our Prototype algorithm received 3.87 % data in first second. After 1 s it shows steady performance i.e. 0.23 % data is received in every 100 ms interval which required 41 s to receive the complete data. This shows better access time.

6 Results and Observations—Part B (Discovery of User Latent Behavior)

We now try to extract the different user latent behavior of the Junior Students, Senior Students, Alumni, Staff, and Faculty. The following figure user categories and their participation in different activities. From these activities we extract their latent behavior by using our prototype.

Figure 5 shows that the junior students are interested in wall post (31 %) and Academic activity (30 %) as compare to the other activities such as Political views (4 %), Contact Information (7 %), Privacy (8 %), and Relationship Status (20 %). Then we extract their behavior using these clusters which is shown by our

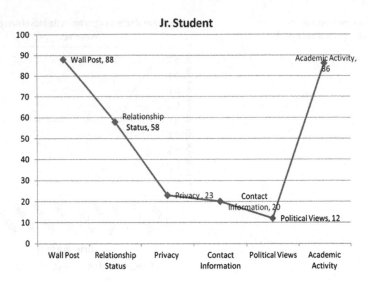

Fig. 5 Junior student's participation

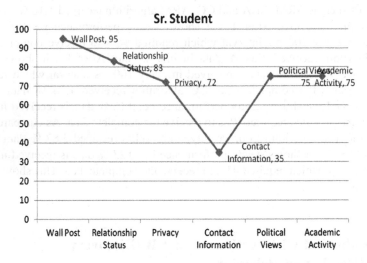

Fig. 6 Senior student participation

prototype. Hence we say that the junior students are more concentrate in Face book on the wall post and academic activity. Figure 6 shows that the senior students are interested in wall post (22 %) and Relationship Status (19 %) as compare to the other activities such as Political views (17 %), Contact Information (8 %), Privacy (17 %), Academic Activity (17 %). Then we extract their behavior using these clusters which is shown by our prototype. Hence we say that the senior students are more concentrate in Face book on the wall post and Relationship status. Figure 7

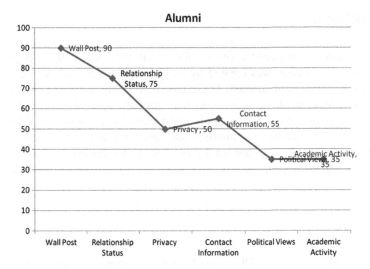

Fig. 7 Alumni participation

shows that the alumni are interested in wall post (27 %) and Relationship Status (22 %) as compare to the other activities such as Political views (10 %), Contact Information (16 %), Privacy (15 %), Academic Activity (10 %). Then we extract their behavior using these clusters which is shown by our prototype. Hence we say that the alumni are more concentrate in Face book on the wall post and Relationship status. The Fig. 8 shows that the staff is interested in Political Views (33 %) and wall post (27 %) as compare to the other activities such as Relationship Status (14 %), Contact Information (6 %), Privacy (10 %), Academic Activity (10 %). Then we extract their behavior using these clusters which is shown by our prototype. Hence we say that the staff is more concentrate in Face book on the Political Views and wall post.

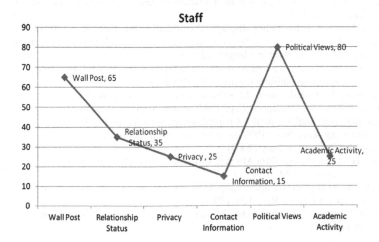

Fig. 8 Staff participation

7   Conclusion

In this research, an empirical approach for Social Network Analysis is presented. A social network dataset was created by using questionnaire and ARM, LDA and CC algorithms were implemented. These algorithms were compared on the basis of error rates; access time, accuracy and data package receive ratios. The second contribution of this paper is to apply clustering for collaborative filtering for computer social networks communities and discovery user latent behavior. This was achieved by suggesting our own prototype. Cross comparisons of our prototypes with selected algorithms was also done.

References

1. Boyd, D.M., Ellison, N.B.: Social network sites: definition, history, and scholarship. J. Comput. Med. Commun. (2007)
2. Muthuselvi, et al.: Information retrieval from social network. In: IJCA Proceedings on E-Governance and Cloud Computing Services, vol. 4. (2012)
3. Thelwall, M.: Social networks, gender and friending: an analysis of MySpace member profiles. J. Am. Soc. Inform. Sci. Technol. **59**(8), 1321–1330 (2008)
4. Tufekci, Z.: Grooming, gossip, face book and MySpace: what can we learn about these sites from those who won't assimilate? Inf. Commun. Soc. **11**(4), 544–564 (2008)
5. Internet Reference At. http://Blog.Sciencenet.Cn/Blog-89075-558555.html
6. Park, H.W.: Hyperlink Network Analysis: A New Method for the Study of Social Structure on the Web. Connections (2003)
7. Sheldon, P.: The relationship between unwillingness-to-communicate and students face book use. J. Media Psychol.: Theor. Methods Appl. **20**(2), 67–75 (2008)
8. Roberts, D.F., Foehr, U.G.: Kids and Media in America. Cambridge University Press, Cambridge (2004)
9. McLuhan, M.: Understanding Media: The Extensions of Man Cambridge. MIT Press (1964)
10. Boase, J., Wellman, B.: Personal relationships: on and off the internet. In: Vangelisti, A., Perlman, D. (eds.) Handbook of Personal Relationships, pp. 709–723. Cambridge University Press, Cambridge (2006)
11. Donath, J., Boyd, D.M.: Public displays of connection. BT Technol. J. (2004)
12. Hangwoo, L.: Privacy, Publicity, and Accountability of Self-presentation in an On-line Discussion Group. Sociological Inquiry (2006)
13. Burns, E.: Global internet adoption slows while involvement deepens. http://www.clickz.com/stats/sectors/demographics/article.php/3596131#table1
14. Wasserman, S., Galaskiewicz, J.: Advances in Social Network Analysis: Research in the Social and Behavioral Sciences. Sage Publications, Thousand Oaks (1994)
15. Social Network Analysis of the Co-Authorship Network of the Australian Conference of Information Systems from 1990 to 2006. In: 17th European Conference on Information Systems, EICS, Verona, Italy (2009)
16. Jiawei, H., Kamber, M.: Data Mining: Concepts and Techniques. Morgan Kauffmann, San Francisco, California (2001)
17. Chawla, A., et al.: Reverse apriori approach—an effective association rule mining algorithm. Int. J. Comput. Complex Res. **4**(3), (2014)
18. Internet Reference At. http://Data-Mining-Tutorials.Blogspot.In/2008/11/Stepdisc-Feature-Selection-For.html

19. Schrock, A.: Examining social media usage: technology clusters and social network site membership. First Monday **14**(1). http://firstmonday.org/htbin/cgiwrap/bin/ojs/index.php/fm/article/viewArticle/2242/2066 (2009). Accessed 3 Feb 2009

20. Boyd, D.: Why youth (heart) social network sites: the role of networked publics in teenage social life. In: Buckingham, D. (ed.) Youth, Identity, and Digital Media, pp. 119–142. MIT Press, Cambridge (2008)

21. Boyd, D., Ellison, N.: Social network sites. J. Comput. Med. Commun. **13**(1). http://jcmc.indiana.edu/vol2013/issue2001/boyd.htl (2009). Accessed 10 Dec 2007

22. Alroldi, E., et al.: Discovery of latent patterns with hierarchical bayesian mixed membership models and the issue of model choice. In: Poncelet, P., Masseglia, F., Teisseire, M. (eds.) Data Mining Patterns: New Methods and Applications, pp. 240–275. (2006)

23. Erdogan, S.Z., et al.: A data mining application in a student database. J. Aeronaut. Space Technol. **2**(2), (2006)

DE Optimized PID Controller with Derivative Filter for AGC of Interconnected Restructured Power System

Tulasichandra Sekhar Gorripotu, Rabindra Kumar Sahu
and Sidhartha Panda

Abstract This present paper focuses on the design of proportional-derivative-integral controller with derivative filter (PIDF) for automatic generation control problem. A two-area, six-unit reheat thermal-hydro restructured power system is considered with nonlinearities such as time delay (TD) and generation rate constraint (GRC). The gain parameters PIDF controllers are optimized by using differential evolution (DE) algorithm employed with integral of time multiplied absolute error (ITAE) as an objective function. The performance of proposed controller is investigated under all the possible scenarios that take place in a restructured power market. From simulation results reveals that DE optimized PIDF controller minimizes the errors effectively.

Keywords Automatic generation control (AGC) · Differential evolution (DE) · Integral of time multiplied absolute error (ITAE) · Proportional-Derivative-Integral controller with derivative filter (PIDF)

1 Introduction

In an interconnected power system, a sudden load perturbation (SLP) in any area causes the deviation of frequencies of all the areas and also in the tie-line powers [1]. This has to be corrected to ensure the generation and distribution of electric

T.S. Gorripotu (✉) · R.K. Sahu · S. Panda
Department of Electrical Engineering, Veer Surendra Sai University
of Technology (VSSUT), Burla, 768018 Odisha, India
e-mail: gtchsekhar@gmail.com

R.K. Sahu
e-mail: rksahu123@gmail.com

S. Panda
e-mail: panda_sidhartha@rediffmail.com

© Springer India 2016
H.S. Behera and D.P. Mohapatra (eds.), *Computational Intelligence
in Data Mining—Volume 2*, Advances in Intelligent Systems
and Computing 411, DOI 10.1007/978-81-322-2731-1_37

power with good quality and which can be achieved by Automatic Generation Control (AGC) [2]. Automatic generation control, as the name signifies, is used to regulate the power flow between different areas while holding the frequency constant. The supply frequency may go down if there is an increase in the load and vice versa. In present days, AGC is becoming popular issue among the researchers in power system design and operation because of its increase in size, change in the structure, emergence of non-conventional energy sources, constraints in environment. In restructured power system, Generating Companies (GENCOs) may or may not involve themselves in AGC problem [3]. On the other side, Distribution Companies (DISCOs) may contract with any generating unit of any area on their interest [4, 5]. Hence, in restructure environment, control is decentralized to a great extent and Independent System Operators (ISOs) are responsible for maintaining frequency and tie-line power flows within the tolerable limits [6].

Many articles are reported in literature regarding power system issues of AGC under restructured scenario [4–11]. Vaibhav et al. [4] demonstrated the concept of AGC under restructured power system and concept of DISCO participation matrix (DPM). Recently, Deepak et al. [5] have proposed load following in deregulated power system. But, they have considered only reheat thermal units without any nonlinearities. KPS. Parmar et al. have presented [6] interconnected power system with multi-source power generation in deregulated power environment. However, they have not considered important any non-linearities in the design model such as Time delay (TD) and Generation rate constraint (GRC). Therefore, this paper presents a comprehensive study on dynamic performance of a more practical restructured two-area, six-unit thermal-thermal-hydro system with physical constraints such as GRC and TD.

2 Power System Description Under Study

A two area, six unit (thermal-thermal-hydro) interconnected power system has been proposed for AGC under deregulation with GRC and TD. The power system investigated widely used in the literature [6] which is given in the Fig. 1. In Fig. 1 area-1 comprises of three GENCOs out of which two GENCOs are thermal power system of reheat turbine and one hydro unit. Area-2 comprises of three GENCOs as same as area-1. Different PIDF controllers are considered for each GENCO. In the present study, GRCs of 3 % per minute for thermal units, GRCs of 270 % per minute for upper generation and 360 % per minute for lower generation are used for hydro units and a 50 ms delay time is considered [7, 8, 12]. The relevant parametric values of different parameters used in Fig. 1 are specified in appendix.

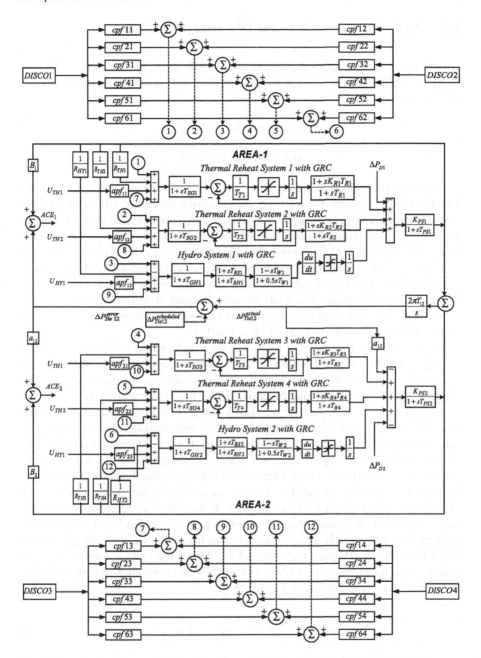

Fig. 1 Transfer function model of multi-source multi area restructured power system

2.1 Design of Controller and Objective Function

To minimize the frequency deviations and deviations in tie-line power, different PIDF controllers are provided at each generating unit and the area control errors (ACE) given by [1]:

$$ACE_1 = B_1\Delta F_1 + \Delta P_{Tie} \tag{1}$$

$$ACE_2 = B_2\Delta F_2 - \Delta P_{Tie} \tag{2}$$

In industrial process control, the proportional-integral-derivative controller (PID) is widely used and the well accepted feedback controller. PID controller can provide better control performance even though there are dynamic characteristics in process plant. PID controllers are used when quick action or a response and constancy are required [13]. However, the derivative term in the PID controller will give unreasonable size control inputs while signal to the input has sharp corners. However, any unwanted signal in the input control signal will leads to large plant input signals. This reason may leads to complications in realistic applications [3]. Keeping this in mind, a filter is incorporated in the derivative term and structure of proposed PIDF controller explained clearly in [8].

In present paper, the ranges of PID gains are considered as [−2 2] and derivative filter gain as [10 300]. The objective function or a cost function can be defined depends up on the essential specifications and constraints, in the design of a modern heuristic optimization technique based controller [8]. Some performance measures such as the Integral of Squared Error (ISE), Integral of Time multiplied Absolute Error (ITAE), Integral of Absolute Error (IAE) and Integral of Time multiplied Squared Error (ITSE) are taken into consideration for the design of control [14]. ITAE not only improve system performance and settling time as well as reduces the peak and under overshoots [15]. Generally, ITSE has a disadvantage of producing undesirable controller output for a rapid variations in set point which is not attractive for the controller design analysis. It has been reported in the literature survey that ITAE is a better objective function in AGC studies [10–14]. Therefore, in the present paper ITAE is chosen as objective function. The controller parameters of PIDF controller (Proportional, Integral and Derivative) and it's derivative filter gain are optimized by using ITAE. The mathematical notation for the objective function (ITAE) is illustrated in Eq. (3) [14].

$$ITAE = \int_0^{t_{sim}} (|\Delta F_1| + |\Delta F_2| + |\Delta P_{Tie}|) \cdot t \cdot dt \tag{3}$$

where, ΔF_1 and ΔF_2 are frequency deviation of the system; ΔP_{Tie} is the change in tie line power; t_{sim} is the time range of simulation.

3 Results and Discussion

Initially, the transfer function model of multi-area, multi-source power system with nonlinearities such as GRCs and TD under restructured environment is shown in Fig. 1 is designed in MATLAB/SIMULINK environment. Different PIDF controllers are considered as the secondary controllers to control the area frequencies and tie-line powers. In area-1 three different PIDF controllers are placed for each generating unit and the same controllers are placed in area-2 as well. The controller gains of PIDF are optimized by using differential Evolution (DE) technique. The concept of DE algorithm and process of tuning the gain parameters is clearly explained by the authors in [8, 16]. In present study, all the possible open market scenarios such as poolco, bilateral and contract violation scenarios are considered. The concepts of above transactions are clearly demonstrated in [4–8] and DE technique optimized PIDF controller gain values are given in Table 1. In poolco based scenario, the DISCOs must contract with GENCOs of same area. The area participation factors (apfs) of 0.4 for each reheat thermal unit and 0.2 for hydro unit are assumed and a particular case of poolco based scenario is simulated based on the following DPM:

$$DPM = \begin{bmatrix} 0.4 & 0.4 & 0 & 0 \\ 0.3 & 0.3 & 0 & 0 \\ 0.3 & 0.3 & 0 & 0 \\ 0 & 0 & 0 & 0 \\ 0 & 0 & 0 & 0 \\ 0 & 0 & 0 & 0 \end{bmatrix}$$

Table 1 Optimized PIDF controller parameters for different power scenarios

Cases/parameters	Poolco based transaction	Bilateral based transaction	Contract violation based transaction
K_{P1}	−0.5559	−1.8479	−1.1032
K_{P2}	0.4811	0.6033	0.2250
K_{P3}	1.2446	1.2655	−1.6243
K_{I1}	−1.9230	−1.6650	−1.6505
K_{I2}	−1.6645	−1.9541	−1.7092
K_{I3}	1.8992	−0.5375	1.8352
K_{D1}	0.6054	−0.3489	−1.8718
K_{D2}	−1.0750	−0.3359	−0.9239
K_{D3}	−0.3860	−0.1465	0.4047
N^{-}_1	45.3860	145.1419	151.3499
N_2	87.8473	168.9660	195.9272
N_3	84.7754	164.4804	154.3227

In this scenario, it is considered that 0.5 % load disturbance occurs in DISCO1 and DISCO2 so that the step load perturbation (SLP) in area-1 becomes 1 %. The SLP in area-2 is zero as DISCO3 and DISCO4 are participated in this particular scenario and steady state power is also zero. In bilateral scenario, the DISCOs can have a business with GENCO of any area and in contract violation scenario the DISCOs may violate the contract of power with GENCOs. The change in scheduled tie-line power is 0.15 % and DPM for bilateral and contract violation scenarios assumed as:

$$DPM = \begin{bmatrix} 0.2 & 0.1 & 0.3 & 0 \\ 0.2 & 0.25 & 0.1 & 0.1666 \\ 0.1 & 0.25 & 0.1 & 0.1666 \\ 0.2 & 0.1 & 0.1 & 0.3666 \\ 0.2 & 0.2 & 0.2 & 0.1666 \\ 0.1 & 0.1 & 0.2 & 0.1666 \end{bmatrix}$$

In bilateral scenario, it is assumed that SLP of 1 % in each area such that all four DISCOs have 0.5 % load disturbance. In contract violation scenario, it is considered that 1 % of uncontracted power demanded in area-1. Hence, in area-1 step disturbance becomes 2 % and in area-2 is 1 %. The dynamic responses of above transactions are given in Figs. 2, 3 and 4 and the performance index values in terms of ITAE, minimum damping ratio (MDR), peak over shoot and under shoot values are given in Table 2. From, Figs. 2, 3 and 4 it is known that the PIDF controller better enough to minimize the oscillations quickly. It is evident from Table 2 that poolco based scenario having minimum objective function value (ITAE = 0.2952) compared to bilateral scenario (ITAE = 0.5300) and contract violation scenario (ITAE = 0.6612).

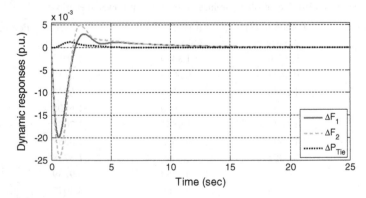

Fig. 2 Dynamic responses of the system for poolco based transaction

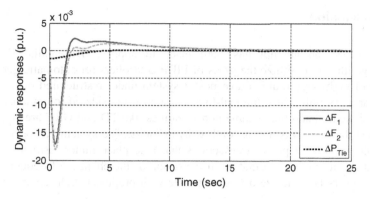

Fig. 3 Dynamic responses of the system for bilateral based transaction

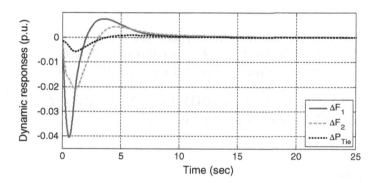

Fig. 4 Dynamic responses of the system for contract violation based transaction

Table 2 Performance index values for different power transactions

Parameters		Poolco based transaction	Bilateral based transaction	Contract violation based transaction
ITAE		0.2952	0.5300	0.6612
MDR		0.2778	0.2977	0.7053
T_S (s)	ΔF_1	14.37	17.43	16.72
	ΔF_2	14.15	17.91	17.76
	ΔP_{Tie}	4.39	3.63	11.33
Peak over shoot	ΔF_1	0.0030	0.0022	0.0074
	ΔF_2	0.0050	0.0012	0.0043
	ΔP_{Tie}	0.0012	0.0001	0.0008
Under shoot	ΔF_1	−0.0199	−0.0170	−0.0406
	ΔF_2	−0.0245	−0.0180	−0.0212
	ΔP_{Tie}	0.0000	−0.0015	−0.0058

4 Conclusion

In this paper, an attempt has been made to apply a Differential Evolution (DE) algorithm to optimize the gains of PIDF controllers for automatic generation control of multi area multi source power system under restructured environment. A two area six unit reheat thermal-hydro system is considered to demonstrate the proposed method and the suitable non-linearities like TD and GRC are considered for reheat thermal units and as well as hydro units. The system has been investigated all possible of power transactions that take place under restructured environment and it is observed that in all the scenarios the changes of frequencies and tie-line powers becomes zero in steady state with proposed PIDF controller.

Appendix

Data for two area six unit thermal-thermal-hydro system [8]
$B_1 = B_2 = 0.425$ p.u. MW/Hz; $K_{PS1} = K_{PS2} = 120$ Hz/p.u. MW; $T_{PS1} = T_{PS2} = 20$ s; $R_{TH1} = R_{TH2} = R_{TH3} = R_{TH4} = R_{HY1} = R_{HY2} = 2.4$ Hz/p.u.; $T_{T1} = T_{T2} = T_{T3}$ $T_{T4} = 0.3$ s; $T_{SG1} = T_{SG2} = T_{SG3} = T_{SG4} = 0.08$ s; $K_{R1} = K_{R2} = K_{R3} = K_{R4} = 0.5$; $T_{R1} = T_{R2} = T_{R3} = T_{R4} = 10$ s; $T_{12} = 0.0433$; $T_{GH1} = T_{GH2} = 48.7$ s; $T_{W1} = T_{W2} = 1$ s; $T_{RS1} = T_{RS2} = 0.513$; $T_{RH1} = T_{RH2} = 10$; $a_{12} = -1$.

References

1. Elgerd, O.I.: Electric Energy Systems Theory–An Introduction. Tata McGraw Hill, New Delhi (2000)
2. Chidambaram, I.A., Paramasivam, B.: Optimized load-frequency simulation in restructured power system with redox flow batteries and interline power flow controller. Int. J. Electr. Power Energy Syst. **50**, 9–24 (2013)
3. Sahu, R.K., Panda, S., Rout, U.K.: DE optimized parallel 2-DOF PID controller for load frequency control of power system with governor dead-band nonlinearity. Int. J. Electr. Power Energy Syst. **49**, 19–33 (2013)
4. Donde, V., Pai, M.A., Hiskens, I.A.: Simulation and optimization in an AGC system after deregulation. IEEE Trans. Power Syst. **16**, 481–489 (2011)
5. Deepak, M., Abraham, R.J.: Load following in a deregulated power system with thyristor controlled series compensator. Int. J. Electr. Power Energy Syst. **65**, 136–145 (2015)
6. Parmar, K.P.S., Majhi, S., Kothari, D.P.: LFC of an interconnected power system with multi-source power generation in deregulated power environment. Int. J. Electr. Power Energy Syst. **57**, 277–286 (2014)
7. Sahu, R.K., Chandra Sekhar, G.T., Panda, S.: DE optimized fuzzy PID controller with derivative filter for LFC of multi source power system in deregulated environment. Ain Shams Eng. J. **6**, 511–530 (2015)
8. Chandra Sekhar, G.T., Sahu, R.K., Panda, S.: AGC of a multi-area power system under deregulated environment using redox flow batteries and interline power flow controller. Eng. Sci. Technol. Int. J. **18**, 555–578 (2015)

9. Gorripotu, T.C.S., Sahu, R.K., Panda, S.: Application of firefly algorithm for AGC under deregulated power system. Comput. Intell. Data Mining **1**, Smart Innovation Syst. Technol. **31**, 677–687 (2015)
10. Chandra Sekhar, G.T., Sahu, R.K., Baliarsingh, A.K., Panda, S.: Load frequency control of power system under deregulated environment using optimal firefly algorithm. Int. J. Electr. Power Energy Syst. **74**, 195–211 (2016)
11. Sahu, R.K., Gorripotu, T.C.S., Panda, S.: A hybrid DE–PS algorithm for load frequency control under deregulated power system with UPFC and RFB. Ain Shams Eng. J. 6, 893–911 (2015)
12. Sahu, R.K., Panda, S., Padhan, S.: A hybrid firefly algorithm and pattern search technique for automatic generation control of multi area power systems. Int. J. Electr. Power Energy Syst. **64**, 9–23 (2015)
13. Panda, S., Yegireddy, N.K.: Automatic generation control of multi-area power system using multi-objective non-dominated sorting genetic algorithm-II. Int. J. Electr. Power Energy Syst. **53**, 54–63 (2013)
14. Sahu, R.K., Panda, S., Chandra Sekhar, G.T.: A novel hybrid PSO-PS optimized fuzzy PI controller for AGC in multi area interconnected power systems. Int. J. Electr. Power Energy Syst. **64**, 880–893 (2015)
15. Sahu, R.K, Panda, S., Yegireddy, N.K.: A novel hybrid DEPS optimized fuzzy PI/PID controller for load frequency control of multi-area interconnected power systems. J. Process Control 24,1596–1608 (2014)
16. Stron, R., Prince, K.: Differential evolution—a simple and efficient adaptive scheme for global optimization over continuous spaces. J Global Optim. **11**, 341–359 (1995)

Opposition-Based GA Learning of Artificial Neural Networks for Financial Time Series Forecasting

Bimal Prasad Kar, Sanjib Kumar Nayak and Sarat Chandra Nayak

Abstract Artificial neural network (ANN) based forecasting models have been established their efficiencies with improved accuracies over conventional models. Evolutionary algorithms (EA) are used most frequently by the researchers to train ANN models. Population initialization of EA can affect the convergence rate as well as the quality of optimal solution. Random population initialization of EAs is the most commonly used technique to generate candidate solutions. This paper presents an opposition-based genetic algorithm (OBGA) learning to generate initial candidate solutions for ANN based forecasting models. The present approach is based on the concept that, it is better to begin with some fitter candidate solutions when no a priori information about the solution is available. In this study both GA and OBGA optimizations are used to optimize the parameters of a multilayer perceptron (MLP) separately. The efficiencies of these methods are evaluated on forecasting the daily closing prices of some fast growing stock indices. Extensive simulation studies reveal that OBGA method outperforms other with better accuracies and convergence speed.

Keywords Artificial neural network · Evolutionary algorithm · Random population initialization · Genetic algorithm · Opposition-based genetic algorithm · Multilayer perceptron · Stock market · Financial time series forecasting

B.P. Kar (✉)
Department of Computer Science & Engineering, GIET, Bhubaneswar,
Odisha, India
e-mail: bimalprasadkar@gmail.com

S.K. Nayak
Department of Computer Application, VSSUT, Burla, Odisha, India
e-mail: fortunatesanjib@gmail.com

S.C. Nayak
Department of Computer Science & Engineering, SIET, Bhubaneswar,
Odisha, India
e-mail: sarat_silicon@yahoo.co.in

H.S. Behera and D.P. Mohapatra (eds.), *Computational Intelligence in Data Mining—Volume 2*, Advances in Intelligent Systems and Computing 411, DOI 10.1007/978-81-322-2731-1_38

1 Introduction

Stock market index prediction has been considered as an important and challenging task for the researchers since last few decades. Stock index forecasting is the process of making prediction about future performance of a stock market, based on present and past stock market behavior. Stock market behaves very much like a random walk process, which suggests that stock prices change randomly, and it is impossible to predict the stock prices. Due to the influence of uncertainties involved in the movement of the market, stock market forecasting is regarded as a challenging and difficult task in financial time-series forecasting. Predicting stock market prices movements is quite difficult also due to its nonlinearities, highly volatile nature, discontinuities, movement of other stock markets, political influences and other many macro-economical factors even individuals psychology. Hence an effective and more accurate forecasting model is necessary in order to predict the stock market behavior. It has been observed that MLP has been adapted as the most frequently used ANN by the researchers. A MLP contains more than one hidden layer, and each layer can contain more than one neurons. The input pattern is applied to the input layer of the network and its effect propagates through the network layer by layer. During the forward phase, the synaptic weights of the networks are fixed. In the backward phase, the weights are adjusted in accordance with the error correction rule. Some applications of MLP include financial time series forecasting [1], market trend analysis [2], macroeconomic data forecasting [3] and stock exchange movement [4]. Some other applications of MLP are found in railway traffic forecasting [5], airline passenger traffic forecasting [6] and maritime traffic forecasting [7] etc. MLP has been successfully applied to forecast the short-term demand of electric load consumption [8] and air pollution [9]. However, suffering from slow convergence, sticking to local minima are the two well known lacuna of a MLP. Also there is no formal method of deriving a MLP network for a given classification task. In order to overcome the local minima, more number of nodes added to the hidden layers. Multiple hidden layers and more number of neurons in each layer also add more computational complexity to the network. Hence, there is no direct method of finding an optimal structure of MLP for solving a problem. Defining optimal architecture and parameters for MLP is a matter of trial and error which are computationally very expensive.

The EAs have been used as popular and efficient approach to optimize the network structure as well as the parameters of neural networks by the researchers. They are more effective than the conventional algorithm in terms of learning accuracy and prediction accuracy [10]. Many researchers have adopted a neural network model, which is trained by genetic algorithm for nonlinear forecasting due to its broad adaptive and learning ability [11, 12]. In the past few decades several nature-inspired optimization techniques, particularly population based algorithms have been applied successfully. Normally these algorithms are motivated by biological evolution and termed as evolutionary algorithms of metaheuristic. Some of the genetic level evolutionary algorithms are Genetic Algorithm (GA) [13],

Memetic Algorithm (MA) [14] and Differential Algorithm (DE) [15]. Similarly, Particle Swarm Optimization (PSO) [16], Bee Colony Optimization (BCO) etc. are some of creature level metaheuristic.

It is very common that all EAs start from scratch known as initial population and gradually try to improve the search performance, hopefully, toward optimal solution. Due to unavailability of a priori information about the solution, the search process starts with a random guess and terminates when satisfied with a predefined criteria. It may happen that the random guess is far from the optimal solution, in worst case it is in the opposite location and move farther in the successive iterations. As a result the computational time will increase and may hamper the quality of solution. Of course it is not possible that we can adopt the best initial guess. In another way the search can be looking in all directions simultaneously, or more simply in the opposite direction of the initial guess. This concept is a well known as opposite base learning (OBL) introduced by Tizhoosh in [17, 18]. The OBL concept is used in this study to generate the better initial population for GA with a hope of achieving higher convergence speed and prediction accuracy. The GA and OBGA are used to search the optimal weight and biases vector of a MLP with single hidden layer.

The rest of the paper is organized as follows. Section 2 describes about the architecture of the ANN based forecasting model and problem formulation. Section 3 explains about the conventional GA and OBGA learning for ANN based forecasting model. Experimental results and analysis have been carried out by Sect. 4. Section 5 concludes the study followed by a list of references.

2 ANN Based Forecasting Model

The multilayer perceptron has been considered to be capable of approximating arbitrary functions in terms of mapping abilities. MLP is one of the most widely implemented neural network topologies. The feed forward neural network model considered here consists of one hidden layer only. The architecture of the MLP model is presented in Fig. 1.

The MLP performs a functional mapping from the input space to the output space. The model considered here contains a single hidden layer and let there are m neurons in this layer. Since there are n input values in an input vector, the number of neurons in the input layer is equals to n. The first layer corresponds to the problem input variables with one node for each input variable. The second layer is useful in capturing non-linear relationships among variables. This model consists of a single output unit to estimate one-day-ahead closing prices. The neurons in the input layer use a linear transfer function, the neurons in the hidden layer and output layer use sigmoid function presented in Eq. (1).

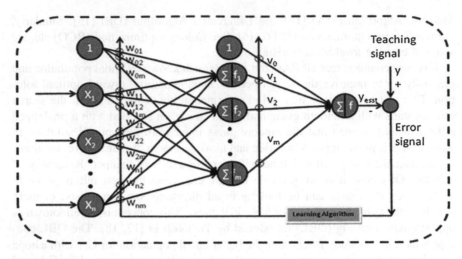

Fig. 1 Architecture of MLP based forecasting model

$$y_{out} = \frac{1}{1 + e^{-\lambda y_{in}}} \tag{1}$$

where y_{out} is the output of the neuron, λ is the sigmoidal gain and y_{in} is the input to the neuron. At each neuron, j in the hidden layer, the weighted output z is calculated as in Eq. 2.

$$z = f\left(B_j + \sum_{i=1}^{n} V_{ij} * X_i\right) \tag{2}$$

where X_i is the ith input vector, V_{ij} is the synaptic weight value between ith input neuron and jth hidden neuron and B_j is the bias value. The output y at the single output neuron is calculated as in Eq. 3.

$$y = f\left(B_0 + \sum_{j=1}^{m} W_j * z\right) \tag{3}$$

where W_j is the synaptic weight from jth hidden neuron to output neuron, z is the weighted sum calculated as in Eq. 2, and B_0 is the output bias. This output y is compared to the desired output and the error is calculated.

$$e(i) = y(i) - d(i) \tag{4}$$

The error signal $e(i)$ and the input vector are employed to the weight update algorithm to compute the optimal weight vector. To overcome the demerits of back propagation, we employed the GA which is a popular global search optimization.

The network has the ability to learn through training by GA. During the training, the network is repeatedly presented with the training vector and the weights as well as biases are adjusted by GA till the desired input-output mapping occurs. The error is calculated by Eq. 4 and our objective is to minimize the error function as in Eq. 5 with an optimal set of weight and biases vectors.

$$E(i) = \sum_{i=1}^{N} e(i) \tag{5}$$

3 Evolutionary Algorithm Based Learning

This section explains about the two learning techniques used to optimize the parameters of the MLP based forecasting model. The Sect. 3.1 explains about the conventional GA with random population initialization method. The Sect. 3.2 describes the OBGA method which adopting the OBL concept.

3.1 GA Based ANN Learning

Genetic algorithms (GA) are based on biological evolutionary theory which can be used to solve the optimization problems. The GA works with encoding parameter instead of parameter itself. They have been considered as popular global search optimization techniques which work on a population of potential solutions in the form of chromosomes, attempting to locate the best solution through the process of artificial evolution. It consists of repeated artificial genetic operations such as evaluation, selection, crossover, and mutation. The genetic evolution process consist the following basic steps as described by Algorithm 1.

Algorithm 1- Basic steps of GA

1. Initialization of the search space.
2. Evaluation of fitness of individuals.
3. Apply selection operator.
4. Apply crossover operator.
5. Apply mutation operator.
6. Repeat of the above steps until convergence.

The basic search mechanism of GA can be achieved through the genetic operators such as crossover and mutation. These two operators used to create new individuals from the existing individuals of the population. The working of GA starts with the random/uniform population of individuals. The individuals/chromosomes are iteratively cycled through the genetic operators. The GA allows the survival of the fittest individual over some continuous generation for solving the problems. As the GA moves from generation to generation, there is chance of producing better solutions. The GA stops by meeting the stopping criteria, which can be the maximum number of generation or population convergence criteria. Genetic algorithms are robust, efficient and not vulnerable to getting stuck at local minima like gradient descent based learning. A complete discussion about GA can be found in [13].

3.2 OBGA Learning of ANN

The basic GA adopts random initialization of search space. The concept of opposition based learning is to generate better initial candidate solutions for the search space of the given problem. In this method another set of solutions opposite to the original solutions is generated so that the probability of choosing fitter solutions can increase. In a simple opposite of a multidimensional vector x_i in an interval of $[a, b]$ can be calculated as in Eq. 6.

$$\breve{x}_i = a_i + b_i - x_i \tag{6}$$

An opposite point of a given point in a search space can be viewed as a point with all of its dimensions opposite to that of the original point. The chromosome of GA represents a potential solution in the search space for a given problem. So it is simple to calculate an opposite of a chromosome which is another potential solution with better fitness. Let $C = (c_1, c_2, c_3, \ldots, c_n)$ be a chromosome in an n-dimensional search space, where $c_i \in [a_i, b_i] \forall i \in \{1, 2, \ldots, n\}$. An opposite chromosome of C can be calculated as $\breve{C} = (\breve{c}_1, \breve{c}_2, \breve{c}_3, \ldots, \breve{c}_n)$. In this manner an opposite population can be generated from the original population. Then instead of evaluating the original population only, its opposite population is evaluated simultaneously in order to continue the search process with better fit solutions. If the fitness of opposite population is found to be superior then the original population can be replaced with the opposite population, otherwise the search process continue with the original. The major steps of OBGA learning ANN are represented by Fig. 2.

4 Experimental Results and Analysis

For experimental studies closing prices of four stock indices such as FTSE100, CNX Nifty, Nikkei 225, LSE and S&P 500 have been considered. The indices are collected from January 01 of 2010 to December 31 of 2014 for a period of 5 years. The raw data are normalized by min-max normalization method. Both of the

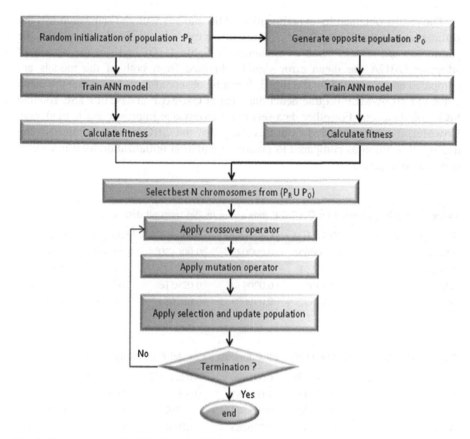

Fig. 2 Flow process of OBGA based ANN learning

learning algorithms, i.e. GA and OBGA are used to search the optimal weight and bias vectors of a MLP shown by Fig. 1. Hence the forecasting models are termed as GA-MLP and OBGA-MLP. To avoid the stochastic biases of ANN based models, the average of 10 simulations are considered for performance evaluation. The optimal parameters of GA and OBGA are presented by Table 1.

Table 1 Simulated parameters of GA and OBGA

Parameter	Value
Population size	60
Cross over probability	0.6
Mutation probability	0.004
Max. number of generation	200
Selection	10 % best + 90 % binary tournament selection

The results obtained by experimenting of the models with the same data sets are summarized in the Tables 2 and 3. For the two evolutionary learning methods the parameters are remain same as explained in Table 1. To find out the benefit of adopting OBGA the mean error signals obtained from both of the models are compared and presented by the Fig. 3. It can be observed from Fig. 3 that the mean value of OBGA-MLP is quite better than that of GA-MLP in case of FTSE 100 and S&P 500 data sets. For other data sets the performance improvement is moderate. However, it can be revealed from this study that the OBGA method improves the quality of solution as compared to random population initialization used in case of conventional GA.

Table 2 Results generated by GA-MLP forecasting model from all data sets

Descriptive statistics	FTSE 100	CNX Nifty	Nikkei 225	S&P 500	LSE
Minimum	0.000065	0.000025	0.000120	0.000075	0.000024
Maximum	0.027501	0.012102	0.012146	0.005833	0.003168
Mean	0.047090	0.020930	0.029516	0.032874	0.036449
Std. deviation	0.174522	0.099603	0.082584	0.305269	0.176500

Table 3 Results generated by OBGA-MLP forecasting model from all data sets

Descriptive statistics	FTSE 100	CNX Nifty	Nikkei 225	S&P 500	LSE
Minimum	0.000043	0.000081	0.000118	0.000109	0.000027
Maximum	0.023130	0.018573	0.017972	0.021829	0.055480
Mean	0.022947	0.015263	0.020015	0.015663	0.028836
Std. deviation	0.002272	0.001604	0.001882	0.001715	0.003612

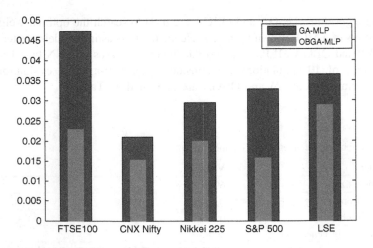

Fig. 3 Performance comparisons of GA-MLP and OBGA-MLP for all data sets

5 Conclusions

Population initialization of evolutionary algorithms can influence the quality of solution and convergence speed. This study considered two EA based ANN models for prediction of next day's closing prices of five real stock indices. The first model was trained with the basic GA with random initial population and the second one with opposition based population initialization called OBGA. For both training algorithms the same MLP architecture was used as the base model. It has been observed that OBGA-MLP performed better than simple GA-MLP, due to its ability to represent the search space unevenly.

The work can be extended further by exploring the OBL concept for other evolutionary algorithms with faster convergence. Attention can be given toward adopting higher order neural networks.

References

1. Yu, L., Lai Wang, S.K.: A neural-network-based nonlinear metamodeling approach to financial time series forecasting. Appl. Soft Comput. **9**, 563–574 (2009)
2. Aiken, M., Bsat, M.: Forecasting market trends with neural networks. Inf. Syst. Manag. **16**, 42–49 (1999)
3. Aminian, F., Suarez, E., Aminian, M., Walz, D.: Forecasting economic data with neural Netw. Comput. Econ. **28**, 71–88 (2006)
4. Mostafa, M.M.: Forecasting stock exchange movements using neural networks: empirical evidence from Kuwait. Expert Syst. Appl. **37**, 6302–6309 (2010)
5. Zhuo, W., Li-Min, J., Yong, Q., Yan-hui, W.: Railway passenger traffic volume prediction based on neural network. Appl. Artif. Intell. **21**, 1–10 (2007)
6. Nam, K., Yi, J.: Predicting airline passenger volume. J. Bus. Forecast. Methods Syst. **16**, 14–17 (1997)
7. Mostafa, M.M.: Forecasting the suez canal traffic: a neural network analysis. Marit. Policy Manag. **31**, 139–156 (2004)
8. Darbellay, G., Slama, M.: Forecasting the short-term demand for electricity: do neural networks stand a better chance. Int. J. Forecast. **16**, 71–83 (2000)
9. Videnova, I., Nedialkova, D., Dimitrova, M., Popova, S.: Neural networks for air pollution forecasting. Appl. Artif. Intell. **20**, 493–506 (2006)
10. Yu, L., Zhang, Y.Q.: Evolutionary fuzzy neural networks for hybrid financial prediction. IEEE Trans. Syst. Man Cybern. Part C: Appl. Rev. **35**(2), (2005)
11. Nayak, S.C., Misra, B.B., Behera, H.S.: Index prediction using neuro-genetic hybrid networks: a comparative analysis of performance. In: International Conference on Computing Communication and Application, IEEEXplore (2012). doi:10.1109/ICCCA.2012.6179215
12. Kwon, Y.K., Moon, B.R.: A hybrid neuro genetic approach for stock forecasting. IEEE Trans. Neural Netw. **18**(3), (2007)
13. Goldberg, D.E.: Genetic Algorithms in Search, Optimization, and Machine Learning. Addison-Wesley, Reading (1989)
14. Chen, X.S., Ong, Y.S., Lim, M.H., Tan, K.C.: A multi-facet survey on memetic computation. IEEE Trans. Evol. Comput. **15**(5), 591–607 (2011)
15. Price, K., Storn, R., Lampinen, J.: Differential Evolution: A Practical Approach to Global Optimization. Springer, Berlin (2005)

16. Kennedy, J., Eberhart, R.C.: Swarm Intelligence. Morgan Kaufmann, San Francisco (2001)
17. Tizhoosh, H.R.: Opposition-based learning: a new scheme for machine intelligence. In: International Conference on Computational Intelligence for Modeling Control and Automation —CIMCA'2005 (2005)
18. Tizhoosh, H.R.: Reinforcement learning based on actions and opposite actions. In: International Conference on Artificial Intelligence and Machine Learning (2005)

Evolving Low Complex Higher Order Neural Network Based Classifiers for Medical Data Classification

Sanjib Kumar Nayak, Sarat Chandra Nayak and H.S. Behera

Abstract Multilayer neural network based classifiers have been proven their better approximation and generalization ability in medical data classification. However they are characterize with both computational and structural complexities. This article proposes an Evolving Functional Link Network (EFLN) for medical data classification. First, the input signals are mapped from lower to higher dimensional feature space applying some trigonometric expansion functions. Then the optimal number of expanded input signals, weight vectors and network parameters are obtained by an evolutionary search technique. Therefore the optimal network structure can be achieved on fly by evolving a set of FLNs during training rather fixing it earlier. The proposed EFLN classifiers are validated with some benchmark data sets from UCI machine learning repository. The performances are compared with that of a gradient descent based FLN (GDFLN), multiple linear regressions (MLR) and a multilayer perceptron (MLP) and found to be superior.

Keywords Multilayer neural network · Evolving functional link network · Random medical data classification

S.K. Nayak (✉)
Department of Computer Application, VSSUT, Burla, Odisha, India
e-mail: fortunatesanjib@gmail.com

S.C. Nayak
Department of Computer Science, Silicon Institute of Technology, Bhubaneswar,
Odisha, India
e-mail: sarat_silicon@yahoo.co.in

H.S. Behera
Department of Computer Science & Engineering, VSSUT, Burla, Odisha, India
e-mail: mailtohsbehera@gmail.com

© Springer India 2016
H.S. Behera and D.P. Mohapatra (eds.), *Computational Intelligence
in Data Mining—Volume 2*, Advances in Intelligent Systems
and Computing 411, DOI 10.1007/978-81-322-2731-1_39

1 Introduction

Artificial Neural Networks (ANNs) are found to be extensively utilized tools which can approximate any continuous function to desired accuracy. It also allows the adaptive adjustment to the model and nonlinear description of the problems. The ANNs have recently been applied to many areas such as data mining, stock market analysis, medical data classification and many other fields. It is observed that Multilayer Perceptron (MLP) has been adopted as the most frequently used ANN by the researchers for the task of classification and prediction. An MLP contains one or more hidden layer, and each layer can contain more than one neurons. Though MLP is the most widely and frequently used technique, but it suffers from slow and non-convergence. Defining a feasible architecture and parameters for MLP is very often a matter of trial and error which is also computationally very expensive. The multilayer neural network approach adds more computational complexity to the network and also suffers from slow convergence, sticking to local minima. To bridge the gap between the linearity in the single layer neural network, an alternative to multilayer approach and highly complex and computation intensive of multilayer approach, functional link neural (FLN) networks have been employed and proved itself as successful as well as popular model in the domain of classification and forecasting.

Y.H. Pao has proposed a simplified single layer ANN model called FLN which maps the nonlinearity of input-output relationship by functional expansion of the input patterns [1]. It is a class of Higher Order Neural Networks (HONN) that utilizes higher combination of inputs [2]. The properties of expanding the input space into a higher dimensional space without hidden units of ANN were introduced. The FLN is basically a single layer network, in which the need of hidden layers has been removed by incorporating functional expansion of the input pattern. The functional link acts on an element by generating a set of linearly independent functions. The input, expanded by a set of linearly independent functions in the functional expansion block causes an increase in the input vector dimensionality, which enables the FLN to solve complex classification problems by generating non-linear decision boundaries. The performance of FLN models have been experimented during several research works, including pattern classification and recognition [3], system identification and control [4], functional approximation [5] and digital communications channel equalization [6]. The trigonometric expansion is chosen in many research works, because such expansion based models have been shown to provide improved performance for various applications [7].

In medical data classification, methods with better classification accuracy will provide more sufficient information to identify the potential patients and to improve the diagnosis accuracy [8]. Medical database classification is a kind of complex optimization problem whose goal is not only to find an optimal solution but also to provide accurate diagnosis for diseases. And therefore, meta-heuristic algorithms such as genetic algorithm, particle swarm optimization etc., soft computing and machine learning tools such as neural networks, decision tree, and fuzzy set theory have been successfully applied in this area and have achieved significant results [9]. ANN based approach aims to provide a filter that distinguishes the cases which do not have disease, therefore reducing the cost of medication and overheads of doctors. Back propagation neural network (BPNN) was used by Floyd et al. for classification of medical data and this work achieved an overall accuracy of 50 % [10]. In another study, rule extraction from ANN has been employed for prediction of breast cancer from Wisconsin dataset [11]. All the above methods used back propagation learning for training the ANN, where solution got trapped in the local minima. Therefore, Fogel et al. attempted to solve the medical database classification problem using evolutionary computation and could achieve higher prediction accuracy than the above techniques [12]. However, this work suffered from higher computational cost in application.

The main objective of this paper is to select optimal number of polynomials as a functional input to the network to construct the classifier. The selection of optimal number of functional links is carried out by genetic algorithm. Since the classifier uses optimal number of functional links, the network becomes less complex in structure and consuming less computational overhead.

The rest of the paper is organized as follows: Sect. 2 discusses about the FLN based classifiers along with the proposed EFLN model used. The experimental results and analysis are carried out by Sect. 3. Section 4 gives the concluding remark followed by a list of references.

2 FLN Based Classifier

The FLN architecture was originally proposed by Pao et al. [2]. They have shown that, their proposed network may be conveniently used for function approximation and pattern classification with faster convergence rate and lesser computational load than the multilayer perceptron (MLP) structure.

The FLN model considered here is a single layer model with on trigonometric expansion. Let each element of the input pattern before expansion be represented as $z(i)$, $1 < i < d$, where each element $z(i)$ is functionally expanded as $fn(z(i))$, $1 < n < N$, where N is the number of expanded points for each input element. For designing the network, at the first instance a functional expansion (FE) unit expands each input attribute of the input data.

The attributes of each input pattern is passed through the FE unit. Here a single output neuron is considered. This model does not contain any hidden layer other than the FE unit. The sum of the output signals of the FE units multiplied with weights is passed on to the sigmoidal activation function of the output unit. This estimated output is compared with the target output and error signal is obtained. This error signal is passed on to the training algorithm as its fitness function and the model having minimum error is considered as the best fit individual.

2.1 Gradient Descent Based FLN: FLN-GD

The FLN introduces higher order effects through nonlinear functional transforms via links rather than at nodes. The FLN architecture uses a single layer feed forward network without hidden layers. The functional expansion effectively increases the dimensionality of the input vector and hence the hyper plane generated by the FLN provides greater discrimination capability in the input pattern space.

For designing the network, at the first instance a functional expansion unit (FE) expands each input attribute of the input data. The simple trigonometric basis functions of *sine* and *cosine* are used here to expand the original input value into higher dimensions. An input value x_i expanded to several terms by using the trigonometric expansion functions such as in Eq. 1.

$$\begin{cases} c_1(x_i) = (x_i) \\ c_2(x_i) = \sin(x_i) \\ c_3(x_i) = \cos(x_i) \\ c_4(x_i) = \sin(\pi x_i) \\ c_5(x_i) = \cos(\pi x_i) \\ c_6(x_i) = \sin(2\pi x_i) \\ c_7(x_i) = \cos(2\pi x_i) \end{cases} \tag{1}$$

The attributes of each input pattern is passed through the FE unit. The sum of the output signals of the FE units multiplied with weights is passed on to the sigmoidal activation function of the output unit. The estimated output is compared with the target output and error signal is obtained. This error signal is used to train the model. Figure 1 shows the architecture of the FLN-GD model. The FLN model

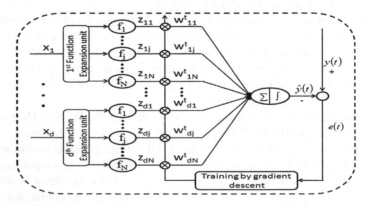

Fig. 1 Architecture of FLN-GD classifier

obtains the optimal weight set W iteratively using gradient descent learning algorithm based on the training samples. The learning process of FLN can be explained as follows:

The estimated output of the FLN forecasting model is calculated as follows:

A fixed number of weight parameters. W and a set of basis functions f are used to represent an approximating function $\varphi_w(X)$. Now the problem can be represented as 'find the optimal weight set W which provide the best possible approximation of f on the set of input-output samples'. The FLN obtains this solution by updating W iteratively.

Let there are k training pattern to be applied to the FLN and denoted by $\langle X_i : Y_i \rangle, 1 \leq i \leq k$

Let at ith instant $(1 \leq i \leq k)$ the N-dimensional input-output pattern are given by $X_i = \langle x_{i1}, x_{i2}, - - -, x_{iD} \rangle$, $\hat{Y}_i = [\hat{y}_t] 1 \leq i \leq k$

The dimension of the input signal is increased from N to N' by a set of basis functions f given by

$$
\begin{aligned}
f(X_i) &= [f_1(x_{i1}), f_2(x_{i2}), \ldots, f_1(x_{i2}), f_2(x_{i2}), \ldots, f_1(x_{iN}), f_2(x_{iN}), \ldots] \\
f(X_i) &= [f_1(x_{i1}), f_2(x_{i2}), \ldots, f_N(x_{iN})]
\end{aligned}
\tag{2}
$$

$W = [W_1 \quad W_2 \quad -- \quad W_k]^T$ is the $k \times N$ dimensional matrix where W_i is the weight vector associated with ith output and is given by $W_1 = [w_{i1} \quad w_{i2} \quad -- \quad w_{iN}]$

The ith output model is calculated by Eq. (3)

$$
\hat{y}_i(t) = \varphi \left(\sum_{j=1}^{N'} f_j(x_{ij}) w_{ij} \right) = \varphi \left(W_i f^T(X_i) \right) \forall i
\tag{3}
$$

The error signal associated with the ith output is calculated by Eq. 4.

$$
e_i(t) = y_i(t) - \hat{y}_t(t)
\tag{4}
$$

The weights of the model can be updated using adaptive learning rule as given by Eq. 5.

$$
w_{ij}(t+1) = w_{ij}(t) + \mu \Delta(t)
\tag{5}
$$

$$
\Delta(t) = \delta(t)[f(x_i)]
\tag{6}
$$

where

$$\delta(t) = [\,\delta_1(t) \quad \delta_2(t) \quad \cdots \quad \delta_k(t)\,]$$
$$\delta_i(t) = \left(1 - \hat{y}_t^2(t)e_i(t)\right)$$

μ is the learning parameter.

2.2 Proposed EFLN Based Classifier

In the proposed approach, first we define a network structure with a fixed number of inputs, and a single output as shown in the Fig. 1. Then the input signals are mapped to higher dimension by using the functional expansion units. The genetic algorithm performs search over the whole solution space, finds the optimal solution relatively easily, and it does not requires continuous differentiable objective functions. The problem of finding an optimal FLN to train the model could be seen as a search problem into the space of all possible parameters. The fitness of the best and average individual (i.e. FLN) in each generation increases towards a global optimum. It can be used as the tool for decision making in order to solve the complex nonlinear problems such as medical data classification. The chromosome set of GA represent a set of potential FLN models. The individual representation in GA is shown by Fig. 2. Suppose the maximum numbers of trigonometric polynomials are set to N, then the structure of the individual is represented as follows:

The fitness function which is used to guide the search process is defined in Eq. (7).

$$E(w; B) = \frac{1}{N} \sum_{i=1}^{N} (y - f(w; B; x_i))^2 \tag{7}$$

where N is the total number of training sample, y_i is the actual output and $f(.)$ is the estimated output of EFLNs.

Input data along with the chromosome values are fed to the set of FLN models. The fitness is obtained from the absolute difference between the target class level and the estimated output class level. The less the fitness value of an individual, GA considers it better fit. The high level architecture of EFLN forecasting model is shown by Fig. 3.

Functional Input Section					Weight Vector					Bias Value
1	0	1	-------	1	W₁	W₂	W₃	---------------	Wₙ	B

Fig. 2 Structure of individual

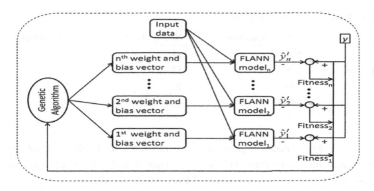

Fig. 3 Architecture of FLN-GD classifier

The major steps of the GA based FLNN models can be summarized and described by Algorithm 1.

Algorithm 1 -	EFLN Training:

1. *Random initialization of search spaces, i.e. populations.*
   Initialize each search space, i.e. chromosome with values from the domain [0, 1].
2. *Mapping of input patterns.*
   Map each pattern from the lower dimension to higher dimension, i.e. expand each feature value according to the polynomial basis functions.
3. *While (termination criteria not met)*
   For each chromosome in the search space
   Calculate the weighted sum and feed as an input to the node of output layer.
   Present the desired output, calculate the error signal and accumulate it.
   Fitness of the chromosome is equal to the accumulated error signal.
   End
   Apply crossover operator.
   Apply mutation operator.
   Select better fit solutions.
   End

3 Experimental Results and Analysis

3.1 Dataset Description

The proposed model has been tested with five medical datasets obtained from UCI machine learning repository [13]. The following table shows a summary of these

Table 1 UCI medical dataset properties

Dataset	No. of instances	No. of attributes	No. of classes
WBC	699	11	2
Heart disease	270	14	2
Hepatitis	155	20	2
ILPD	583	10	2

datasets. It should be noted that number of attributes in Table 1 is the number of input attributes plus one class attribute. Further details and properties of all the mentioned five datasets can be obtained from UCI repository.

3.2   Experimental Setup and Simulation Parameters

All the simulations are carried out in MATLAB R2010a which is installed in a PC having Windows 7 OS and 2 GB main memory. The processor is Intel dual core and each processor has an equal computation speed of 2 GHz (approx.). We adopted the binary encoding for GA. Each weight and bias value constitute of 15 binary bits. For calculation of weighted sum at output neuron, the decimal equivalent of the binary chromosome is considered. A randomly initialized population with 80 genotypes is considered. GA was run for maximum 250 generations with the same population size. Parents are selected from the population by elitism method in which first 10 % of the mating pools are selected from the best parents and the rest are selected by binary tournament selection method. A new offspring is generated from these parents using uniform crossover followed by mutation operator. In this way the new population generated replaces the current population and the process continues until convergence occurs. The fitness of the best and average individuals in each generation increases towards a global optimum. The uniformity of the individuals increases gradually leading to convergence. The simulated parameters of GA are shown in Table 2.

Table 2 Simulation parameters of GA

Parameter	Value
Population size	80
Gene size	15
Crossover probability (C_p)	0.7
Mutation probability (C_m)	0.002
Selection i elitism ii binary tournament	10 % 90 %
Max. no. of generation	250

For all the three models, FLN-GD, MLR, MLP and EFLN, RMSE is noted against each epoch during training. RMSE is calculated using the formula as in Eq. 8.

$$RMSE = \sqrt{\frac{\sum_{i=1}^{N}(y_i - \hat{y}_i)^2}{N}} \tag{8}$$

Testing results are classification accuracy, simulation time. Following tables shows the simulation results for all the four tested datasets. All the four models are compared for the above three parameters. It should be noted that each simulation is run for twenty times to avoid any biasness towards results.

From Table 3 it can be observed that the EFLN model generating significantly lower MSE value as compared to other classifiers. Table 4 presents the classification accuracies of all the models. It can be observed that the EFLN has better accuracies compared to others. Similarly, Table 5 presents the computational time of all the models in second. Here, it can be observed that the simulation time of MLP is quite higher than others and the FLN based models have lower computational overhead.

Table 3 RMSE (mean ± std. deviation) for four tested medical datasets

Dataset	MLR	MLP	FLN-GD	EFFLN
WBC	0.4282 ± 0.0198	0.2718 ± 0.0105	0.2600 ± 0.0275	0.1356 ± 0.0023
Heart disease	0.2182 ± 0.0118	0.1928 ± 0.0216	0.1811 ± 0.0124	0.1375 ± 0.0114
Hepatitis	0.2520 ± 0.0248	0.2314 ± 0.0141	0.2322 ± 0.0115	0.2015 ± 0.0088
ILPD	0.4033 ± 0.0322	0.3057 ± 0.0225	0.3013 ± 0.0272	0.1565 ± 0.00116

Table 4 Classification accuracy (mean ± std. deviation) for four tested medical datasets

Dataset	MLR	MLP	FLN-GD	EFFLN
WBC	70.69	72.60	75.52	82.56
Heart disease	74.29	76.87	76.93	76.97
Hepatitis	56.28	56.85	57.41	66.30
ILPD	61.65	67.37	67.58	80.77

Table 5 Simulation time (training time + testing time) for four tested medical datasets

Dataset	MLR	MLP	FLN-GD	EFFLN
WBC	218.940 + 0.097	276.940 + 0.082	138.990 + 0.072	124.280 + 0.046
Heart disease	256.250 + 0.105	275.131 + 0.095	117.660 + 0.083	87.650 + 0.092
Hepatitis	144.270 + 0.068	194.750 + 0.048	112.280 + 0.033	104.320 + 0.088
ILPD	110.260 + 0.072	134.601 + 0.084	103.261 + 0.033	101.500 + 0.025

4 Conclusion and Further Works

In order to achieve high classification accuracy, an evolving functional link network has been presented in this article. The higher order neural network; particularly FLN has been adopted as the base model. The higher order terms are able to improving generalization capability of the network while the single layer architecture is able to overcome the structural complexities as in case of multilayer approach. A functional link network is represented as an individual in GA and a set of such FLNs can be seen as a possible search space for the classification problem. Therefore the optimal FLN structure is chosen by the evolutionary search technique during training process and not fixed earlier. The performance of the proposed EFLN models is evaluated on four benchmark data sets used for classification problem. Also, the performance is compared with that of FLN-GD, MLR, MLP based classifiers and found to be superior in terms of RMSE, accuracy and simulation time.

The work can be extended by adopting other higher order networks as well as other evolutionary search techniques.

References

1. Pao, Y.H.: Adaptive Pattern Recognition and Neural Networks. Addison-Wesley, Reading (1989)
2. Pao, Y.H., Takefuji, Y.: Functional-link net computing: thory, system architecture, and functionalities. Computer 25, 76–79 (1992)
3. Mishra, B.B., Dehuri, S.: Functional link artificial neural network for classification task in data mining. J. Comput. Sci. 3, 948–955 (2007)
4. Majhi, R., Majhi, B., Panda, G.: Development and performance evaluation of neural network classifiers for Indian internet shoppers. Expert Syst. Appl. 39, 2112–2118 (2012)
5. Patra, J.C., Pal, R.N., Chatterji, B.N., Panda, G.: Identification of nonlinear dynamic systems using functional link artificial neural networks. IEEE Trans. Syst. Man Cybern. B Cybern. 29 (2), 254–262 (1999)
6. Yang, S.S.: Tseng, C.S: An orthonormal neural network for function approximation. IEEE Trans. Syst. Man Cybern. 26, 779–784 (1996)
7. Majhi, R., Panda, G., Sahoo, G.: Development and performance evaluation of FLN based model for forecasting of stock markets. Expert Syst. Appl. 36, 6800–6808 (2009)
8. Fan, C.-Y., Chang, P.-C., Lin, J.-J., Hsieh, J.C.: A hybrid model combining case-based reasoning and fuzzy decision tree for medical data classification. Appl. Soft Comput. 11, 632–644 (2011)
9. Bojarczuk, C.-C., Lopes, H.-S., Freitas, A.-A.: Genetic programming for knowledge discovery in chest-pain diagnosis. IEEE Eng. Med. Biol. Mag. 19(4), 38–44 (2000)
10. Floyd, C.E., Lo, J.Y., Yun, A.J., Sullivan, D.C., Kornguth, P.J.: Prediction of breast cancer malignancy using an artificial neural network. Cancer 74, 2944–2998 (1994)
11. Setiono, R., Huan, L.: Understanding neural networks via rule extraction. In: Proceedings of the International Joint Conference on Artificial Intelligence, Morgan Kauffman, San Mateo, CA, pp. 480-487. (1995)

12. Fogel, D.B., Wasson, E.C., Boughton, E.M.: Evolving neural networks for detecting breast cancer. Cancer Lett. **96**(1), 49–53 (1995)
13. Bache, K., Lichman, M.: UCI machine learning repository. Irvine, CA. University of California, School of Information and Computer Science. http://archive.ics.uci.edu/ml (2013)

Analysis of ECG Signals Using Advanced Wavelet Filtering Approach

G. Sahu, B. Biswal and A. Choubey

Abstract Electrocardiogram signal is principally used for the interpretation and assessment of heart's condition. The main criteria in ECG signal analysis is interpretation of QRS complex and obtaining its feature information. R wave is the most significant segment of this QRS complex, which has a prominent role in finding HRV (Heart Rate Variability) features and in determining its characteristic features. This paper intends to propose a novel approach for the analysis of ECG signals. The ECG signal is preprocessed using stationary wavelet transform (SWT) with interval dependent thresholding integrated with the wiener filter and is then subjected to Hilbert transform along with a window to enhance the presence of QRS complexes, to detect R-Peaks by setting a threshold. The proposed algorithm is validated with different parameters like Sensitivity, +Predictivity and Accuracy. The proposed method yields promising results with 99.94 % Sensitivity, 99.92 % +Predictivity, 99.87 % Accuracy. Finally the proposed method is compared with other methods to show the efficiency of the proposed technique for the analysis of ECG Signal.

Keywords Electrocardiogram · Stationary wavelet transform · Wiener filtering · Hilbert transform · Thresholding

G. Sahu (✉) · B. Biswal
Department of Electronics and Communication Engineering,
GMR Institute of Technology, Rajam 532127, India
e-mail: gupteswar.sahu@gmail.com

B. Biswal
e-mail: birendra_biswal1@yahoo.co.in

A. Choubey
Department of Electronics and Communication Engineering,
National Institute of Technology, Jamshedpur 831014, India
e-mail: achoubey@nitjsr.ac.in

© Springer India 2016   427
H.S. Behera and D.P. Mohapatra (eds.), *Computational Intelligence in Data Mining—Volume 2*, Advances in Intelligent Systems and Computing 411, DOI 10.1007/978-81-322-2731-1_40

1   Introduction

Electrocardiogram is a diagnostic measure, which reports the electrical activity of the heart. It is a record of the bio-electric potentials caused due to polarization and depolarization of cardiac muscles. The ECG signal is a sequence of electric events labelled with the letters P, Q, R, S, and T as shown in Fig. 1. The amplitude and the timing of the wave patterns in ECG give the pertinent information about the heart's condition. Any variation in the actual rhythm of the heart or heart rate or any variation in its morphological pattern reports cardiac arrhythmia [1]. Signal accuracy is of extreme important for feature extraction and analysis about heart condition. But, during the acquisition of the ECG signal, it is contaminated by different sources of noises like base line wandering, motion artifacts, power line interference, electrode contact noise, instrumentation noise generated by electronic devices, electromyography (EMG) noise etc. These noises greatly lower the signal quality, which will affect the morphology of the ECG signal containing significant information. The ECG being a non-stationary and low frequency signal, normal linear filters cannot be effective for removing the noise to a greater extent [2]. Even the normal linear filters cannot be able to remove all the noises present in the ECG. In literature, different methods of QRS complex detection algorithms have been suggested. Kohler et al. [3] have classified and compared some significant approaches on QRS detection. Methodology based on ECG signal derivative has already been carried out in reference [4]. ECG signal analysis based Hilbert transform [5, 6] have been used most frequently. QRS detection [7] based on wavelet transform has been profusely used. Some other methods based on more intricate approaches using artificial neural networks [8] have already gained some attention. In this work, an advanced approach of de-noising the ECG signal using dyadic stationary wavelet transform (SWT) with interval dependent thresholding (IDT) integrated with the wiener filter is used for initial preprocessing stage and then Hilbert transform is applied on denoised ECG signal to improve the presence of QRS complexes thereby detecting the exact position of the R-Peaks by setting a threshold of about 30 % of the maximum peak.

Fig. 1 ECG waveform

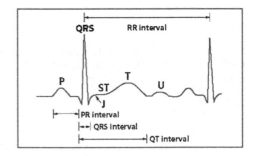

2 Methodology

Figure 2 shows the block diagram for the proposed QRS detection methodology. The algorithm processes the ECG signals in three different stages like pre-processing, post-processing and parameter calculation. The detailed explanations of each of the stages are discussed below.

2.1 Pre-processing

The ECG signals that are available in MIT-BIH database are mixed up with different types of artefacts. These artefacts have to be removed for further analysis and interpretation. For this an advanced way of filtering the ECG signals is followed. The procedure for de-noising follows two steps and is shown in Fig. 3.

Step 1: For this, first decompose the signal using stationary wavelet transform (SWT) [9] with the interval dependent thresholding at different intervals up to third level, this can be represented by the expression as given below

$$y_m g(n) = a_m(n) + b_m(n) \tag{1}$$

where $a_m(n)$ represents noise-free signal coefficients, $b_m(n)$ gives the coefficients of the noise, m gives decomposition level and then reconstructing the signal for further processing. In the interval dependent thresholding (IDT), the detailed coefficients at a particular level of decomposition are divided into certain intervals depending on

Fig. 2 Block diagram for the proposed methodology

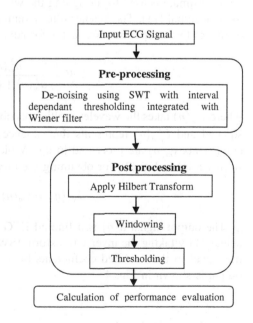

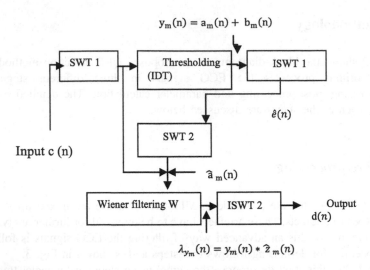

$$y_m(n) = a_m(n) + b_m(n)$$

SWT 1 → Thresholding (IDT) → ISWT 1

$\hat{e}(n)$

SWT 2

Input c (n)

$\hat{a}_m(n)$

Wiener filtering W → ISWT 2 → Output $d(n)$

$$\lambda_{y_m}(n) = y_m(n) * \hat{2}_m(n)$$

Fig. 3 Stationary wavelet transforms with interval dependent thresholding (IDT) and the Wiener filtering used for pre-processing of ECG Signal

the distribution of detail amplitudes. With IDT, the threshold values are taken separately for each level of transformation as the high frequency and low frequency parts of signals have different values of mean and standard deviation [10].

Step 2: The second step consists of SWT2, wiener filter (W) and ISWT2, which is the inverse of the wavelet transform of SWT2. By implementing ISWT1, the estimate $\hat{e}(n)$ is obtained, which approximates the noise-free signal $e(n)$. This above estimate is used for designing the wiener filter (W), which can be applied to the real signal [11]. The Wiener filter requires an estimate of a clean signal, which estimates the error correction factor for adjusting of transform coefficients.

$$W = \frac{\hat{a}_m^2(n)}{\hat{a}_m^2(n) + \hat{\sigma}_{bm}^2(n)} \tag{2}$$

where, $\hat{a}_m^2(n)$ takes the wavelet coefficients obtained from the estimate $\hat{e}(n)$ which is squared and $\hat{\sigma}_{bm}^2(n)$ represents the variance of the noise coefficients. The noisy coefficients $y_m(n)$ are processed in the W block, using the correction factor of the wiener filter in Eq. (2), for obtaining the modified coefficients

$$\lambda_{y_m}(n) = y_m(n) * W(n) \tag{3}$$

The output signal $d(n)$ is a filtered ECG signal used for further assessment is achieved by taking the inverse transform ISWT2 representing the reconstruction of the signal in the modified coefficients by $\lambda_{y_m}(n)$ as taken in Eq. (3). The filtered output is shown in Fig. 4.

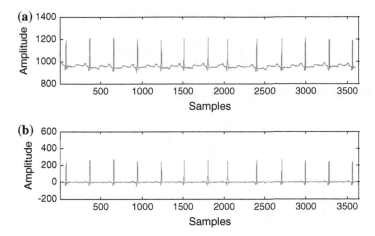

Fig. 4 **a** Input ECG from MIT-BIH database. **b** Filtered ECG after applying advanced wavelet filtering approach

2.2 Post-processing

The fundamental step in ECG signal processing is the precise detection of the QRS complexes, especially R peaks. After pre-processing, the resultant filtered signal is post processed for the QRS complex extraction. In this paper, the R-Peak detection can be best enhanced by using Hilbert transform along with a window which gives the exact location of R-Peaks [5].

2.2.1 Hilbert Transform

For a given function $q(t)$, the Hilbert transform (HT) is defined as

$$\hat{q}(t) = \mathcal{H}[q(t)] = \frac{1}{\pi} \int\limits_{-\infty}^{\infty} q(\tau) \frac{1}{t-\tau} d\tau \tag{4}$$

where, '$\mathcal{H}$' represents the Hilbert transform operator and $\hat{q}(t)$ is the HT of signal $q(t)$. The HT provides a time varying and a linear function of $u(t)$ and equation can be rewritten as the convolution between the signal and $1/\pi t$, i.e.

$$\hat{q}(t) = \frac{1}{\pi t} * q(t) \tag{5}$$

Re-arranging the Eq. (5) and applying FT, we have

$$F\{\hat{q}(t)\} = \frac{1}{\pi} F\left\{\frac{1}{t}\right\} F\{q(t)\} \tag{6}$$

The Fourier transform of $\frac{1}{t}$ is simplified to

$$F\left\{\frac{1}{t}\right\} = \int_{-\infty}^{\infty} \frac{1}{y} e^{-j2\pi fy} dy = -j\pi * \mathrm{sgn}\, f \tag{7}$$

where $sgn f = \begin{cases} +1; f > 0 \\ 0; f = 0 \\ -1; f < 0 \end{cases}$

Then the Fourier Transform of the HT of $q(t)$ is set by (6) as

$$F\{(\hat{q}(t))\} = -j\, sgn f\, F\{q(t)\} \tag{8}$$

The results are obtained in the frequency domain by multiplying the spectrum of $q(t)$ with $j(+90°)$ for the negative frequencies and $-j(-90°)$ for positive frequencies. Further by applying IFT, the time domain results can be achieved. Hence, the HT of the original function $q(t)$ represents its harmonic conjugate.

The analytic signal of the original signal $q(t)$ is expressed as

$$v(t) = q(t) + j\hat{q}(t) \tag{9}$$

The envelope $A(t)$ of $v(t)$ can be obtained by

$$A(t) = \sqrt{q^2(t) + \hat{q}^2(t)} \tag{10}$$

The instantaneous phase angle is given by

$$\theta(t) = arctan\left(\frac{\hat{q}(t)}{q(t)}\right) \tag{11}$$

The Hilbert transform has the advantage of nature of odd functionality. That means whenever there is a change in inflection of the input signal (i.e. when the signal slope changes from positive to negative, or from negative to positive) the Hilbert transformed signal will cross the horizontal axis [6]. This property is favorable for QRS detection as the R-wave in the complex is obeying the property of change in slope.

2.2.2 Windowing

Windows are commonly used to reduce the effect of leakage but cannot eliminate leakage entirely. In effect, they only change the shape of the leakage. In this work Triangular window is taken, which reduces the effects of final samples. Filters designed using this window have wider transition region. As a result, it is necessary to have a higher order filter in order to keep the same transition region. In process the technique should maintain the normal ECG specifications to get the real R-R intervals, to this the large sampled signals such as ECG signals are need to maintain window length.

2.2.3 Thresholding

Thresholds are dynamically calculated with regard to estimates of signal noise. Basically noise is any non-QRS ECG signal shapes, which is approximated by taking the normal ECG characteristic values of the result of the envelope over a time window to decide the reliable QRS complex. A threshold of about 35 % of the maximum value over the window is selected and points which exceed this value are selected as QRS points where as the others are treated as the non QRS points. If two peaks are detected within 200 ms, one of the peaks must be eliminated upon review of both amplitudes and relative position of both peaks to the previous QRS peak. By this process, accurate QRS complexes are detected by the Hilbert transform with a window by setting a threshold value (Fig. 5).

3 Results

The proposed methodology is tested using MIT-BIH Arrhythmia database [12]. The database contains 48 half-hour of two channel ECG records each having 30 min of duration at a sampling rate of 360 Hz with 11-bit resolution over a 10 mV range. This database includes different categories of ECG signals with acceptable quality, sharp, tall or wide P and T waves, muscle noise, multiform premature complex ventricular junction, supraventricular arrhythmias, baseline drift, QRS morphology variations, conduction abnormalities, and irregular heart rhythms. The accuracy of the proposed technique to detect the QRS complexes is quiet high.

3.1 Performance Assessment

The actual performance of the proposed algorithm is validated by the quantitative results of True-Positive, False-Positive and False-Negative [13].

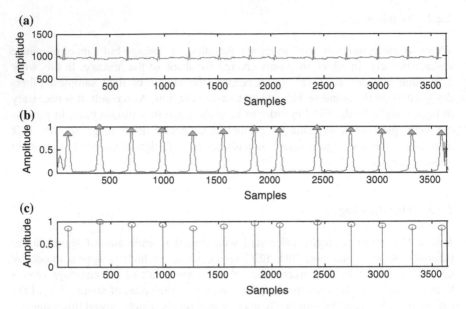

Fig. 5 **a** Input ECG from MIT-BIH database. **b** Signal envelop with detected R-peaks. **c** R-peak positions of the input

True-positive (TP): The number of correctly detected actual heart beats; *False-positive (FP)*: The number of incorrectly detected heart beats labelled as heart beats which are actually not heart beats; *False-negative (FN)*: The number of heart beats that are actually heart beats but were not labelled as heart beats.

The performance criteria of this QRS detection methodology is validated using the sensitivity (S), the positive predictivity (+P) and Accuracy (Acc) values which are calculated as given below:

$$Sensitivity = \frac{TP}{TP + FN} * 100 \tag{12}$$

$$+ Predicivity = \frac{TP}{TP + FP} * 100 \tag{13}$$

The entire performance of the method is validated through detection accuracy. The accuracy is commonly expressed as:

$$Accuracy = \frac{TP}{TP + FP + FN} * 100 \tag{14}$$

Table 1 shows the results obtained from the signals in MIT-BIH arrhythmia database. The proposed algorithm resulted 31 false negatives (FN), and 26 false positives (FP), overall sensitivity of 99.94 %, Positive Predictivity of 99.92 % and

Table 1 Performance evaluation of the methodology for the signals in MIT-BIH Database

MIT-BIH data base	TP	FP	FN	S (%)	P (%)	A (%)
100	2273	0	0	100	100	100
101	1866	0	0	100	100	100
102	2186	3	2	99.9	99.86	99.77
103	2084	0	0	100	100	100
104	2228	9	1	99.95	99.6	99.55
105	2561	3	10	99.61	99.88	99.49
106	2025	2	2	99.9	99.9	99.8
107	2135	2	2	99.9	99.9	99.81
109	2530	0	2	99.92	100	99.92
111	2123	3	1	99.95	99.86	99.81
112	2539	0	0	100	100	100
113	1793	1	2	99.93	99.95	99.93
114	1878	3	1	99.94	99.84	99.79
115	1953	0	0	100	100	100
117	1534	2	1	99.93	99.86	99.80
118	2275	0	0	100	99.86	99.86
119	1986	2	1	99.95	99.89	99.84
121	1863	0	0	100	100	100
122	2475	1	1	99.96	99.96	99.91
123	1518	0	0	100	100	100
124	1619	0	0	100	100	100
Total	**43,444**	**31**	**26**	**99.94**	**99.92**	**99.87**

S Sensitivity, *P* Predictivity, *A* Accuracy

an overall accuracy of 99.87 % for all annotated beats. The results obtained here are compared with the other existing methodologies.

Table 2 shows a comparison on results of performance evaluation among Pan-Tompkins algorithm and Moving average method. On the whole, the proposed method produces better performance.

Table 2 Comparison of the proposed algorithm with existing methods

S. no	Method	Sensitivity	Predictivity
1	The proposed method	99.94	99.92
2	Pan-Tompkins method [13]	99.75	99.54
3	Moving average method [14]	99.87	99.27

4 Conclusion

The proposed method provides a robust and proficient method of QRS detection based on Hilbert transform incorporating with advanced wavelet filtering approach. The pre-processing stage in this method is based on interval dependent thresholding in stationary wavelets integrated with the wiener filter. This noise reduction is an important initial step to improve the results of later processing. The post-processing stage uses Hilbert transform and a windowing technique to identify the exact positions of true R-peaks of the ECG signal. The proposed algorithm is performing well with a Sensitivity of 99.94 %, Positive Predictivity of 99.92 % and accuracy with 99.87 %. The results, when compared with the earlier techniques like Pan-Tompkins algorithm and Moving average method, are showing that the proposed method is working efficiently for QRS complex detection.

References

1. Afonso, X., Tompkins, W.J., Nguyen, T., Luo, S.: ECG beat detection using filter banks. IEEE Trans. Biomed. Eng. **46**, 230–236 (1999)
2. Al Mahmdy, M., Bryan Riley, H.: Performance study of different denoising methods for ECG signals, Science Direct. Procedia Comput. Sci. **37**, 325–332 (2014)
3. Kohler, B.U., et al.: The principles of software QRS detection. IEEE Eng. Medicine Biol. Mag. **21**, 42–57 (2002)
4. Arzeno, N.M., Deng, Z.D., Poon, C.S.: Analysis of first-derivative based QRS detection algorithms. IEEE Trans. Biomed. Eng. **55**(2), 478–484 (2008)
5. Thulasi Prasad, S., Varadarajan, S.: Heart rate detection using Hilbert transform. Int. J. Res. Eng. Technol. **2**, 508–513 (2013)
6. Benitez, D., Gaydecki, P.A., Zaidi, A.: The use of the Hilbert transform in ECG signal analysis. Comput. Biol. Med. **31**, 399–406 (2001)
7. Ruchita, G., Sharma, A.K.: Detection of QRS complexes of ECG recording based on wavelet transform using Matlab. Int. J. Eng. Sci. **7**(2), 3038–3034 (2010)
8. Arbateni, Khaled, Bennia, Abdelhak: Sigmoidal radial basis function ANN for QRS complex detection. Neurocomputing, Elsevier **145**, 438–450 (2014)
9. Shivappriya, S.N., Shanthaselvakumari, R., Gowrishankar, T.:ECG delineation using stationary wavelet transform. In: International Conference on Advanced Computing and Communications (ADCOM), IEEE, pp. 271–274 (2006)
10. Li, G.: Noise removal of Raman spectra using interval thresholing method. In: Second International Symposium on intelligent Information Technology Application, IITA'08. Shanghai, China, vol. 1, pp. 535–539 (2008)
11. Nikolev, N., Nikolov, Z., Gotchev, A., Egiazarian, K.: Wavelet domain Wiener filtering for ECG denoising using improved signal estimate. In: IEEE International Conference on Acoustics, Speech and Signal Processing (ICASSP), vol. 6, pp. 3578–3581 (2000)
12. Mark, R., Moody, G.: MIT-BIH arrhythmia database. http://www.physionet.org/physiobank/database/mitdb
13. Pan, J., Tompkins, W.J.: A real time QRS detection algorithm. IEEE Trans. Biomed. Eng. **BME-32**(3) March (1985)
14. Chen, S.W., Chen, H.C., Chan, H.L.: A real-time QRS detection method based on moving-averaging incorporating with wavelet denoising. Comput. Methods Programs Biomed. **82**, 187–195 (2006)

Fast Multiplication with Partial Products Using Quaternary Signed Digit Number System

P. Hareesh, Ch. Kalyan Chakravathi and D. Tirumala Rao

Abstract The computation speed is restricted with the binary number system by generation and transmission particularly for the carry, as the bit size increments. Using a quaternary Signed Digit number system one may achieve an adder, subtraction which can be of carry free, borrow free respectively for multiplication. However QSD number system necessitates a different set of major modulo based logic aspects for every arithmetic operation. QSD is a superior radix number system, which is used for every arithmetic operation with generate carry free operation and the number scope for QSD is −3 to 3. Proposed QSD multipliers with generating partial products which perform very high speed operation as compared to QSD adders and also used for number of times addition process increasing. The addition process is a carry free operation and other processes for the digits of large numbers like 126,256 or further can be exerted with persistent delay and low difficulty. Tools Modelsim 6.0, Microwind are used.

Keywords Quaternary signed digit multiplier · Quaternary signed digit addition · Intermediate carry · Intermediate sum

1 Introduction

These high recital adders [1] are necessary then the digital computer efficiency [2] greatly determined by the accuracy of adders used in the system. Digital signal processing in which adders are used to achieve a variety of algorithms like FIR, IIR

P. Hareesh (✉) · Ch.Kalyan Chakravathi · D.T. Rao
Department of Electronics and Communication Engineering, GMR Institute
of Technology, Rajam, Andhra Pradesh, India
e-mail: hareeshpudi@gmail.com

Ch.Kalyan Chakravathi
e-mail: challachakravarthi@gmail.com

D.T. Rao
e-mail: tirumalarao.d@gmrit.org

© Springer India 2016
437
H.S. Behera and D.P. Mohapatra (eds.), *Computational Intelligence
in Data Mining—Volume 2*, Advances in Intelligent Systems
and Computing 411, DOI 10.1007/978-81-322-2731-1_41

etc. In earlier period the main confront for VLSI designer is to decrease area of chip by using proficient optimization methods. The addition carry free, which is possible due to redundancy in QSD numbers. The representation of signed-digit which allocates for fast subtraction and fast addition So the subtraction or addition digit is merely purpose of the digits in two adjoining digit locations of the operands for a radix greater than 2, and 3 adjoining digit locations for a radix of 2. Therefore, the add time for two redundant signed-digit numbers is a persistent self-reliant of the word size of the operands, which is the key to high quickness computation. An Algorithm for plan of QSD adder and multiplication is projected. Simulation and synthesis processes are done and the timing report is generated for QSD adder by using VHDL codes. QSD numbers have saving storage of 25 % as compared to BCD [3, 4]. As scope of QSD number is from $\{\bar{3}, \bar{2}, \bar{1}, 0, 1, 2, 3\}$ which is equal to $\{-3, -2, -1, 0, 1, 2, 3\}$. QSD number can be converted into decimal number $D = \sum_{i}^{n-1} x_i 4^i$. QSD number is 3 bit 2's complement notation. x_i is any value from set.

2 Design Algorithm for n-Digit QSD Adder

The number scope of QSD is -3 to 3, the addition outcome of two QSD digits differs number scope -6 to $+6$. Table 1 represents output for all feasible combinations of two numbers. In two digit QSD result the digit LSB stands for the sum bit and the MSB digit stands for the bit carry. QSD digit illustration is used for avoid this bit carry to transmit from lower digit LSB location to MSB location. To achieve carry free addition [2], the addition follows 2 steps [5].
Step 1

Table 1 IC and IS among -6 to $+6$

Sum	QSD number representation	QSD number coding
-6	$\bar{2}2, \bar{1}\bar{2}$	$\bar{1}\bar{2}$
-5	$\bar{2}3, \bar{1}\bar{1}$	$\bar{1}\bar{1}$
-4	$\bar{1}0$	$\bar{1}0$
-3	$\bar{1}1, 0\bar{3}$	$\bar{1}1$
-2	$\bar{1}2, 0\bar{2}$	$0\bar{2}$
-1	$\bar{1}3, 0\bar{1}$	$0\bar{1}$
0	00	00
1	$01, 1\bar{3}$	01
2	$02, 1\bar{2}$	02
3	$03, 1\bar{1}$	$1\bar{1}$
4	10	10
5	$11, 2\bar{3}$	11
6	$12, 2\bar{2}$	12

An Intermediate carry (IC) and Intermediate sum (IS) are produced by this primary stage from the input QSD numbers i.e. addend or augend

Step 2   Next stage unites intermediate sum of present digit with the intermediate carry of the minor significant digit

After these two steps, for eliminate the advance moving of carry there are two laws are executed.

Rule 1   Primary rule declares that the magnitude of intermediate sum must be in the number scope is −2 to +2

Rule 2   Next one declares that magnitude of the intermediate carry must be in the number scope is −1 to +1. It can be represented in 2 bits but functionally it can be represented in 3 bits. This n-digit QSD adder is explained by one example

Example: Perform 4 digit QSD addition for A = 56, B = 23.

Here A, B are decimal numbers. So A, B should be converted into their equivalent QSD.

$$A = (56)_{10} = 0 \times 4^3 + 3 \times 4^2 + 2 \times 4^1 + 0 \times 4^0 = (0320)_{QSD}$$

$B = (23)_{10} = 0 \times 4^3 + 1 \times 4^2 + 2 \times 4^1 + \bar{1} \times 4^0 = (012\bar{1})_{QSD}$. Now addition of two QSD numbers

A = (56)$_{10}$=	0	3	2	0
B = (23)$_{10}$=	0	1	2	$\bar{1}$
Decimal sum	0	4	4	$\bar{1}$
IC 0	1	1	0	
IS	0	0	0	$\bar{1}$
Sum	1	1	0	$\bar{1}$
C$_{out}$ 0				

The sum output result is $(110\bar{1})_{QSD}$. If it is converted into decimal number. Then got $(80)_{10}$. Here above example shows QSD numbers added one time. Here one important issue is if suppose for example 56 is added 56 times. Then what happened is the time delay is increased to get the resultant output of QSD addition. So QSD addition is complicated for addition of repeated times and also large time is taken to get addition result for addition of repeated times. So QSD multiplication process is used to get these results [6].

3 Algorithm for n-Digit QSD Multiplier

The multiplication result M_i is partial product among an n-number input. $A_{n-1} - A_0$ is a single number input, B_i, where $i = 0$ to $n - 1$. The number scope of each QSD is -3 to 3. The multiplication result number scope is -9 to $+9$. The multiplication contains following 2 steps [6, 7].

Step 1 An Intermediate carry (IC) and Intermediate sum (IS) are produced by this primary stage from the input QSD numbers

Step 2 Next stage unites intermediate sum of present digit with the intermediate carry of the minor significant digit. After these two steps, for eliminate the advance moving of carry there law is executed

Rule: The intermediate sum and intermediate carry must be in number scope -2 to $+2$.

Hence the resultant QSD multiplication digits having scope between -4 and $+4$. This multiplication procedure can be explained below example.

Example: To perform QSD multiplication for A = 56; B = 23

This n-digit QSD adder is explained by one example.

Example: Perform 4 digit QSD addition for A = 56, B = 23.

Here A, B are decimal numbers. So A, B should be converted into their equivalent QSD numbers.

$$A = (56)_{10} = 0 \times 4^3 + 3 \times 4^2 + 2 \times 4^1 + 0 \times 4^0 = (0320)_{QSD}$$

$$B = (23)_{10} = 0 \times 4^3 + 1 \times 4^2 + 2 \times 4^1 + \bar{1} \times 4^0 = (012\bar{1})_{QSD}$$

Now addition of two QSD numbers

$$0\,3\,2\,0 \times 0\,1\,2\,\bar{1}$$

$$
\begin{array}{l}
0\,3\,5\,8\,1\,\bar{2}\,0 \\
\text{IC } 0\,1\,1\,2\,0\,0\,0 \\
\quad\text{IS} \quad 0\,\bar{1}\,1\,0\,1\,\bar{2}\,0
\end{array}
$$

$$0\,1\,0\;3\,0\,1\,\bar{2}\,0$$

This is QSD multiplier output result. So $56 \times 23 = (010301\bar{2}0)_{QSD} = (1288)_{10}$ (Table 2, Fig. 1).

Table 2 IC and IS of single number QSD multiplication

Multi	QSD number represent	QSD number coding
−9	$\bar{2}\bar{1}, \bar{3}3$	$\bar{2}\bar{1}$
−8	$\bar{2}0$	$\bar{2}0$
−4	$\bar{1}0$	$\bar{1}0$
−3	$\bar{1}1, 0\bar{3}$	$\bar{1}1$
−2	$\bar{1}2, 0\bar{2}$	$0\bar{2}$
−1	$\bar{1}3, 0\bar{1}$	$0\bar{1}$
0	00	00
1	$01, 1\bar{3}$	01
2	$02, 1\bar{2}$	02
3	$03, 1\bar{1}$	$1\bar{1}$
4	10	10
5	$11, 2\bar{3}$	11
6	$12, 2\bar{2}$	12
7	$13, 2\bar{1}$	$2\bar{1}$
8	20	20
9	$21, 3\bar{3}$	21

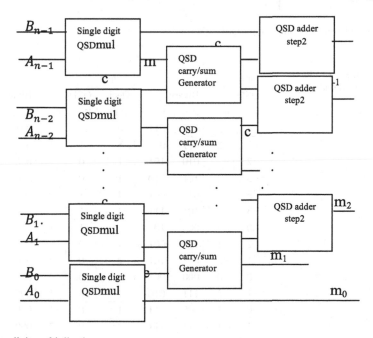

Fig. 1 n-digit multiplication

4 Simulation Results

(a) Simulation results for QSD addition

QSD adders are perform high speed operations as compared to binary number adders. In binary adders carry free additions are not performed. Because propagation delay is increased. QSD adders perform carry free addition operations. So overall propagation delay is decreased as compared to binary additions. Verilog HDL code is compiled and simulated producing tool Modelsim. The simulation results are shown below for single-digit QSD addition. By using this single digit addition n-digit QSD additions or operations may perform. Example of n-digit QSD adder is shown in Design algorithm for n-digit QSD adder (Fig. 2).

Here a[2:0] and b[2:0] is 3 bit inputs and IC[2:0], IS[2:0] are Intermediate Carry, Intermediate Sum respectively. For dissimilar or different value for a, b the output values are shown below Table 3.

Name	Value	999,995 ps	999,996 ps	999,997 ps	999,998 ps	999,999 ps
▶ IS[2:0]	000			000		
▶ IC[2:0]	001			001		
▶ a[2:0]	011			011		
▶ b[2:0]	001			001		

Fig. 2 Output results of single digit QSD adder

Table 3 Output results of single digit QSD adder

Input				Output				
QSD		Binary		Deci	QSD		Binary	
a	b	a	b	Sum	IC	IS	IC	IS
3	3	011	011	6	1	2	001	010
3	2	011	010	5	1	1	001	001
2	−3	010	101	−1	0	−1	000	111
−3	2	101	010	−1	0	−1	000	111
3	1	011	001	4	1	0	001	000
−3	−1	101	111	−4	−1	0	111	000
0	−3	000	101	−3	−1	1	111	001
−2	1	110	001	−1	0	−1	000	111
1	−3	001	101	−2	0	−2	000	110
0	0	000	000	0	0	0	000	000
−1	2	111	010	1	0	1	000	001
0	−1	000	111	−1	0	−1	000	111
−1	−1	111	111	−2	0	−2	000	110
0	1	000	001	1	0	1	000	001
1	−1	001	111	0	0	0	000	000
−2	1	110	001	−1	0	−1	000	111

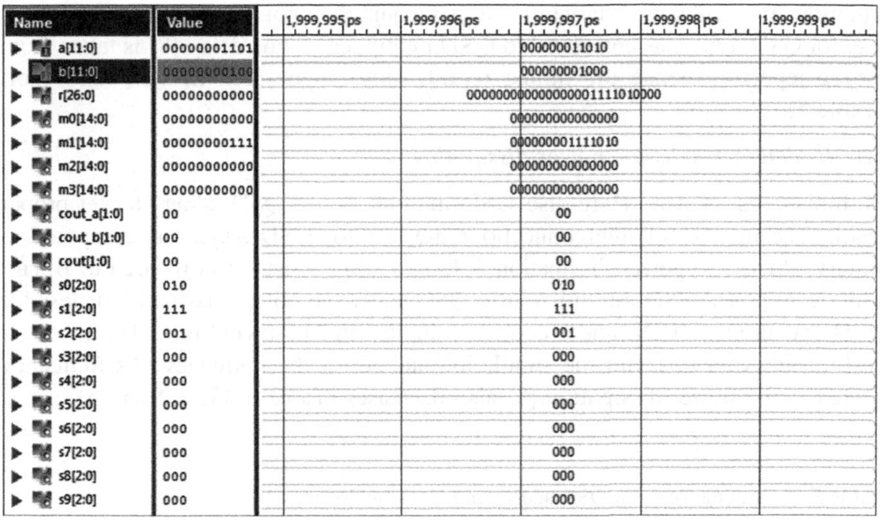

Name	Value	1,999,995 ps	1,999,996 ps	1,999,997 ps	1,999,998 ps	1,999,999 ps
▶ a[11:0]	00000001101			000000011010		
▶ b[11:0]	00000000100			000000001000		
▶ r[26:0]	00000000000		000000000000000001111010000			
▶ m0[14:0]	00000000000		000000000000000			
▶ m1[14:0]	00000000111		000000001111010			
▶ m2[14:0]	00000000000		000000000000000			
▶ m3[14:0]	00000000000		000000000000000			
▶ cout_a[1:0]	00		00			
▶ cout_b[1:0]	00		00			
▶ cout[1:0]	00		00			
▶ s0[2:0]	010		010			
▶ s1[2:0]	111		111			
▶ s2[2:0]	001		001			
▶ s3[2:0]	000		000			
▶ s4[2:0]	000		000			
▶ s5[2:0]	000		000			
▶ s6[2:0]	000		000			
▶ s7[2:0]	000		000			
▶ s8[2:0]	000		000			
▶ s9[2:0]	000		000			

Fig. 3 Simulation result using 4-digit QSD multiplication

(b) Simulation results for QSD multiplication

Proposed QSD multipliers perform very high speed operation as compared to QSD adders for large numbers and also used for number of times addition process increasing. QSD multipliers are formed by QSD adders. The n-digit QSD multiplication with partial products is performed by using VHDL code. Verilog HDL code is compiled and simulated producing tool Modelsim. Here 4-digit QSD multiplication with partial products is performed by using VHDL code. The simulation results are shown below (Fig. 3).

Here Table 4 shows results for 4-digit QSD multiplication. Each input having 4-QSD digits. Each QSD digit having 3 bits. So for 4-digit each input having 12 bits. The resultant output having maximum 27 bits. So a[11:0], b[11:0] are inputs, r[26:0] is output.

(c) Comparison delay time between QSD adders and QSD multipliers

Proposed QSD multipliers perform very high speed operation as compared to QSD adders for also for used number of times addition process increasing. So at this

Table 4 QSD multiplication results for 4-Digit QSD multiplication

Input		Output
a[11:0]	b[11:0]	r[26:0]
000000011010	000000000010	000000000000000001111010000
000011010000	000001010111	000000001000011000001110000
000001000000	000000001000	000000000000000001000000000
000000001000	000000001000	000000000000000000001000000

situation QSD adder takes large time to get required output. So speed of operation is less for QSD adders as compared to QSD multipliers. Time delay [8] is taken from timing report. Generate the timing report. This can be seen in synthesis report (Table 5).

(d) Simulation results for power dissipation

Multipliers use adders. Here also QSD multipliers use QSD adders to get proper result. The number of interconnections are less for QSD adders as compared to Binary adders. So power dissipation is less in QSD adders as compared to Binary adders. Proposed QSD multipliers use QSD adders so power dissipation is less for QSD operations. VHDL adder code is compiled in Microwind tool. The layout is performed. After this, run the simulation and adjust the frequency. As frequency decreases the power dissipation [8] also decreases (Table 6, Figs. 4 and 5).

Table 5 Delay time between QSD adders and QSD multipliers

Time delay	QSD adder	QSD multiplication
Number of times addition process increasing	5.854 ns	4.635 ns

Table 6 Comparison between frequency versus power dissipation

Frequency (MHz)	Power dissipation (µW)
20	34.929
10	34.353
5	33.492

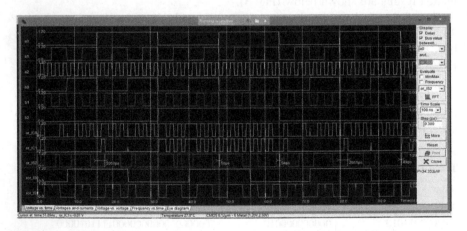

Fig. 4 Simulation result for power dissipation 34.929 µW using Microwind (20 Mz)

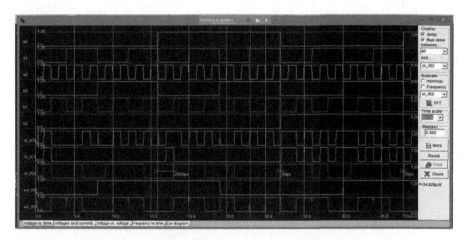

Fig. 5 Simulation result for power dissipation 33.492 μW using Microwind (f = 5 Mz)

5 Conclusion

Proposed QSD multipliers with generating partial products which perform very high speed operation as compared to QSD adders and also used for number of times addition process increasing. The single digit QSD multipliers and single digit QSD adders are used for design n-digit multiplication. These circuits have less power dissipation. The dynamic power dissipation is 33.492 μW at 5 MHz frequency due to using QSD adders and better performance is shown. The delay of planed scheme is 4.635 ns. Tool Modelsim6.0 used for simulating process of this plan and power dissipation is observed by using Microwind.

References

1. Vasundara Patel, K.S., Gurumurthy, K.S.: Design of high performance quaternary adders. Int. J. Comput. Theory Eng. **2**(6), 944–952 (2010)
2. Moskal, J., Oruklu, E., Saniie, J.: Design and synthesis of a carry-free signed-digit decimal adder. In: IEEE International Symposium on Circuits and Systems, pp. 1089–1092 (2007)
3. Parhami, Behrooz: Carry-free addition of recoded binary signed-digit numbers. IEEE Trans. Comput. **37**(11), 1470–1476 (1988)
4. Ahmed, J.U., Awwal, A.A.S.: Multiplier design using RBSD number system. In: Proceedings of the 1993 National Aerospace and Electronics Conference, vol. 1, pp. 180–184 (1993)
5. Dubey, S., Rani, R., Kumari, S., Sharma, N.: VLSI implementation of fast addition using quaternary signed digit number system. In: ICECCN, pp. 655–659 (2013)
6. Ongwattanakul, S., Chewputtanagul, P., Jackson, D.J., Ricks, K.G.: Quaternary arithmetic logic unit on a programmable logic device. Electrical and Computer Engineering The University of Alabama Tuscaloosa, 35487-0286 (2013)

7. Ishizuka, O., Ohta, A., Tannno, K., Tang, Z., Handoko, D.: VLSI design of a quaternary multiplier with direct generation of partial products. In: Proceedings of the 27th International Symposium on Multiple-Valued Logic, pp. 169–174 (1997)
8. Pucknal, A.: For power dissipation and delay time purpose go through Hard copy. In: VLSI Design

Mathematical Model for Optimization of Perishable Resources with Uniform Decay

Prabhujit Mohapatra and Santanu Roy

Abstract Waste stemmed from inappropriate management is a major challenge for perishable resources. Improvement of the inappropriate management has great potential to improve the efficiency of the resources. This research aims to maximize profit and reduce resource spoilage through a fitness value approach based on the decay rate of the perishable resources. A particular type of resource whose decay rate is uniform with time is considered here and is defined as uniform perishable resource. But here in this paper it is shown that the best way to utilize those resources is to follow the first method (i.e. to pick up the best resource first).

Keywords Optimization · Resource management · Uniform decay

1 Introduction

A resource is a supplier from which profit can be generated. The various types of resources may include natural resources such as materials, consumable resource such as energy, or other inputs that are utilized in the process of production. Profits of the resource utilization can be proper system functioning, fulfillment of needs, and increased assets. From the perspective of human being, a natural resource is the source obtained from the nature to satisfy various human needs [1]. The concept of resource has been applied in different fields, including with respect to computer science [2], economics [3–5], biology and ecology [6], management, and human resources [5]. Various factors of human society need allocation of resource through proper management. Resource management is an efficient and effective way for resource utilization [7, 8]. Resources have three primary features: limited stock,

P. Mohapatra (✉) · S. Roy
National Institute of Technology, Silchar 788010, Assam, India
e-mail: prabhujit.mohapatra@gmail.com

S. Roy
e-mail: santanuroy1000@gmail.com

© Springer India 2016

447

H.S. Behera and D.P. Mohapatra (eds.), *Computational Intelligence in Data Mining—Volume 2*, Advances in Intelligent Systems and Computing 411, DOI 10.1007/978-81-322-2731-1_42

utility, and tend to degrade. The fitness value of the resource is defined as the measure of utility of the resource. It represents the amount of output or satisfaction the resource can produce. The fitness value of those resources may get increased, decreased or may remain fixed with time depending on their nature. The resources whose fitness value gets decreased with time are only considered here and they are defined as perishable resources. The examples of such type of resources are stocks like chemicals, fruits and vegetables, drugs, blood cells which are perishable in nature [9, 10]. If the perishable nature is uniformly proportional to time, then it is being defined as uniform perishable resources.

2 Background Theory

The concept of perishable resources arises when in an industry or organization the resources have a short span of life. Some of these resources can be stored for considerably longer time but decomposes eventually. They begin to deteriorate until eventually they become unfit for utilization. This deterioration is known as decay and leads to resource spoilage. These gradual changes that cause deterioration and decay in resources are due to certain organisms and chemicals present in the resources and outside the resources. Particular examples of such types of resources are perishables foods such as fish, meat, milk, bread and vegetables. Most natural food resources have a limited life.

The *fitness value* is defined as the measure of utility that the resource produces and it is represented in real positive values. Fitness value is used only for comparison purpose among all the resources of same kind. The resource which produces maximum utility among resources of same kind is defined as best resource, and the resource which has minimum fitness value is defined as the worst resource and all other's fitness values lie in between them. Some of the resources may have the same fitness values. As mentioned above the resources are perishable in nature, so their utility also starts to decrease as time passes. Hence, their fitness value decreases. The rate at which it decreases is defined as *decay rate*. The minimum utility fitness value can be denoted as zero. If any resource's utility fitness value reaches zero, then it can't fall further. The resource which is picked up first will have full utility fitness value. Another thing, the fitness value of these resources can decrease uniformly or non-uniformly depending on time. But here for simplicity the *uniform decay rate* is considered only.

The next thing is to optimize the utility fitness values of the resources, so that the maximum profit or gain can be reached. One way is to pick up the worst one among all the resources of same kind, so that its utility fitness value will not fall further. Then other resources would be picked up one by one in ascending order. The best resource would be picked up at the end. Another method is the reverse order i.e. to pick up first the best resource, then the other resources in descending order. One can also pick up the resources randomly. To give an example of this situation, one can consider of a bunch of fruits, where the person has to eat only one at a time. In the bunch of fruits all may look differently. Some may have ripened very much, some may not. So, depending on their condition the one which can stay longer or the best among them is given maximum fitness value and the one whose condition is very bad or the worst among them is given the minimum fitness value. The fitness values of remaining fruits lie in between them. Now, the question arises—which one to eat first? In general, people do select first the one which is very much in bad condition as it may not last longer and at last the best one. Alternately one may select first the best among them and at last the worst one. The mathematical modeling of these methods gives a clear picture in the next section.

3　Mathematical Modeling

Let $(f_1, f_2, f_3, \ldots \ldots, f_n)$ be fitness values of n resources $(x_1, x_2, x_3, \ldots \ldots, x_n)$ of one kind. Assume that they have been sorted in descending order such that $f_i \geq f_j, \forall x_i > x_j; i, j \in N$. As the resources are perishable in nature, their fitness values are likely to decrease uniformly. Let the fitness values decrease at the decay rate of $p\%$ per unit of time. For our convenience the unit of time may be considered as 1 day. So, for each day the utility fitness value decreases with decay rate of $p\%$ of the total value initially present for each resource. If a resource x_i is chosen on mth day, then the fitness value f_i decreases m times of $p\%$ of f_i. Instead of $p\%$, a constant amount of p can also be considered to decrease for each day in some specific problems. These problems are independent of initial fitness values and are defined as constant decay rate.

Two methods forward ($\uparrow$) and backward ($\downarrow$) are defined here, where the forward moves from best (f_1) to worst (f_n) where as backward moves from worst (f_n) to best (f_1). In forward method f_1 is picked up first, so it will receive full fitness value. Similarly in backward method f_n is picked up first, so backward method will receive complete fitness values. If any f_i is received on nth step, then its fitness values will decrease $(n - 1)$ times of their total fitness value present initially. Both f_n and f_1 are picked up in forward and backward methods respectively. The modified fitness value of each variable is given in utility fitness Table 1.

Claim 1 The forward method is always better than or equal to the backward method.

Proof we have to show that the sum of all fitness values of the forward method is greater than or equal to that of backward method. From the utility Table 1 it is to prove that

$$f_1 + (f_2 - f_2 * p\%) + \cdots + (f_{n-1} - f_{n-1} * (n - 2)p\%) + (f_n - f_n * (n - 1)p\%)$$
$$\geq$$
$$f_n + (f_{n-1} - f_{n-1} * p\%) + \cdots + (f_2 - f_2 * (n - 2)p\%) + (f_1 - f_1 * (n - 1)p\%).$$

$$(1)$$

Table 1 Utility fitness Table 1

Resources	Fitness values	Forward method (↑)	Backward method (↓)
x_1	f_1	f_1	$f_1 - f_1 * (n-1)p\%$
x_2	f_2	$f_2 - f_2 * p\%$	$f_2 - f_2 * (n-2)p\%$
x_3	f_3	$f_3 - f_3 * 2p\%$	$f_3 - f_3 * (n-3)p\%$
x_4	f_4	$f_4 - f_4 * 3p\%$	$f_4 - f_4 * (n-4)p\%$
.	.	.	.
.	.	.	.
.	.	.	.
x_{n-3}	f_{n-3}	$f_{n-3} - f_{n-3} * (n-4)p\%$	$f_{n-3} - f_{n-3} * 3p\%$
x_{n-2}	f_{n-2}	$f_{n-2} - f_{n-2} * (n-3)p\%$	$f_{n-2} - f_{n-2} * 2p\%$
x_{n-1}	f_{n-1}	$f_{n-1} - f_{n-1} * (n-2)p\%$	$f_{n-1} - f_{n-1} * p\%$
x_n	f_n	$f_n - f_n * (n-1)p\%$	f_n

Simplifying by subtracting the common terms, the final equation is to show that

$$(f_2 * p\%) + (f_3 * 2p\%) + \cdots + (f_{n-1} * (n-2)p\%) + (f_n * (n-1)p\%)$$
$$\leq$$
$$(f_1 * (n-1)p\%) + (f_2 * (n-2)p\%) + \cdots + (f_{n-2} * 2p\%) + (f_{n-1} * p\%). \tag{2}$$

If n = odd, then Eq. (2) becomes

$$(f_{\frac{n+3}{2}} * 2p\%) + (f_{\frac{n+5}{2}} * 4p\%) + \cdots + (f_{n-1} * (n-3)p\%) + (f_n * (n-1)p\%)$$
$$\leq$$
$$\left(f_{\frac{n-1}{2}} * 2p\%\right) + \left(f_{\frac{n-3}{2}} * 4p\%\right) + \cdots + (f_2 * (n-3)p\%) + (f_1 * (n-1)p\%). \tag{3}$$

But $\left(f_{\frac{n+3}{2}}\right) \leq \left(f_{\frac{n-1}{2}}\right)$, $\left(f_{\frac{n+5}{2}}\right) \leq \left(f_{\frac{n-3}{2}}\right), \ldots, (f_{n-1}) \leq (f_2), (f_n) \leq (f_1)$.
Hence Eq. (1) concludes.
Similarly when n = even, then Eq. (2) becomes

$$(f_{\frac{n+2}{2}} * p\%) + (f_{\frac{n+4}{2}} * 3p\%) + \cdots + (f_{n-1} * (n-3)p\%) + (f_n * (n-1)p\%)$$
$$\leq$$
$$\left(f_{\frac{n}{2}} * p\%\right) + \left(f_{\frac{n-2}{2}} * 3p\%\right) + \cdots + (f_2 * (n-3)p\%) + (f_1 * (n-1)p\%). \tag{4}$$

But $\left(f_{\frac{n+2}{2}}\right) \leq \left(f_{\frac{n}{2}}\right)$, $\left(f_{\frac{n+4}{2}}\right) \leq \left(f_{\frac{n-2}{2}}\right), \ldots, (f_{n-1}) \leq (f_2), (f_n) \leq (f_1)$.
Hence Eq. (1) concludes.

Equation (1) is satisfied in both the cases. Hence the claim is proved.

Claim 2 The forward method is equal to the backward method in case of constant decay rate, if all the fitness values $(f_i - mp)$ are non-zero positive or strictly positive.

Table 2 Utility fitness Table 2

Resources	Fitness values	Forward method (↑)	Backward method (↓)
x_1	f_1	f_1	$f_1 - (n-1)p$
x_2	f_2	$f_2 - p$	$f_2 - (n-2)p$
x_3	f_3	$f_3 - 2p$	$f_3 - (n-3)p$
x_4	f_4	$f_4 - 3p$	$f_4 - (n-4)p$
.	.	.	.
.	.	.	.
.	.	.	.
x_{n-3}	f_{n-3}	$f_{n-3} - (n-4)p$	$f_{n-3} - 3p$
x_{n-2}	f_{n-2}	$f_{n-2} - (n-3)p$	$f_{n-2} - 2p$
x_{n-1}	f_{n-1}	$f_{n-1} - (n-2)p$	$f_{n-1} - p$
x_n	f_n	$f_n - (n-1)p$	f_n

Proof Two methods forward (↑) and backward (↓) are also defined here as mentioned before. The only difference is that if any f_i is received on nth step, then its fitness values will decrease $(n-1)$ times of a fixed value p instead of $p\%$. The utility fitness table is given (Table 2).

Let the fitness value of f_i at any step is non-zero. Then the total fitness value of the forward method is given by

$$f_1 + (f_2 - p) + (f_3 - 2p) + \cdots + (f_{n-1} - (n-2)p) + (f_n - (n-1)p).$$

It can be rearranged as

$$(f_1 - (n-1)p) + (f_2 - (n-2)p) + \cdots + (f_{n-2} - 2p) + (f_{n-1} - p) + f_n.$$

This is nothing but the total fitness value of the backward method. Hence the second claim is proved.

4 Examples and Applications

4.1 Problem 1

A person purchases one dozen of bananas from market to consume it over 12 days (one each per day). Consider three cases, where in first case maximum bananas are in very good condition, in second case maximum are in average condition and in third case maximum are in very bad condition. Based on their appearance and condition, they may be given fitness values in between zero to hundred as follows.

Case 1:   98,95,90,90,87,85,81,79,75,70,65,60.
Case 2:   95,90,75,75,65,65,45,40,35,30,25,15.
Case 3:   60,55,50,45,40,35,30,25,20,15,10,05.

Due to perishable nature, let these values decrease 5 % per day. Then we have to find the order in which the person has to consume the bananas in each case so that maximum benefit can be gained and check whether the hypothesis is satisfied in each case or not.

Solution

Let the bananas be consumed using both forward and backward methods.

Case 1   Maximum bananas are in good condition.
        The corresponding utility fitness of each banana is given in utility fitness Table 3.
        From the table, it is found that the average of forward method is 60.72083, whereas the average of backward method is 56.9666.

Case 2   Maximum bananas are in average condition.
        The corresponding utility fitness of each banana is given in utility fitness Table 4.
        From the utility fitness Table 4, it is found that the average of forward method is 43.8958, whereas the average of backward method is 35.25.

Case 3   Maximum bananas are in bad condition and the corresponding utility fitness of each banana is given in utility fitness Table 5.
        From the utility fitness Table 5, it is found that the average of forward method is 26.54166, whereas the average of backward method is 20.5833.
        From the tables and results, one can easily conclude that—*forward method is always better than backward method.*

Table 3 Utility fitness Table 3

Resources	Fitness values	Forward method (↑)	Backward method (↓)
x_1	98	98.00	44.10
x_2	95	90.25	47.50
x_3	90	81.00	49.50
x_4	90	76.50	54.00
x_5	87	69.60	56.55
x_6	85	63.75	59.50
x_7	81	56.70	60.75
x_8	79	51.35	63.20
x_9	75	45.00	63.75
x_{10}	70	38.50	63.00
x_{11}	65	31.00	61.75
x_{12}	60	27.00	60.00
Average	81.25	60.72083	56.9666

Table 4 Utility fitness Table 4

Resources	Fitness values	Forward method ($\uparrow$)	Backward method ($\downarrow$)
x_1	95	95.00	42.75
x_2	90	85.50	45.00
x_3	75	67.50	41.25
x_4	75	63.75	45.00
x_5	65	52.00	42.25
x_6	65	48.75	45.50
x_7	45	31.50	33.75
x_8	40	26.00	32.00
x_9	35	21.00	29.75
x_{10}	30	16.50	27.00
x_{11}	25	12.50	23.75
x_{12}	15	06.75	15.00
Average	54.58	43.8958	35.25

Table 5 Utility fitness Table 5

Resources	Fitness values	Forward method ($\uparrow$)	Backward method ($\downarrow$)
x_1	60	60.00	27.00
x_2	55	52.25	27.50
x_3	50	45.00	27.50
x_4	45	38.25	27.00
x_5	40	32.00	26.00
x_6	35	26.25	24.50
x_7	30	21.00	22.50
x_8	25	16.25	20.00
x_9	20	12.00	17.00
x_{10}	15	08.25	13.50
x_{11}	10	05.00	09.50
x_{12}	05	02.25	05.00
Average	32.50	26.54166	20.5833

5 Conclusions

The proposed work is particularly applicable only to uniform perishable resources and it has significant importance to some industrial problems where we have to keep stock of perishable resources like vegetables, fish, milk, chemicals etc. The same work may be applicable to non-uniform perishable resources also. The only drawback of the proposed work is that if the number of resources is large, then some resources' utility functional values may vanish before the algorithm reaches the resource. For example in the given problem with 5 % decay rate, if there are 50 resources then the utility fitness value of each resource becomes zero after 21 resources. So, the utility fitness values of remaining 29 resources have no effect to the method. The proposed work is good when the number of resources is less or the rate of decay is very slow. The same work can also be extended to the durable resources whose fitness values increase periodically with time.

References

1. Miller, G.T., Spoolsman, S.: Living in the Environment: Principles, Connections, and Solutions, 17th edn. Brooks Cole (2011)
2. Morley, D.: Understanding Computers: Today and Tomorrow, 13th edn. Stamford, Course Technology (2010)
3. McConnell, C.R., Brue, S.L., Flynn, S.M.: Economics: Principles, Problems, and Policies, 19th edn. New York, McGraw-Hill (2011)
4. Mankiw, N.G.: Principles of Economics, 5th edn. South-Western College Publishing, Boston (2008)
5. Samuelson, P.A., Nordhaus, W.D.: Economics, 18th edn. Boston, McGraw-Hill (2004)
6. Ricklefs, R.E.: The Economy of Nature, 6th edn. WH Freeman, New York (2005)
7. Project Management Institute (PMI), A Guide to the Project Management Body of Knowledge, 3rd edn. Newtown Square, Pennsylvania, Project Management Institute (2004)
8. Project Management Institute (PMI), A Guide to the Project Management Body of Knowledge, 4th edn. Newtown Square, Pennsylvania, Project Management Institute (2008)
9. Wang, X., Li, D.: A dynamic product quality evaluation based pricing model for perishable food supply chains. Omega **40**(6), 906–917 (2012)
10. Ahumada, O., Vilalobos, J.R.: Operational model for planning the harvest and distribution of perishable agricultural products. Int. J. Prod. Econ. **133**(2), 677–687 (2011)

Server Consolidation with Minimal SLA Violations

Chirag A. Patel and J.S. Shah

Abstract Cloud computing is very promising technology to deliver computing resources like infrastructure (IaaS), platform (PaaS), software (SaaS) etc. in form of services over the Internet. Server consolidation is the mechanism to minimize active (running) physical servers in the data center of cloud service provider. Consolidation helps the service provider to make efficient usage data center resources and in turn to minimize running cost of data center. Users have to pay as per service usage. SLA is like contract between cloud service provider and users. Service Level Agreement (SLA) is the mechanism of providing guaranteed services to cloud users. SLA Violation may cause high penalties to service provider. In this paper it is shown that if consolidation is maximized, it may cause more number of SLA violation. A balanced approach of server consolidation with minimal SLA violations is suggested in this paper.

Keywords Server consolidation · Datacenter · VM migration · Cloudsim

1 Introduction

Cloud computing is the Internet based paradigm where computing resources are provided in form of services. Cloud service provider (CSP) maintains one or more data centers to provide services. Each cloud data center may contain large number of computing servers and networking equipments. The concept of Service level agreement (SLA) is used to guarantee client services. SLA considers various parameters like service availability, response time, elasticity, scalability and many

C.A. Patel (✉)
Computer Engineering Department, Government Engineering College,
Modasa, Gujarat, India
e-mail: chirag.email@yahoo.com

J.S. Shah
Gujarat Technological University, Ahmedabad, India
e-mail: jssld@yahoo.com

© Springer India 2016   455
H.S. Behera and D.P. Mohapatra (eds.), *Computational Intelligence
in Data Mining—Volume 2*, Advances in Intelligent Systems
and Computing 411, DOI 10.1007/978-81-322-2731-1_43

more. Data center contains large number of physical servers which are used to deploy virtual machines. One physical server hosts more than one virtual machines using virtualization technology. Energy consumption of datacenter is directly proportional to number of active physical servers. Server consolidation is the process of minimizing active physical servers in data center.

Server consolidation helps CSP to make efficient use of datacenter resources and in turn reduces energy consumption. Virtual machine migration is the process of moving VM from one physical machine to other physical machine. This concept of VM migration is used to achieve efficient consolidation. VM migration may cause performance degradation and SLA violations. In this work, limit on maximum allowed migration of each VM is kept while doing server consolidation. Limit on maximum allowed migrations for each VM helps to achieve balanced VM migrations i.e. Even numbers of migrations are distributed across all VMs. Experiment results show that, limit on migrations of individual VM increases energy consumption of datacenter but reduces overall SLA violations. In Sect. 2, related work is discussed. Problem formulation is discussed in Sect. 3. Section 4 represents proposed methodology. Section 5 discusses about experiments and results. At last, Sect. 6 contains conclusion and future work.

2 Related Work

Cloud computing is latest paradigm to deliver each computing resource in form of services over the medium of Internet. Many authors have worked in areas like VM placement, VM migration, Server consolidation, SLA management etc.

In [1], power aware best ft decreasing (PABFD) technique is used for VM placement. Algorithms are suggested form host overload and under load detection. Host overload is detected by methods like utilization threshold. For consolidation of VMs, various selection methods are suggested like minimum migration time. In [2] authors have concentrated on various resource optimization parameters like reactivity, robustness, resource optimization and SLA violations. Bin packing approach [3] is used to solve optimization problem. Main idea is to divide items (VMs) into different types like B-item, L-item, S-item etc. similarly bins are also classified. Each bin can hold specific type of item. This approach is applied to minimize unnecessary migrations. In paper [4] concept of computational geometry is used to manage SLA based advance reservation of resources. By using concept of AR, IaaS providers are able to lock infrastructure resources to guarantee that customers can invoke services during agreed time interval. Geometric representation (graph of capacity over time) helps to record, manage and trace AR requests. In [5, 6], work related to cloud SLA management is done [5]. Proposes a language SYBL for SLA management [7]. Considered elasticity as SLA parameter and compared private cloud with public cloud like Amazon.

In [8], multi-dimensional SLA constraint model is proposed to consider different types of hardware resources. Resource outage event occurs when aggregate resource consumption from VMs hosted by same server exceeds hardware capacity of server.

3 Problem Formulation

In this work, we have considered two SLA parameters. First, number of migrations for individual VM during consolidation and second, overall host utilization. If same VM is migrated multiple times, its performance will degrade and SLA violation may occur. Similarly if host utilization value is above defined threshold, VMs on that host may not get required resources and may lead to SLA violation. Our goal is to minimize active servers with minimal SLA violations. Same can be represented as bellow.

$P = \{P_1, P_2,....., Pm\}$ is set of m physical machines
P_i^C = Total Capacity of PM Pi in terms of tuple (RAM, CPU, N/W)
P_i^U = Utilized Capacity of PM P_i in terms of tuple (RAM, CPU, N/W)
$V = \{V1, V2,....., Vn\}$ set of n virtual machines in datacenter
V_j^R = required capacity of VM V_j terms of tuple (RAM, CPU, N/W)\
$M = m \times n$ matrix where
Mij = 1 if VM V_j is assigned on PM P_i

  0 otherwise
$S = \{S1, S2,..... Ss\}$ is a set of SLAs
th = Threshold defined by user e.g. 80 % (to identify over utilization of host)
mig_th = Threshold defined by user e.g. 3 (to specify maximum number of allowed migrations for each VM)

Optimization function = Use minimum number of active PMs
Minimize $\sum Yi$ where i $\varepsilon\{1....m\}$
  $Y_i = 1$ if $\sum M_{ij} \geq 1$ where j $\varepsilon\{1....n\}$
    0 Otherwise
Here function, $\sum Yi$ represents total number of active servers. Each Yi is one, if number of hosted VMs on that server is greater than or equal to 1.

Constraints:
  C1: For all VM $V_j, \sum M_{ij} = 1$ where i $\varepsilon\{1....m\}$
  C2: For each VM V_j, migrationCount(V_j) <= mig_th
  Here constraint C1 ensures that VM cannot be hosted partially on two different servers. C2 ensures that same performance of same VM is not degraded all time.

4 Proposed Methodology

During consolidation phase, servers which are over utilized and underutilized are identified and migration map is prepared. Migration map contains VMs to be migrated. The selection of VM for migration is done based on criteria MMT (Minimum migration time). Migration time of VM is related to its size and network speed. VM with minimum migration time is added into map. For identifying over utilized hosts, threshold value is checked. Based on migration map two operations are performed, first VMs are migrated and unused servers are powered off and second load of each over utilized host is kept bellow threshold value. Consolidation is done periodically. Algorithms for scheduling VMs and consolidation are as bellow.

Schedule_VM (Fist fit based approach)

   Input: set of VM Requests V, Set of Hosts (PMs) H

   Output: Mapping of requested VMs to Hosts

   Begin

   1. For each VM request V_i, in V,

   2. For each host H_i in H

   3.   If current load of H_i < threshold && H_i has enough resources for V_i then

   4.   Deploy V_i on H_i

   5.   Repeat step 2, 3 till V_i is deployed

   6. If no suitable host found for vi, then display message

   7. Repeat step 2–6 till VM request set is empty.

   End

Optimize_DC (MMT as selection criteria)

   Input: Set of currently deployed VMs VM_LIST   Set of Hosts HOST_LIST
           Current mapping of VMs and PMs

   Output:
           Optimized mapping of VMs and PMs as MIGRATION_MAP

   Begin

   1. Find over utilized hosts (hosts_overutilized) based on pre defined threshold value

   2.   for each over utilized host,

   2.1   find VM to migrate based on selection criteria (MMT)

   2.2   Add VM to MIGRATION _MAP if migration count of that VM is less than mig_th

   2.3   Repeat 2.1 and 2.2 till host is not over utilized

   3. Repeat above step till no host is over utilized

   4. For each VM in MIGRATION_MAP, find suitable host and complete migration

   5. Find under-utilized hosts (hosts_underutilized) based on pre defined threshold value

   6. for each under-utilized host,

           6.1 Add each currently running VM to MIGRATION _MAP if migration count of that VM is less than mig_th
           6.2 Repeat 6.1 till host is empty

   7. Migrate all VMs in MIGRATION_MAP to suitable hosts

   8. Repeat step 1–7 after each pre defined interval.

5 Experiments and Results

CloudSim simulator [9] is used to perform experiments. Three parameters were considered in experiments namely: energy consumption of datacenter, SLA performance degradation of VMs and SLA time per active host. Experiment scenario is as bellow.

VM allocation Policy: Best Fit decreasing

VM selection Policy: Minimum migration time

Overload threshold th=0.8

Max migration per VM mig_th=3

Workload: Planetlab

We have taken above parameters with different number of hosts and VMs like (20 hosts, 50 VMs), (50 hosts, 100 VMs), (100 hosts, 200 VMs) and (200 hosts, 400 VMs). Experiment results are tabulated as bellow. Here the assumption is made that performance of VM degrades by 10 % during migration. SLA time per active host defines the amount of time active host runs with load more than th.

We have considered three evaluation parameters namely: Energy consumption of data center, %SLA performance degradation due to migrations and %SLA time per active host.

Energy consumption in datacenter is mainly because of cpu, memory, disk storage, networking equipments. It is linearly proportional to number of active PMs. Server consolidation reduces number of active PMs and hence energy consumption of data center.

Other parameter is %SLA performance degradation. When VM is migrated from one PM to another, during migration process performance of VM degrades. It is assumed here that 10 % CPU performance will degrade while VM is in migration stage. If more migrations will occur, it may cause more SLA violation.

Third parameter we have considered is %SLA time per active host. percentage of time, during which active hosts have experienced the CPU utilization of 100 % (Tables 1, 2 and 3).

Based on tabular results, graphs are plotted as bellow. From results we can find that restriction of maximum migrations for individual VM (Figs. 1, 2 and 3).

1. Decreases SLA performance degradation due to migrations.
2. Decreases SLA time per active host
3. Increases energy consumption of datacenter

Table 1 Energy consumption of datacenter (kWh)

VMs	PMs	W/o migration limit	With migration limit
50 VMs	20 PMs	0.71	0.98
100 VMs	50 PMs	1.52	1.99
200 VMs	100 PMs	3.22	4.55
400 VMs	200 PMs	6.75	10.03

Table 2 % SLA performance degradation due to migration

VMs	PMs	W/o migration limit	With migration limit
50 VMs	20 PMs	0.03	0.03
100 VMs	50 PMs	0.06	0.04
200 VMs	100 PMs	0.05	0.04
400 VMs	200 PMs	0.07	0.05

Table 3 % SLA time per active host

VMs	PMs	W/o migration limit	With migration limit
50 VMs	20 PMs	0.03	0.03
100 VMs	50 PMs	0.06	0.04
200 VMs	100 PMs	0.05	0.04
400 VMs	200 PMs	0.07	0.05

Fig. 1 Energy consumption of datacenter kWh

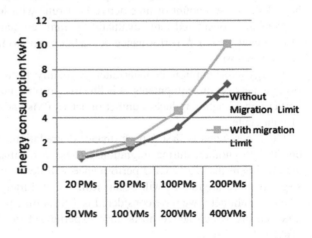

Fig. 2 %SLA performance degradation due to migrations

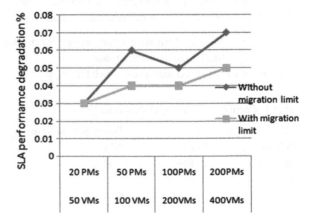

Fig. 3 %SLA per active host

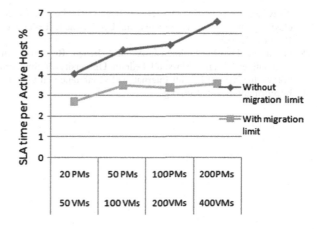

6 Conclusion and Future Work

A balanced approach for server consolidation is discussed in this paper. Form the experiments it is shown that keeping limit on maximum migration count of individual VM increases datacenter energy consumption slightly but causes reduction in SLA violations. Overall numbers of migrations are also reduced. In our work we have taken static threshold value for overload/under load detection and minimum migration time (MMT) as VM selection technique for migration.

In future some other methods for overload and under load detection can be considered. Same way selection of VM for migration can be done based on some other parameters.

References

1. Lionel Eyraud-Dubois, H.L.: Optimizing resource allocation while handling SLA violations in cloud computing platforms. In: IEEE 27th International Symposium on Parallel & Distributed Processing (2013)
2. Buyya, A.B.: Optimal Online Deterministic Algorithms and Adaptive heuristics for energy and Performance Efficient Dynamic Consolidation of Virtual Machines in Cloud Data Centers. Wiley InterScience (2012)
3. Song, W., Xiao, Z.: Adaptive resource provisioning for the cloud using online bin packing. In: IEEE 2013 Transactions on Computers
4. Kuan Lu, T.R.: QoS-aware SLA-based advanced reservation of infrastructure as a service. In: Third IEEE International Conference on Cloud Computing Technology and Science (2011)
5. Georgiana Copil, D.M.-L.: SYBL: an extensible language for controlling elasticity in cloud applications. In: 13th IEEE/ACM International Symposium on Cluster, Cloud, and Grid Computing (2013)
6. Kouki, Y., de Oliveira, F.A.: A language support for cloud elasticity management. In: 14th IEEE/ACM International Symposium on Cluster, Cloud and Grid Computing (2014)

7. Azevêdo, E., Dias, C., Dantas, R., Sadok, D., Fernandes, S.: Profiling core operations for elasticity in cloud environments. In: Universidade Federal de Pernambuco (UFPE) Recife, Brazil 2012 IEEE
8. Goudarzi, H.: Multi-dimensional SLA-Based Resource Allocation for Multi-tier Cloud Computing Systems. Massoud Pedram University of Southern California, Los Angeles, CA 90089. In: 2011 IEEE 4th International Conference on Cloud Computing
9. http://cloudbus.cloudsim.org

Predicting Consumer Loads for Improved Power Scheduling in Smart Homes

Snehasree Behera, Bhawani Shankar Pattnaik, Motahar Reza
and D.S. Roy

Abstract Smart homes form one of the major components leveraging demand response within the smart grid paradigm. Flexible pricing policies along with the capability of scheduling power among many homes form the crux of a wide variety of smart home power management controllers. However leveraging power scheduling for smart homes while keeping user costs minimal is a challenging proposition and involves complex multistage, stochastic, non-linear optimization techniques. For ease of computation, heuristic algorithms can be employed that require consumer load corresponding to smart homes which are not available a priori. The efficiency of power scheduling heuristics, however depend on the accuracy of the consumer loads forecasted. In this paper, we focus on developing a technique that can efficiently forecast consumer loads and thereafter the predicted load is fed to a GA heuristic based power scheduling algorithm for smart homes. Detailed procedure for the aforementioned forecasting has been presented and the results obtained are analyzed.

Keywords Demand response · Smart home · Genetic algorithm power scheduling · Time series forecasting · ARIMA

S. Behera (✉) · B.S. Pattnaik · M. Reza · D.S. Roy
Department of Computer Science & Engineering, National Institute of Science & Technology, Berhampur 761008, Odisha, India
e-mail: b.snehasree@gmail.com

B.S. Pattnaik
e-mail: bhawani.pattnaik@gmail.com

M. Reza
e-mail: tonikr@gmail.com

D.S. Roy
e-mail: diptendu.sr@gmail.com

© Springer India 2016
H.S. Behera and D.P. Mohapatra (eds.), *Computational Intelligence in Data Mining—Volume 2*, Advances in Intelligent Systems and Computing 411, DOI 10.1007/978-81-322-2731-1_44

1 Introduction

Demand response has been hinted as one of the key elements of smart grid parlance that promises operating the grid as its optimal [1]. Several variation of implementing DR has been proposed in literature such as Direct Load Control (DLC) that is very effective in maintaining energy balance [2] yet suffer from being inflexible and intruding from customers point of view; Dynamic pricing that allows customers a way of participating the Demand Response scheme through incentive-penalty mechanism [3] and so on. Power scheduling refers to the scheme where Demand Response is employed by altering customers' original power usage pattern to achieve the higher goal of sustaining the grid optima. Power schedules for every customer needs to be prepared for each and every customer participating in such Demand Response scheme within the Smart Grid parlance. In order to prepare valid and feasible schedules for customer, the Smart Grid regulates this task to a controller, which has been referred to as the Home Energy Management Systems (HEMS) in literature [4, 5]. It has to be born in mind that in order to generate optimal power schedules, the HEMS require information regarding the customer consumption load, in an hour-wise denomination. Since the HEMS generate power schedules for every house within the set of smart homes in a day ahead fashion at the granularity of 1 h slots, such load data is not available a prior to the HEMS. This necessitates the use of appropriate techniques for predicting the hourly load for every customer.

In one of our recent works [6], we have presented a metaheuristic formulation using Genetic Algorithm that minimizes customers' overall energy cost by means of an intelligent power scheduling scheme. However, in that work the proposed scheme was tested with very limited data pertaining to customers' hourly demand. In order to test the efficacy of such metaheuristic, it is necessary to its employ much more realistic predicted customers load. To the end, this paper presents comprehensive details of how such data can be forecasted using a well known dataset, namely the UCI Machine Learning Repository [7] that provides the electric consumption for one household.

The remainder of this paper is organized as follows: Sect. 2 presents the related work. Section 3 summarizes the Genetic Algorithm based optimal power scheduling problem in a HEM. Section 4 presents details of how realistic dataset is formed for UCI dataset. In Sect. 5, the simulation setup and simulation parameters have been presented. Section 6 introduces the experiments conducted. Section 7 concludes the paper.

2 Related Work

Demand response (DR) [1] in its multifarious forms including market DR & physical DR; when employed combinedly has been hinted as a key for attaining sustained grid optimal. Besides, DR scheme can range from as invasive and inflexible schemes like direct load control (DLC) [2] to much more customer-premise oriented and non-invasive, like dynamic pricing [3]. In the former case, the grid directly takes initiative to turn residential appliances ON or OFF to maintain grid energy balance; whereas the latter employs a varying pricing strategy with incentive-penalty scheme for inducing customers to shift their load in a harmonious way.

Within the purview of residential energy management, there has been a variety of research areas, ranging from issues related to design of smart homes and their implementation [8] to a host of DR programs. A few researchers have also delved into optimal energy management problem by means of certain non-linear optimization techniques as in [4], extending these to account for possible rebound peak [6] and so forth. Reference [9] presents coordinated management strategies that accounts for multimode appliances by means of Markov Decision Process.

3 Genetic Algorithm Based Optimal Power Scheduling Problem in a HEM

3.1 Objective Function

For the power scheduling problem, the optimization criteria considered fulfills the demand of the customer by minimizing the cost of user u. Mathematically the problem is defined as:

$$\min_{S_i \in Sched} \sum_{\substack{u \in User \\ s \in Slot}} EPC[u][s] * PR[s] \tag{1}$$

Subject to:

$$\sum_{\substack{u \in User \\ s \in Slot}} EPC[u][s] = TOTDEM[u] \tag{2}$$

$$\sum_{\substack{u \in User \\ s \in Slot}} EPC[s][u] \leq TOTPOW[s] \tag{3}$$

The Eq. (1) depicts the objective function that minimizes cost for all users. Here $EPC[u][s]$ and $PR[s]$ denotes the power required by user u in slot s and its corresponding prices given by retailer in advance. *Sched, User, Slot* indicates the sets of all possible schedules, participating users and number of slots in a day respectively. The constraint in (2) and (3) specifies the real time scheduling constraints. *TOTDEM[u]* and *TOTPOW[s]* denotes the total power demand provided by every user in a day-ahead fashion and total hour wise dispensable power for a retailer (for slot *s*) respectively. For easy computation of the function, a cost vector is considered of size *Sched* and the cost is calculated for each schedule of *Sched* storing the result in increasing order.

3.2 Mapping the Power Scheduling Problem with Genetic Algorithm

In reality, scheduling power among a large number of residences is a challenging problem. For this paper, we propose a simulation model that uses the Expected Power Consumption (EPC) model has been proposed in [6]. This model has been formulated by estimating total power demand of individual users and also the hour wise power demand for each user. In [6], the *EPC* matrix was formed based on limited information. In this paper, however we generated the *EPC* matrix for every house based on realistic dataset through time series analysis. An entry in the *EPC* matrix, for instance *EPC[u][s]* denote the power demanded by user *u* in slot *s*. Entries corresponding *EPC[u][s]* are the predicted values generated from the time series analysis of the UCI dataset for a user. Rest of the EPC matrices will be generated based on the forecasted EPC matrix using normal distribution.

Encoding of chromosomes is one of the challenges while solving a problem with GA. We have considered the direct representation of encoding in the proposed power scheduling algorithm. All feasible solutions are stored in a vector of size *no_chromosomes(n)*. An initial population of chromosomes is generated at the beginning to generate *n* chromosome. To do so, we generated random chromosomes with normal distribution. The fitness considers two parameters namely user cost and demand fulfillment. On retailers' side, it has to provide slot-wise price information and also balance load among users such that overall load (demand) is within available power with the grid overall time slots In this paper, we have considered two fitness functions: in the first case (FF1), at every iteration the best chromosomes are chosen whereas for the other case (FF2), 0.9 % chromosomes are chosen from the best chromosomes and the remaining 0.1 % are those with worst fitness values.

We have considered single-point crossover for our simulation purposes. For our problem we have considered two mutation operators, namely move and swap based on their fitness. The set of selected chromosomes, *selected[i]*, is denoted with 1 if the chromosome is selected or is set to 0 otherwise.

4 Dataset Formation

4.1 Forming the EPC Matrix by Time Series Analysis

Time series are the series of data over a time period and analysis of the time series means using the past and current observation we can find the future values. Time series analyses are used in various fields such as statistics, signal processing, econometrics and mathematical finance [10]. In time series normally forecasting models such as ARIMA, ARMA, MA, SRIMA are found by using which we can predict future values for better estimation of anything (i.e. stock, weather report etc.). In our paper we are using this time series analysis for predicting consumers load so that we can minimize the overall cost of HEM and can balance the grid from blackouts.

The objective of this research is to decide upon an appropriate model for predicting day ahead electricity consumption in a household effectively. For that purpose, we have employed Box and Jenkins method [11]. This predicted data will be used in the HEM for optimizing the electricity consumption and cost for each individuals.

The proposed load prediction comprises of a time series process and the steps involved have been summarized in the schematic depicted Fig. 1.

This paper uses the UCI dataset [7] for household power consumption in a single residence at a sampling rate of 1 s between 2006-12-26 to 2010-11-26. We used R [12] and R Studio [13] for building the model. The first step in using the aforementioned dataset is to fill the missing values since it can negatively affect the

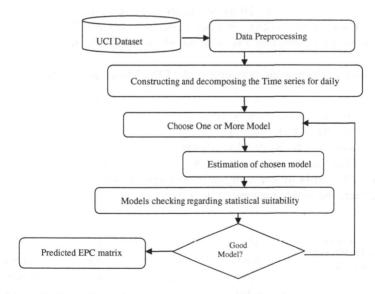

Fig. 1 Schematic diagram depicting the steps to generate EPC matrix

prediction efficacy and thus lead to poor forecasted model. The UCI dataset's resolution is every second, i.e., power consumption for every second is noted. However, for residential energy management, consumption data required is on hourly or in some cases on a daily basis. Thus we had to modify the data format to suit the forecasted user consumptions..

Auto regressive integrated moving average (ARIMA) [10] is one of the time series models which is a mixed version of autoregressive moving average (ARMA) model and ease to find mixed model parameters. ARIMA model uses three parameters named p, d and q for auto regression, integration and moving average of the dataset. This model is represented as ARIMA (p, d, q) where p, d, q are non negative integer values. For understanding and predicting time series, ARIMA model is mostly used. For a particular datasets initially an appropriate ARIMA model is found out with the smallest possible value for the parameters so that analysis and prediction of data can be appropriate.

The Akaike Information Criteria (AIC) is basically used for statistical model. While comparing two or more time series models, it is considered that that the lowest AIC should be chosen for the realistic data analysis. According to Box-Jenkins method, the value of p and q of ARIMA model should be *two* or less or the total number of parameters should be less than *three* [10].

The model with the smallest AICc (Akaike Information Criterion) and RMSE (Root Mean Square Error) value is selected. In R we have used the *auto.arima()* function to get the best suitable model for our dataset. After analysis we got ARIMA (0, 0, 1) as the suitable model for our dataset. Extensive details pertaining to model selection has been adopted based on [14].

Here the prediction is done on daily basis for an individual user which is represented in a matrix i.e. EPC matrix. As the predicted realistic data is only for one user it is a limitation for the HEM optimization using Genetic algorithm. So, we have to extrapolate this matrix for other users. Here Table 1 show the parameters which are used for predicting a suitable model for the dataset for each category of user.

EPC matrix which is an extension of the EPC Model we had presented in [6] includes:

  i. A set of users demanding power.
 ii. A set of time slots (periodic).
iii. Daily total power demand for all consumers.
 iv. User-wise total power consumption.
  v. The EPC matrix of size *num_users* × *num_slots*, where a position *EPC[u][s]* indicates the power required by user u in slot s.

Table 1 Parameter estimation for different prediction models

ARIMA	AIC	AICc	RMSE
(0,0,1)	8.56	10.56	0.1786
(1,0,1)	10.87	14.96	0.2895
(1,0,0)	11.54	13.01	0.2109

4.2　Extrapolating the EPC Matrix for All Users

In our analysis we have predicted the daily demand of one user on hourly basis from the UCI dataset. Further this realistic demand is extrapolated for 50 users by generating the random variables under Gaussian assumption for the realistic *EPC[u][s]* matrix which is the input for our HEMs assuming that the consumer loads across a smart homes within a HEMS. The parameters of this Gaussian distribution which predicts realistic demand for the users are estimated by 95 % confidence interval around zero using UCI dataset.

5　Experimental Simulation & Parameter Setup

In this paper to simulate power scheduling for residential consumers certain assumptions are made with respect to customers' involvement and energy retailer. As per our assumption retailer is providing day ahead hourly energy prices to the customers in a dynamic pricing fashion on or before 12:00 am. To get a better cost optimization we have predicted every customers hourly load demand.

5.1　Simulation Setup

For the purpose of simulation, dynamic base prices of electric power has been assumed to be at three levels, namely at Rs. 4, Rs. 8 and Rs. 10 respectively for off-peak, normal and peak load hours. We have derived the hourly prices with an assumption that these hourly prices fluctuate within 20 % of the above mentioned base prices and they are normally distributed. To account for variations in consumers' power demands, we have considered three categories of customers based on their power demand. Category 1 (U1) customers are characterized by a maximum daily demand of 50 Wh, whereas category 2 customers (U2) possess a maximum demand of 100 Wh and category 3 (U3) customers have a maximum demand is 200 Wh. Also we have assumed that each category as 50 users each in the residential area. These values depend on the actual problem at hand and thus once the formulation is done, these parameters can be used.

5.2　Parameters for Genetic Algorithm (GA) and HEMS

This section presents a summary of the simulation parameters used in this paper. Parameters pertaining to smart home energy management and genetic algorithm have been provided in Tables 2 and 3 respectively.

Table 2 Simulation parameters for HEMS

Demand response type	Day ahead dynamic pricing and loads
Load information	Day-ahead forecasted load

Table 3 Genetic algorithm parameters

Number of generation steps	1000
Crossover operator	Single-point
Population size	100
Intermediate population size	90
Selection operator	Rank based
Fitness is based on two categories	Selection
FF1(Fitness Function 1)	Best based on minimum cost
FF2(Fitness Function 2)	90 % best and 10 % worst

6 Results and Discussion

6.1 Simulation Environment

For reckoning the simulation model we used the system with following hardware specifications: Intel Core i5-2400 @ 3.10 GHz X4, RAM: 3.38 GB, HDD: 226.2 GB, and some of the software specification are Windows (32-bit) operating system, C++ as programming language and R3.2, R Studio as statistical tool [12, 13].

As an input for our GA we have an EPC matrix with 50 rows (users) and 24 columns (hourly slots) of 100 matrices for each user category and a cost vector which provides price for 24 h.

6.2 Result Obtained and Observation

Table 4 depicts cost comparison among the two approaches, namely using GA for optimization using fitness function (FF1) and without using GA. Table 5 depicts the same, but considering the other fitness function, namely FF2. Such comparison captures the importance of employing GA.

Table 4 Cost comparison between without GA and with GA (FF1)

User category	Cost without GA	Cost with GA (FF1)
U1	4595.819824	3821.177002
U2	6063.348145	5327.931152
U3	12200.94921	11151.125977

Table 5 Cost comparison between without GA and with GA (FF2)

User category	Cost without GA	Cost with GA (FF2)
U1	4595.819824	3745.321777
U2	6063.348145	4949.039063
U3	12200.94921	10648.509766

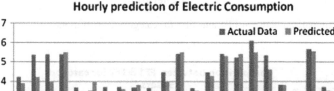

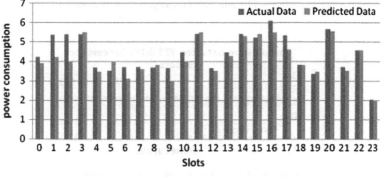

Fig. 2 Hourly predicted data of a user

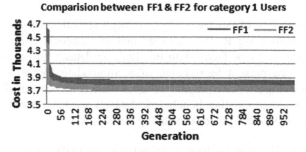

Fig. 3 Cost comparison between FF1 and FF2 for category 1 (U1) user

Figure 2 presents the hourly predicted data of an individual house. Figures 3, 4 and 5 present the cost comparison between Category 1 (U1) users, Category 2 (U2) and Category 3 (U3) for both the fitness functions, namely FF1 and FF2 respectively.

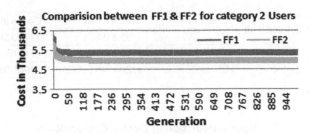

Fig. 4 Cost comparison between FF1 and FF2 for category 2 (U2) users

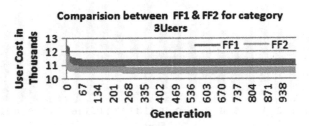

Fig. 5 Cost comparison between FF1 and FF2 for category 3 (U3) users

7   Conclusion

The major contribution of this paper is generating a realistic consumer load data by time series analysis using R. We observed that the genetic algorithm which was developed in [6] works perfectly with this realistic dataset. In our next work we plan to extend the present formulation to accommodate appliance level power schedule, which is much more pragmatic. Since there is no benchmark datasets, we also envision building existing benchmark datasets so that we can have a comparative study of different power scheduling heuristics.

Acknowledgments This work has been carried out using the facilities prevailing at the Smart Grid Analytics Group at National Institute of Technology, Berhampur, India. The authors gratefully acknowledge the facilities provided.

References

1. Palensky, P., Dietrich, D.: Demand side management: demand response, intelligent energy systems, and smart loads. IEEE Trans. Ind. Inf. **7**(3), 381–388 (2011)
2. Chu, W.-C., Chen, B.-K., Chun-Kuei, F.: Scheduling of direct load control to minimize load reduction for a utility suffering from generation shortage. IEEE Trans. Power Syst. **8**(4), 1525–1530 (1993)
3. Borenstein, S.: The long-run efficiency of real-time electricity pricing. Energy J. 93–116 (2005)

4. Chang, T.-H., Alizadeh, M., Scaglione, A.: Coordinated home energy management for real-time power balancing. In: Power and Energy Society General Meeting, 2012 IEEE, San Diego, CA, USA
5. Chang, T.-H., Alizadeh, M., Scaglione, A.: Real-time power balancing via decentralized coordinated home energy scheduling. IEEE Trans. Smart Grid 4(3), 1490–1504 (2013)
6. Polaki, S.K., Reza, M., Roy, D.S.: A genetic algorithm for optimal power scheduling for residential energy management. In: Environment and Electrical Engineering (EEEIC), 2015 IEEE 15th International Conference on. IEEE, 2015, Rome, Italy
7. UCI repository on machine learning database. https://archive.ics.uci.edu/ml/datasets/Individual+household+electric+power+consumption
8. Han, D.-M., Lim, J.-H.: Design and implementation of smart home energy management systems based on zigbee. IEEE Trans. Consum. Electron. 56(3), 1417–1425 (2010)
9. Chang, T.-H., Alizadeh, M., Scaglione, A.: Real-time power balancing via decentralized coordinated home energy scheduling. IEEE Trans. Smart Grid 4(3), 1490–1504 (2013)
10. Li, G., et al.: Day-ahead electricity price forecasting in a grid environment. IEEE Trans. Power Syst. 22(1), 266–274 (2007)
11. Box, G.E.P., Jenkins, G.M., Reinsel, G.C.: Time series analysis: forecasting and control, vol. 734. Wiley (2011)
12. http://www.rstudio.com/. Accessed 25 July 2015
13. http://cran.us.r-project.org/. Accessed 25 July 2015
14. Chujai, P., Kerdprasop, N., Kerdprasop, K.: Time series analysis of household electric consumption with ARIMA and ARMA models. In: Proceedings of the IMECS Conference on, Hong Kong. (2013)

Prevention of Wormhole Attack Using Identity Based Signature Scheme in MANET

Dhruvi Sharma, Vimal Kumar and Rakesh Kumar

Abstract Mobile ad hoc network (MANET) has attracted many security attacks due to its characteristics of dynamic topology, limited resources and decentralize monitoring. One of these vulnerable attack is wormhole in which two or more malicious nodes form a tunnel like structure to relay packets among themselves. This type of attack may cause selective forwarding, fabrication and alteration of packets being sent. In this paper, we have proposed a way to protect network from wormhole attack by using identity based signature scheme on cluster based ad hoc network. Proposed scheme does not require distribution of any certificate among the nodes so it decreases computation overhead. We have also discussed existing work that either require certifcates or does not accomplish all the security requirements of network. Our simulation results show that attack is prevented successfully and it outperforms other schemes.

Keywords Wormhole attack · ID based signature · MANET · Secure AODV

1 Introduction

MANET is an autonomous system of nodes that are free to move at their will [1]. Transmission of data is directly if they lie in radio range of each other or through multi hop transmission if they are far away. So a node can behave as source or destination or router depending upon the scenario. In cluster-based framework

D. Sharma (✉) · V. Kumar · R. Kumar
Department of Computer Science & Engineering, Madan Mohan Malaviya
University of Technology, Gorakhpur 273010, U.P., India
e-mail: dhruvisharma000@gmail.com

V. Kumar
e-mail: vimalmnnit16@gmail.com

R. Kumar
e-mail: rkiitr@gmail.com

© Springer India 2016 475
H.S. Behera and D.P. Mohapatra (eds.), *Computational Intelligence
in Data Mining—Volume 2*, Advances in Intelligent Systems
and Computing 411, DOI 10.1007/978-81-322-2731-1_45

nodes are divided into groups, these groups are called cluster. Each cluster have cluster head who is the central monitoring body of these cluster, they also act as gateway for inter-cluster transmission. An Ad hoc On Demand Distance Vector (AODV [2–5]) routing protocol has been used in this paper for communication between all participating nodes. Wormhole is one of those security attacks that are difficult to detect as they often do no harm to the nodes so no encryption applied on these nodes can stop the attacker. In this attack there is a tunnel like structure present between two or more malicious nodes, which acts as a passage, when packets arrive at one such node they are relayed to another one which is present in same or different network [6]. Such attacks either disturb the functioning of routing protocols or may affect the packets being transferred. Figure 1 shows that Wormhole attack by malicious nodes F and G. Wormhole attack can be launched in either of the following way [7, 8]:

By encapsulation: The packets are encapsulated at one end of tunnel and are decapsulated at another. Nodes lying in between cannot decode it and so hop count cannot be incremented hence malicious nodes relay the packets without any difficulty.

By high power transmission: An attacker node on getting a route request (RREQ) packet forward it at high power level, any normal node hearing it rebroadcasts it toward destination, therefore attacker node increases its chance of being in route selected in route discovery phase of AODV.

By high-quality/out-of-band transmission: Two attacker nodes transfer the packet by single hop link like long range directional wireless or wired link. Specialized hardware is required to launch such attack.

By packet relay: One or more attacker node relays packet between two different nodes to convince them that they are neighbours.

By using protocol distortion: The main aim of attacker is to attract the network traffic. Suppose a routing protocol that is based on shortest time can be made target of this attack by using mechanism like, normally nodes wait for random amount of time before forwarding RREQ to avoid collision but malicious node can broadcast the packet directly to make it reach the destination first.

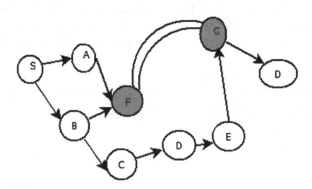

Fig. 1 Wormhole attack by malicious nodes F and G

Digital signature is an application of asymmetric cryptography. Firstly we apply one-way hash function on message to convert it into shorter fixed length string. Then it is encrypted by a private key of sender and sent it along with other information to receiver. Receiver on getting the message decrypts it with public key of sender. Hashing algorithm is again applied on message and computed hashed message is compared with received, if same, it depicts received message has its integrity undamaged. In identity based digital scheme [9] a well-known string is used as public key like email address, IP address etc. A trusted third party called private key generator (PKG), which generates private key to given identity.

This paper has been further organised with Sect. 2 discuss related works. Section 3 describes proposed ID based scheme do and then its performance evaluation has been discussed in Sect. 4. Finally we end our discussion with conclusion and future scope in Sect. 5.

2 Related Work

Zapata et al. [10] introduced a secure AODV (SAODV) routing protocol. It is uses a digital signature scheme for authentication purpose in mobile ad hoc network. The drawback of this approach is key distribution problem between nodes in MANET. Another approach uses packet leaches by Hu et al. [11]. There are two types of packet leaches- geographical leaches that require every node to have GPS device. It ensures communicating nodes are within range. Temporal leaches that require nodes to be tightly synchronized. It enforces time bound up to which packet will be considered valid. Round Trip Time (RTT) by Zhen et al. [12] was used later to avoid the use of special hardware. RTT is the time elapsed since the packet has been sent by a node till it gets the reply. So RTT of fake neighbour would be larger as compared with the actual so it can be used in hidden attacks. However, it cannot detect exposed attack. Directional antennas by Hu and Evans [13] were used too. In this, every node share a secret key to other participating nodes. Announcer broadcast HELLO message to its neighbours by using directional antennas. Nodes send their identity and an encrypted message having identity of announcer and challenge nonce in reply. Before adding the nodes to their neighbours list, announcer checks for authentication. Another lightweight technique is LITEWORP by Khalil et al. [14]. Guard node are common neighbour between two nodes, they perform local monitoring of traffic to detect the malicious node. However, it is difficult to find guard node in sparse network. Transmission time based mechanism was introduced for AODV routing protocol and can be extended for others by Tran el al. [15]. In this method, transmission time between every two node is calculated along a route, it is based on the fact that transmission time between real neighbours will be much smaller than fake neighbours form by wormhole. Sharma et al. [16]

proposed an algorithm that was based on use of digital signature to prevent wormhole attack. This mechanism assumes that only legitimate nodes will have valid digital signature.

3 Proposed ID Based Scheme

In our proposed scheme, we have assumed that network follow cluster based architecture and cluster head of each cluster are chosen in a way that they can't be malicious, so they can work as private key generator (PKI). TTL (Time to Live) has been considered as an average time taken be time taken by a packet to reach any node in network area taken into consideration. This method presents a novel solution to prevent a wormhole attack done through packet replay. Proposed scheme is consists of three phases:

- Setup phase
- Communication phase
- Secure data transfer phase

3.1 Setup Phase

- Cluster head (CH) broadcasts the cluster parameters (AG, MG, pri, e, H, p, q, pub) to nodes present in cluster N = (N1, N2...Nn). The cluster parameter is given in Table 1.
- Each node on receiving parameters sends their ID to their CH.
- **Public key generation**:
  $\{pub_1, pub_2, pub_3,..., pub_n\} = \{H(ID_1), H(ID_2), H(ID_3),..., H(ID_n)\}$
- **Private key generation**:
- $\{pri_1, pri_2, pri_3,..., pri_n.\} = \{pub_1 * PRI, pub_2 * PRI, pub_3 * PRI,..., pub_n * PRI.\}$
- CH sends the private key to corresponding node using secure channel as in Fig. 2.

Table 1 System parameters

S. n.	Term	Definition
1.	AG	Additive cyclic group of prime order p
2.	MG	Multiplicative cyclic group of prime order q
3.	PRI	Cluster heads private key
4.	e	Bilinear mapping e: AG * MG -> MG
5.	H	Hash function H: 0,1* -> Zq+
6.	p	Generator of group AG
7.	q	Generator of group MG
8.	PUB	Cluster heads public key

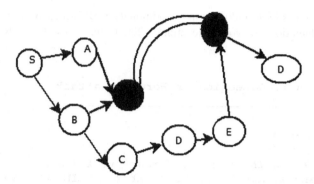

Fig. 2 Distribution of public and private key by cluster head

3.2 Communication Phase

Intra cluster communication (Lying within the same cluster): Direct communication (source and destination are within the range of each other): Node are neighbours in such scenario. So, this type of wormhole attack being considered in this paper does not work in this scenario.

Indirect communication (source and destination do not lie in range of each other): Nodes have to communicate through cluster head which is also a monitoring body as in Fig. 3.

Inter cluster communication (Source and destination belongs to different cluster): Malicious nodes create illusion among nodes in different clusters making them believe they are situated near each other as in Fig. 4.

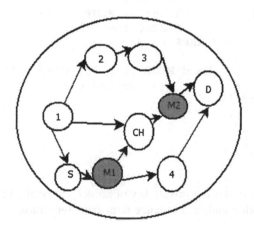

Fig. 3 Intra cluster communication in the presence of malicious node M1 and M2

This phase has slight modification in traditional AODV protocols and Algorithm 1 is used for detection of wormhole attack. Details terms used in Algorithm 1 in Table 2:

Algorithm 1: Detection of Wormhole attack

```
S (RREQ)====> D
for i=1 to N do
  if IN!=S and IN! has valid route to D then
    if destseqnum in routing table of IN > = destseqnum
in RREQ then
      D(RREP)====> S
    else
      Forward RREQ to neighbors
      if RREP is received by S then
      Create RCNF consisting of destseqnum, hopcount,
TTL, ID received in RREP
      Encapsulate RCNF by pri S and then by pub D
      Create new RREQ having D=1 and reqID=2
      Send new RREQ + RCNF to D
      if TTL expires then
        S sends new RREQ + RCNF₁ more time and if TTL
expires again RERR message would be generated to discard
the routes through nodes mentioned in RREP.
      else
        for i=1 to N, if IN!= D, every IN increments
hopcount by 1 and forward toward D do
        When D receives RCNF+ new RREQ, it
decapsulate by pri D and pub S
        if field values of RCNF== field values of new
                   RREQ then
          Send RREP to S
        else
            Send RERR to S for discarding nodes in
                   suggested path
        end if
      end if
    end if
  end if
end if
```

Now data is transferred using private key of sender and public key of receiver, that get decrypted at other end, as in **secure data transfer phase**.

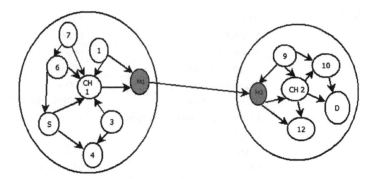

Fig. 4 Inter cluster communication in the presence of malicious node M1 and M2

Table 2 Details of terms used in algorithm used in communication phase

Term	Details
RREQ	Route request packet
N	Number of nodes
S	Source node
D	Destination node
destseqnum	Destination sequence number
RREP	Route reply packet
RCNF	Route confirm packet
TTL	Time to live
RERR	Route error packet
IN	Intermediate node

4 Performance Evaluation

In this section, we apply ID based signature to detect wormhole attack, which is simulated using Qualnet. We have used AODV with certain modifications as routing protocol. The parameters used in simulation are as in Table 3.

4.1 Performance Metrics

The metrics that have been used to evaluate performance of performance of proposed scheme are:

Packet Delivery Ratio (PDR): It is the ratio of total number of packets being sent by a source node and total number of packets that were successfully received at destination node.

End-to-end delay: It is the average time taken by a data packet that were successfully delivered to a destination end.

Table 3 Simulation parameters

Parameter	Value
Simulator	Qualnet
Simulation area	500 × 500
Number of nodes	10–80
Simulation time	600 s
Routing protocol	AODV, SAODV
Maximum speed traffic agent	15 m/s TCP
Pause time	6 s
Node speed	2–10 m/s
Packet size	512 bytes
Transmission range	250 m
Mobility model	Random waypoint
No. of malicious nodes	2

Throughput: It is the ratio of total number of data packets successfully transferred to a destination and total simulation time.

4.2 Simulation Results

Figure 5 shows that comparison graph between AODV with wormhole attack, SAODV and proposed scheme using packet delivery ratio (PDR) as a performance metric. The simulation result shows that proposed scheme having 95.11 % while AODV with wormhole attack and SAODV having 89.23 and 94.4 % for average packet delivery ratio. The simulation result shows that proposed scheme is better than other two schemes. Figure 6 shows that comparison graph for above three schemes with respect to average end to end delay.

The output of simulation results are 171.46 ms for AODV with wormhole attack, 160.21 ms for SAODV and 140.13 ms for proposed scheme. Hence proposed scheme performs better than other two existing schemes. The outcome of simulation is calculated for throughput for all existing schemes in Fig. 7. AODV with

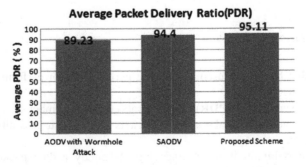

Fig. 5 Packet delivery ratio

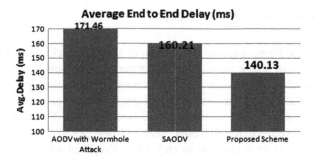

Fig. 6 Average end to end delay

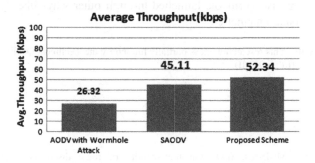

Fig. 7 Average throughput

wormhole attack has 26.32 Kbps, SAODV has 45.11 Kbps while proposed scheme has 52.34 Kbps for throughput parameter.

4.3 Comparison of Security Goals

Wormhole attack has proven to be fatal for many security measures. Several researches have been done in this field. The comparison between proposed scheme and existing scheme is given in Table 4.

Table 4 Comparison of security features among proposed solution and other schemes

S. n. parameters	AODV	SAODV	Proposed scheme
1. Authentication	No	Yes	Yes
2. Secrecy	No	No	Yes
3. Data integrity	No	Yes	Yes
4. Non-repudiation	No	Yes	Yes
5. Forward secrecy	No	No	Yes
6. Backward secrecy	No	No	Yes
7. Group key secrecy	No	No	Yes

5 Conclusion and Future Scope

Wormhole attack forms a tunnel like structure to relay packets among malicious nodes. This type of attack may cause disturbances in communication by causing selective forwarding, fabrication and alteration of packets being sent. In this paper, we have proposed a way to protect network from wormhole attack by using identity based signature scheme on cluster based ad hoc network. The simulation results show that proposed scheme is better than existing scheme in terms of packet delivery ratio, throughput and end to end delay. Proposed scheme protects wormhole attack that are launched by packet replay but still there are scope of enhancement in it such that it can be upgraded to avoid wormhole and it can be made to work against wormhole launched through other ways like by encapsulation, by out-of-band channel etc.

Acknowledgments This research work is partially funded by the Technical Quality Improvement Programme Phase II (TEQIP-II).

References

1. Burmeste, M., Medeiros, B.D.: On the security of route discovery in MANETs. In: Proceedings of IEEE Transactions on Mobile Computing, vol. 8, pp. 1180–1188. IEEE Press (2009)
2. Howarthy, N.M.P.: A survey of MANET intrusion detection prevention approaches for network layer attacks. In: Proceedings of IEEE Communications Surveys Tutorials, pp. 1–19 (2013)
3. Bar, R.K., Mandal, J.K., Singh, M.M.: QoS of MANET through trust based AODV routing protocol by exclusion of blak hole attack. In: Proceedings of Elsevier Technology Elsevier, pp. 530–537 (2013)
4. Argyroudis, P.G., Mahony, D.O.: Secure routing for mobile ad hoc networks. In: Proceedings of IEEE Communications Surveys and Tutorials. Third Quarter, vol. 7, pp. 2–21. IEEE Press (2005)
5. Hu, Y.-C., Perrig, A.: A survey of secure wireless ad hoc routing. In: Proceedings of IEEE Security and Privacy, p. 2839. IEEE Press (2004)
6. Phuong, T.V., Trong Canh, N., Lee, Y.-K., Lee, S., Lee, H.: Transmission time-based mechanism to detect wormhole attack. In: Proceedings of IEEE Asia-Paci_c Service Computing Conference (2007)
7. Sharma, P., Trivedi, A.: An approach to defend against wormhole attacks in ad hoc network using digital signature. In: Proceeding of IEEE (2011)
8. Maulik, R., Chaki, N.: A study on wormhole attacks in MANET. In: Proceeding of International Journal of Computer Information Systems and Industrial Management Applications. ISSN 2150-7988, vol. 3, pp. 271–279 (2011)
9. Shamir, A.: Identity-base cryptosystems and signature schemes. Proc. LNCS **196**, 47–53 (1985)
10. Guerrero Zapata, M., Asokan, N.: Securing ad hoc routing protocols. In: Proceedings ACM Workshop on Wireless Security (WiSe), pp. 1–10. ACM Press (2002)
11. Hu, Y.C., Perrig, A., Johnson, D.B.: Wormhole detection in wireless ad-hoc networks. In: Department of Computer Science, Rice University, Technical Report TR01-384, June (2002)

12. Zhen, J., Srinivas, S.: Preventing replay attacks for secure routing in ad hoc networks. In: Proceeding of Ad Hoc Networks Wireless (ADHOCNOW' 03), pp. 140–150 (2003)
13. Hu, L., Evans, D.: Using directional antennas to prevent wormhole attacks. In: Proceedings of Network and Distributed System Security Symposium, p. 13141 (2004)
14. Khalil, I., Bagchi, S., Shro, N.B.: LITEWORP: a lightweight countermeasure for the wormhole attack in multi-hop wireless networks. In: Proceedings of the International Conference on Dependable Systems and Networks, pp. 612–621 (2005)
15. Tran, P.V., Hung, L.X., Lee, Y.K., Lee, S., Lee, H.: TTM: an efficient mechanism to detect wormhole attacks in wireless ad-hoc networks. In: Proceedings of 4th IEEE Consumer Communication and Networking Conference (CCNC07), pp. 593–598 (2007)
16. Sharma, P., Trivedi, A.: An approach to defend against wormhole attack in ad hoc network using digital signature. In: Proceedings of 3rd IEEE International Conference on Communication Soft-ware and Networks (ICCSN)

Surface Grinding Process Optimization Using Jaya Algorithm

R. Venkata Rao, Dhiraj P. Rai and Joze Balic

Abstract Optimization problem of an important traditional machining process namely surface grinding is considered in this work. The performance of machining processes in terms of cost, quality of the products and sustainability of the process is largely influenced by its process parameters. Thus, choice of the best (optimal) combination machining parameters is vital for any machining process. Hence, in present work a new algorithm is used for solving the considered optimization problem. The Jaya algorithm is a simple yet powerful algorithm and is a algorithm-specific parameter-less algorithm. The comparison of results of optimization show that the results of Jaya algorithm are better than the results reported by previous researchers using GA, SA, ABC, HS, PSO, ACO and TLBO.

Keywords Surface grinding · Optimization · Jaya algorithm

1 Introduction

To minimize the machining time, improve quality of the product and reduce the cost of production determining the optimal setting of parameters of a process is important. Researchers in the past had applied various optimization algorithms [1] to determine the best setting of parameters for machining processes.

R.V. Rao (✉) · D.P. Rai
Department of Mechanical Engineering, S.V. National Institute of Technology,
Surat 395007, Gujarat, India
e-mail: ravipudirao@gmail.com

D.P. Rai
e-mail: dhiraj.p.rai@gmail.com

J. Balic
Faculty of Mechanical Engineering, Institute for Production Engineering,
University of Maribor, Maribor, Slovenia
e-mail: joze.balic@uni-mb.si

© Springer India 2016   487
H.S. Behera and D.P. Mohapatra (eds.), *Computational Intelligence
in Data Mining—Volume 2*, Advances in Intelligent Systems
and Computing 411, DOI 10.1007/978-81-322-2731-1_46

The working of any evolutionary or swarm based optimization algorithms is governed by its control parameters which can be classified as common control parameters and algorithm specific parameters. Selection of control parameters is a critical task, as it influences the performance of the algorithm. Rao et al. [2] proposed the TLBO algorithm. Algorithm-specific parameters are not required for the working of TLBO algorithm. The TLBO algorithm has gained wide acceptance among the optimization researchers [2].

Keeping in view of the success of the TLBO, another algorithm-specific parameter-less algorithm is proposed very recently by Rao [3]. However, unlike two phases of the TLBO algorithm, the proposed algorithm has only one phase and it is comparatively simpler to apply. The working of the proposed algorithm is much different from that of the TLBO algorithm. The Jaya algorithm has also proved its effectiveness in solving a number of constrained and unconstrained benchmark functions [3].

Thus, the optimization problem considered in this work is solved using the Jaya algorithm. The Jaya algorithm is described in Sect. 2. A computer program for Jaya algorithm is prepared using MATLAB r2009a. A computer system with a 2.93 GHz processor and 4 GB random access memory is used for execution of the program.

2 The Jaya Algorithm

Let us suppose that a function $y(x)$ is to be minimized (or maximized). At any iteration i, if the no. of design variables are 'm', no. of candidate solutions are 'n' ($k = 1, 2,\ldots,n$) then $best$ is the candidate solution which provides the best objective function value (i.e. $y(x)^{best}$) among all the candidate solutions and $worst$ is the candidate solution which provides the worst objective function value (i.e. $y(x)^{worst}$) among all the candidate solutions. The jth variable for the kth candidate during the ith iteration (i.e. $x^{j,k,i}$) is modified as per Eq. (1).

$$x_{new}^{j,k,i} = x^{j,k,i} + r_1^{j,i}\left(x^{j,best,i} - abs(x^{j,k,i})\right) - r_2^{j,i}\left(x^{j,worst,i} - abs\left(x^{j,k,i}\right)\right) \qquad (1)$$

In Eq. (1), $x^{j,best,i}$ represents the value of the jth variable corresponding to the $best$ candidate; $x^{j,worst,i}$ represents the value of the jth variable corresponding to the $worst$ candidate; $x_{new}^{j,k,i}$ represents the modified $x^{j,k,i}$; $r_1^{j,i}$ and $r_2^{j,i}$ are the two random numbers for the jth variable during the ith iteration in the range [0, 1]. The term "$r_1^{j,i}((x^{j,best,i} - abs(x^{j,k,i}))$)" indicates the tendency of the solution to move closer to the best solution and the term "$-r_2^{j,i}(x^{j,worst,i} - abs(x^{j,k,i}))$" indicates the tendency of the solution to avoid the worst solution. If the function value given by $x_{new}^{j,k,i}$ is better than the function value given by $x^{j,k,i}$ then $x_{new}^{j,k,i}$ replaces $x^{j,k,i}$. At the end of iteration all the function values that are accepted are given as input to the next iteration.

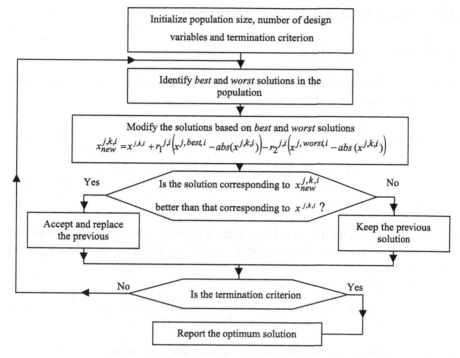

Fig. 1 Flow chart for the Jaya algorithm

The Jaya algorithm is explained with a help of a flowchart shown in Fig. 1. The algorithm always tries to get closer to success (i.e. reaching the best solution) and tries to avoid failure (i.e. moving away from the worst solution). More details and the demonstration of the algorithm can be obtained from: https://sites.google.com/site/Jayaalgorithm/.

3 Optimization of Rough and Finish Grinding Processes

In this work the analytical models developed by Wen et al. [4] for cost of production 'C_T' (\$/pc.), workpiece removal parameter 'WRP' (mm³/min.N) and surface roughness 'R_a' (μm) are considered to formulate the optimization problem. The process parameters considered in this work are same as those considered by Wen et al. [4].

3.1 Rough Grinding Process Optimization

3.1.1 Objective Functions

The objective functions for cost of production and workpiece removal parameter are
expressed by Eqs. (2) and (3).

$$
\begin{aligned}
\text{minimize } C_T ={}& \frac{M_c}{60p}\left(\frac{L_w+L_e}{V_w 1000}\right)\left(\frac{b_w+b_e}{f_b}\right)\left(\frac{a_w}{a_p}+S_p+\frac{a_w b_w L_w}{\pi D_e b_s a_p G}\right)+\frac{M_c}{60p}\left(\frac{S_d}{V_r}+t_1\right) \\
&+\frac{M_c t_{ch}}{60N_t}+\frac{M_c \pi b_s D_e}{60p N_d L V_s 1000}+C_s\left(\frac{a_w b_w L_w}{pG}+\frac{\pi(doc)b_s D_e}{pN_d}\right)+\frac{C_d}{pN_{td}}
\end{aligned}
\tag{2}
$$

Due to space limitation the details of all the constants considered in this work are
not provided in this paper. However, readers may refer to [4] for the description and
values of the constants.

$$
\text{maximize } \quad WRP = 94.4\,\frac{(1+(2doc/3L))L^{11/19}(V_w/V_s)^{3/19}V_s}{D_e^{43/304}VOL^{0.47}d_g^{5/38}R_c^{27/19}}
\tag{3}
$$

In order to satisfy both the objectives simultaneously, a combined objective function (COF) is formed by assigning equal importance to both the objectives. The
COF for rough grinding process is expressed by Eq. (4)

$$
\text{minimize } COF = w1\,\frac{CT}{CT^*}-w2\,\frac{WRP}{WRP^*}
\tag{4}
$$

Where, $w1 = w2 = 0.5$; $CT^* = 10$ ($/pc); $WRP^* = 20$ (mm^3/min. N)

3.1.2 Constraints

(a) Thermal damage constraint [4]

$$
\begin{aligned}
U ={}& 13.8+\frac{9.64\times10^{-4}V_s}{a_p V_w}+\left(6.9\times10^{-3}\frac{2102.4V_w}{D_e V_s}\right) \\
&\times\left(A_0+\frac{K_u V_s L_w a_w}{V_w D_e^{1/2} a_p^{1/2}}\right)\frac{V_s D_e^{1/2}}{V_w a_p^{1/2}}
\end{aligned}
\tag{5}
$$

$$
U^* = 6.2+1.76\left(\frac{D_e^{1/4}}{a_p^{3/4}V_w^{1/2}}\right)
\tag{6}
$$

The thermal damage constraint is expressed by Eq. (7)

$$U^* - U \geq 0 \tag{7}$$

(b) Wheel wear parameter (WWP) constraint [4]

$$WWP = \left(\frac{k_p a_p d_g^{5/38} R_c^{27/29}}{D_c^{1.2/VOL-43/304} VOL^{0.38}} \right) \times \frac{(1 + (doc/L))L^{27/19}(V_s/V_w)^{3/19}V_w}{(1 + (2doc/3L))} \tag{8}$$

The WWP constraint is expressed as follows

$$\frac{WRP}{WWP} - G \geq 0 \tag{9}$$

(c) Machine tool stiffness constraint (MSC) [4]

$$K_c = \frac{1000V_w f_b}{WRP} \tag{10}$$

$$K_s = \frac{1000V_s f_b}{WWP} \tag{11}$$

$$MSC - \frac{|R_{em}|}{K_m} \geq 0 \tag{12}$$

where,

$$MSC = \frac{1}{2K_c}\left(1 + \frac{V_w}{V_s G}\right) + \frac{1}{K_s} \tag{13}$$

(d) Surface roughness constraint [4]
The surface roughness constraint is as given by Eq. (17).

$$R_a = 0.4587 \, T_{ave}^{0.30} \text{ for } 0 < T_{avg} < 0.254 \text{ else,} \tag{14}$$

$$R_a = 0.7866 \, T_{ave}^{0.72} \text{ for } 0.254 < T_{avg} < 2.54 \tag{15}$$

$$T_{avg} = 12.5 \times 10^3 \frac{d_g^{16/27} a_p^{19/27}}{D_e^{8/27}}\left(1 + \frac{doc}{L}\right)L^{16/27}\left(\frac{V_w}{V_s}\right)^{16/27} \tag{16}$$

$$R_a \leq 1.8 \, \mu m \tag{17}$$

3.1.3 Parameter Bounds

$$1000 \leq V_s \leq 2023 \tag{18}$$

$$10 \leq V_w \leq 27.7 \tag{19}$$

$$0.01 \leq doc \leq 0.137 \tag{20}$$

$$0.01 \leq L \leq 0.137 \tag{21}$$

An initial-population of 20 and maximum no. of generations equal to 100 are chosen for the Jaya algorithm in the present work. This combination of common control parameters is chosen based on rigorous experimental trials by considering initial-population sizes of 10, 20, 30, 40, 50, 70, 100 but maintaining the same no. of function evaluations for each experimental trial. The combination of common control parameters which yields the best result in less no. of generations is then chosen.

The solution obtained using the Jaya algorithm is reported in Table 1 along with the solution obtained by previous researchers using other advanced optimization algorithms. Figure 2 shows the convergence graph of Jaya algorithm for rough grinding process. It is observed that the Jaya algorithm required only 40 function evaluations to converge at the minimum value of *COF* (i.e. -0.672) while the other algorithm such as genetic algorithm (GA) [5] required a initial-population size of 20 and no. of generations equal to 100, ants colony optimization (ACO) [6] considered a population size of 20 but the no. of generations required by ACO was not reported in [6], particle swarm optimization (PSO) [7] required 30 to 40 iterations for its convergence, simulated annealing (SA) [8], artificial bee colony (ABC) [8] and harmony search (HS) [8] required 75, 65 and 62 generations, respectively for convergence and teaching-learning-based optimization algorithm (TLBO) [9] required 30 generations to converge at the minimum value of combined objective

Table 1 Results of Jaya algorithm and those other techniques (rough grinding)

Technique	V_s	V_w	doc	L	C_T ($/pc)	WRP (mm^3/N)	COF
QP [4]	2000	19.96	0.055	0.044	6.2	17.47	-0.127
GA [5]	1998	11.30	0.101	0.065	7.1	21.68	-0.187
ACO [6]	2010	10.19	0.118	0.081	7.5	24.20	-0.230
PSO [7]	2023	10.00	0.110	0.137	8.33	25.63	-0.224
SA [8]	2023	11.48	0.089	0.137	7.755	24.45	-0.223
HS [8]	2019.35	12.455	0.079	0.136	7.455	23.89	-0.225
ABC [8]	2023	10.973	0.097	0.137	7.942	25.00	-0.226
TLBO [9]	2023	11.537	0.0899	0.137	7.742	24.551	-0.226
Jaya	2023	22.7	0.137	0.01	**7.689**	**42.284**	**-0.672**

Fig. 2 Convergence graph of Jaya algorithm for rough grinding process

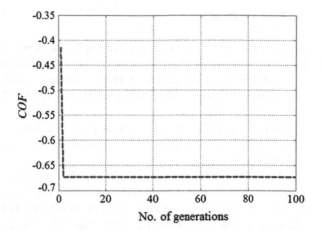

function. In order to reproduce the convergence graph of GA, ACO, PSO, SA, ABC, HS and TLBO in the same graph along with the convergence graph of Jaya algorithm, the information related to value of *COF* obtained by GA, ACO, PSO, SA, ABC, HS and TLBO after the end of every generation is required which is not available in literature [5–9]. For convergence graphs of these algorithms readers may refer to [5–9].

Figure 2 shows the convergence graph of Jaya algorithm for rough grinding process It is observed that convergence graph reduces continuously without getting trapped into local optima until it reaches the minimum value of combined objective function. This shows that the Jaya algorithm is robust, has a good exploration and exploitation capability and is effective in solving multi-objective optimization problems.

3.2 Finish Grinding Process Optimization

3.2.1 Objective Functions

The objective functions for *CT* and R_a for finish grinding process are same as those expressed by Eq. (2) and Eqs. (14)–(16), respectively. In order to satisfy both the objectives simultaneously a *COF* for finish grinding process is formulated and is expressed by Eq. (22).

$$\text{minimize } COF = w1 \ \frac{CT}{CT^*} + w2 \ \frac{R_a}{R_a^*} \tag{22}$$

Where, $w1 = 0.3$ and $w2 = 0.7$; $CT^* = 10$ ($/pc); $R_a^* = 1.8$ (µm)

3.2.2 Constraints

The thermal damage constraint, *WWP* constraint, *MSC* is same as those described in Sect. 3.1. In addition, the constraint on the *WRP* is considered i.e. *WRP* ≥ 20. Due to space limitation the details of all the constants considered in this work are not provided in this paper. However, readers may refer to [4] for the description and values of the constants.

An initial-population of 10 is considered. Maximum no. of function evaluations equal to 2000 is considered. The solution obtained using the Jaya algorithm is reported in Table 2.

The solutions obtained using QP and GA are also reported in Table 2. Jaya algorithm achieved a lower value of *COF* (i.e. 0.522) compared to the other techniques. The Jaya algorithm required on 470 function evaluations (i.e. 47 generations) to converge at the minimum value of *COF*. Figure 3 shows the convergence graph of Jaya algorithm.

Table 2 Results of Jaya algorithm and those other techniques (finish grinding)

Technique	V_s	V_w	*doc*	*L*	C_T	R_a	*COF*
QP [4]	2000	19.99	0.052	0.091	7.7	0.83	0.554
GA [5]	1986	21.40	0.024	0.136	7.37	0.827	0.542
Jaya	2023	22.7	0.011	0.137	**7.13**	**0.7937**	**0.522**

Fig. 3 Convergence graph of Jaya algorithm for finish grinding process

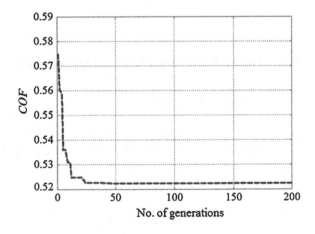

4 Conclusions

In this work the surface grinding optimization problem is solved using Jaya algorithm. The Jaya algorithm produced better results than other techniques such as QP, GA, ACO, PSO, SA, ABC, HS and TLBO in terms of *COF* value with a high convergence rate. Thus, the Jaya algorithm is effective in solving optimization problems in which multiple objectives are considered simultaneously.

Future work may focus on development of a posteriori version of Jaya algorithm and the same may be applied to solve the optimization problems of other traditional and modern machining processes, considering multiple objectives simultaneously. The Jaya algorithm may also be used to solve the optimization problems of other manufacturing processes such as casting, welding, forming, etc.

Acknowledgments The Authors are thankful to the Ministry of Science and Technology of India and the Slovenian Research Agency (ARRS), Ministry of Education, Science and Sport of Slovenia for providing the financial support for the project entitled "Optimization of Sustainable Advanced Manufacturing Processes".

References

1. Rao, R.V., Kalyankar, V.D.: Optimization of modern machining processes using advanced optimization techniques: a review. Int. J. Adv. Manuf. Technol. **73**, 1159–1188 (2014)
2. Rao, R.V.: Teaching-learning-based optimization (TLBO) algorithm and its engineering applications. Springer, London (2015)
3. Rao, R.V.: Jaya-a simple and new optimization algorithm for solving constrained and unconstrained optimization problems. Int. J. Indus. Eng. Comput. **7**(1), 19–34 (2016)
4. Wen, X.M., Tay, A.A.O., Nee, A.Y.C.: Microcomputer based optimization of the surface grinding process. J. Mater. Process. Tech. **29**, 75–90 (1992)
5. Saravanan, R., Asokan, P., Sachidanandam, M.: A multiobjective genetic algorithm approach for optimization of surface grinding operations. Int. J. Mach. Tool. Manuf. **42**, 1327–1334 (2002)
6. Baskar, N., Saravanan, R., Asokan, P., Prabhaharan, G.: Ants colony algorithm approach for multi-objective optimization of surface grinding operations. Int. J. Adv. Manuf. Technol. **23**, 311–317 (2004)
7. Pawar, P.J., Rao, R.V., Davim, J.P.: Multi-objective optimization of grinding process parameters using particle swarm optimization algorithm. Mater. Manuf. Process. **25**(6), 424–431 (2010)
8. Rao, R.V., Pawar, P.J.: Grinding process parameter optimization using non-traditional optimization algorithms. J. Eng. Manuf. **224**(6), 887–898 (2010)
9. Rao, R.V., Pawar, P.J.: Parameter optimization of machining processes using teaching learning based optimization algorithm. Int. J. Adv. Manuf. Technol. **67**, 995–1006 (2013)

BITSMSSC: Brain Image Tomography Using SOM with Multi SVM Sigmoid Classifier

B. Venkateswara Reddy, A. Sateesh Reddy and P. Bhaskara Reddy

Abstract Image segmentation is a process of elevating the objects by partitioning a digital image into multiple segments. To analyze an image, segmentation is the best process to follow. Especially, for detecting tumours from medical images such as brain, skin and breast in the field of medicine. To improve the results of brain images PSNR, ENTROPY image fusion technique is applied. Here the segmentation process is carried out by k-means clustered model algorithm. Then the entire data base is subjecting to classification mode under multi class svm sigmoid classifier. This generates a descent output of 96 % accurate results using various texture features they are contrast, energy, area of the tumour detecting by the cluster model and entropy. These parameters helps in identifying the tumour detection from brain MRI, CT scanned images.

Keywords Image segmentation · MRI · CTSCAN · SVM · Sigmoid

1 Introduction

Contour extraction and edge detection is performed using Kohonen self organising map which is a neural network process. Kohonen algorithm is applied on medical images in this process to prove the convergence of the algorithm in [1, 11] i.e., limiting in the computational calculation and amount of neurons for the network.

B. Venkateswara Reddy (✉) · A. Sateesh Reddy
Department of Electronics and Communication Engineering, Vikas College of Engineering and Technology, Nunna, Vijayawada Rural 521212, AP, India
e-mail: bheemireddyv@gmail.com

A. Sateesh Reddy
e-mail: sateeshreddy.eldt@gmail.com

P. Bhaskara Reddy
Department of Electronics and Communication Engineering, MLR Institutions, Dundigal Village, Hyderabad 500043, AP, India
e-mail: pbhaskarreddy@rediffmail.com

© Springer India 2016
H.S. Behera and D.P. Mohapatra (eds.), *Computational Intelligence in Data Mining—Volume 2*, Advances in Intelligent Systems and Computing 411, DOI 10.1007/978-81-322-2731-1_47

Fig. 1 Output of a 20 × 20
network, **a** and **b** different
distributions of neurons

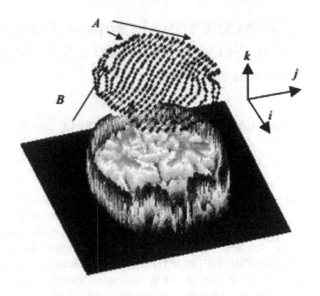

These implementations were done by using (1) and (2) and the extraction can be
explained using Fig. 1.

$$\|x(t) - m_c(t)\| = \min_i \|x(t) - m_i(t)\|$$ (1)

$$m_i(t+1) = m_i(t) + \propto (t)[x(t) - m_i(t)]i \in N_e$$

$$m_i(t+1) = m_i(t)i \notin N_e,$$ (2)

where, for time t: x is the input, m_i is any node, m_c is the winner, α is the gain
sequence and N_e is the neighbourhood of the winner.

The color image segmentation is followed in the proposed work of [2–4] is SOM
K-means clustering in combination of saliency mapped and helps the image segmented output and compared with brain image data base. Figure 2 indicates the
flow chart with saliency mapped output.

The brain images were subjected to the fusion process to increase the entropy
and psnr values of the image, which are the texture feature values. Entropy is
defined as the total information present in the data. PSNR indicates the ratio
between received signal and noise in the signal. A wavelet fusion process is
approached for fusion technique which improves the texture values of the image
and the results were shown and explained in the Table 2. The entire process is
explained in the block diagram of schematic. The fusion is done on the same image.

$$PSNR = 20 \log_{10} \frac{I - I_0}{MSE}$$ (3)

$$ENTROPY(H) = \log_2 2^N$$ (4)

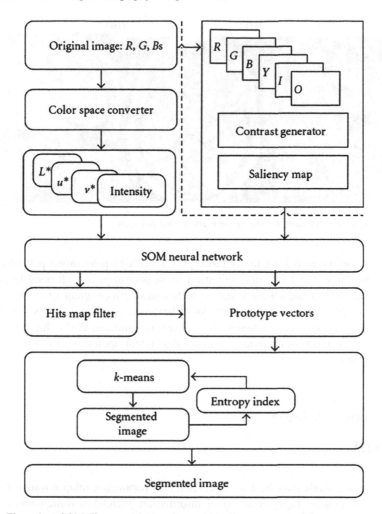

Fig. 2 Flow chart SOM–K means cluster along with saliency map module

A hybridised multi objective optimization for feature selection is introduced in [3, 7, 10, 11]. Due to this the average metric values obtained were 0.32, 0.75 and 0.69.

A mind mri will be those examine of the head which aides for picture test. Mri is those short structure for attractive reverberation imaging.

The MRI image consists of three modes of views shown in Fig. 3 [4]: a Saggital view, b Pivotal view and c Coronol view

The organised sectional flow in rest of the paper is as follows: Wavelets fusion, SOM, Multi SVM classifier, Work flow and Conclusion.

(i)   (ii)   (iii)

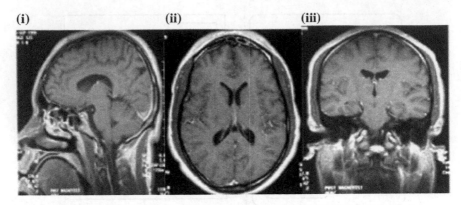

Fig. 3 [4]: **a** Saggital view, **b** Pivotal view and **c** Coronol view

Picture combination is blending those gray-level high-positioning panchromatic picture and the colored low-determination multispectral picture. It need been discovered that those standard combination techniques perform great spatially Be that as generally present ghastly twisting. With beat this problem, various multiscale convert built combination schemes have been recommended. In this paper, we concentrate on those combination strategies dependent upon those discrete wavelet convert (DWT).

2 Proposed Scheme

2.1 SOM

Self Arranging guide may be a unsupervised non-parametric relapse transform that speaks to nonlinear, high-dimensional information ahead low-dimensional yield. The data information focuses would mapped will som units Typically around a absolute alternately two-dimensional grid. The mapping may be used to gain from those preparation information tests Toward An straightforward stochastic Taking in process, the place the som units need aid balanced Toward utilizing little steps to get those characteristic vectors that were concentrated from the information Furthermore exhibited one after an alternate (in a irregular order). Starting with mathematical statement (1) Also comparison (2) som will conveyed crazy. Figure 4 clarifies the pictorial representational of som and may be begun and Johnson had proceeded Eventually Tom's perusing som calculation (Table 1).

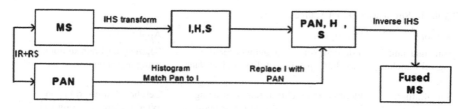

Fig. 4 Standard IHS fusion scheme

ALGORITHM FOR SOM

SOM ALGORITHM

Step 1: Select the extent Furthermore kind of the map. The shape camwood make hexagon alternately square, relying upon the state of the hubs Concerning illustration for every your prerequisite. Typically, hexagon grids would favored since every hub need 6 quick neighbours.

Step 2: Initialize the sum hub weighted vectors haphazardly.

Step 3: Decide irregular information focuses starting with preparation dataset Also submit it of the som.

Step 4: "Best matching Unit" (BMU)

-It is a comparative hub similitude is ascertained eventually Tom's perusing utilizing EDF (Euclidean separation formula).

Step 5: Focus the hubs inside the "neighborhood" of the BMU.

-The extent of the neighborhood diminished for each cycle.

STEP 6: Alter weights from claiming hubs in the BMU neighborhood towards the decided information point.

-The Taking in rate declines for each cycle.

-The extent of the conformity is proportional of the vicinity of the hub of the BMU.

STEP 7: Repeatable 2-5 steps for n iterations.

2.2 Multi SVM Classifier

The SVM will be characterizes Likewise those paradigm on be gazed for a choice surface that might have been maximally a wide margin from any information point inside the information set. This separation from choice surface of the closest information point starting with information set draws those edge of the classifier. This system for development edge by that choice capacity to an SVM will be

Table 1 Summary of other sources

Author	Description	Application
Shirsagar and Jagruti Panchal [5]	4 diverse sorts arrangement features encoding spatial anatolian dialect data	Segmenting brain tumours using alignment
Selvakumar, Lakshmi and Arivoli [6]	Identifying cerebrum tumor utilizing K-mean grouping division Furthermore arranged by fluffy C-means calculation	Cerebrum tumor division What's more Its ae figuring clinched alongside mind mr pictures utilizing K-Mean grouping What's more fluffy C-mean algorithm
Pham and Hopkins [7]	With distinguish edges of tissues Previously, cerebrum pictures fluffy C-means utilizing an destination capacity	Unsupervised tissue arrangement On therapeutic pictures utilizing edge versatile grouping
Kavith [8]	Utilizing both bolster ahead neural system RBF neural system with altered developing	An productive methodology to cerebrum tumor identification dependent upon changed area developing and neural system over mri pictures

completely specified by that little subset of information that characterizes those positions of the separator. These focuses would alluded with make the help vectors for a vector space, a perspective cam wood make considered perfect Similarly as An vector between the sources toward that perspective [9].

SVM classifier relies looking into an expansive edge around the choice limit. Contrasted with an unequivocal hyperplane What's more should spot a even separator between classes with fewer decisions for the place it could a chance to be held. Concerning illustration an aftereffect for this, the memory limit of the model need been decreased, What's more Consequently its capacity should effectively sum up the test information may be expanded.

To examine SVM classifier part capacities were utilized. Assuming that the portion capacities need aid not furnished At that point the order portion is known as SVM classifier mode. SVM classifier utilizing part capacities are LINEAR, POLYNOMIAL, RBF What's more SIGMOID (Figs. 5, 6, 7).

$$Linear \; : \; K(x_i, x_j) = x_i \cdot x_j \tag{5}$$

$$Polynomial \; : \; K(x_i, x_j) = (\gamma x_i \cdot x_i + C)^\alpha \tag{6}$$

$$RBF \; : \; K(x_i, x_j) = \exp(-\gamma |x_i - x_j|^2) \tag{7}$$

$$SIGMOID \; : \; K(x_i, x_j) = \tanh(\gamma x_i \cdot x_j + C) \tag{8}$$

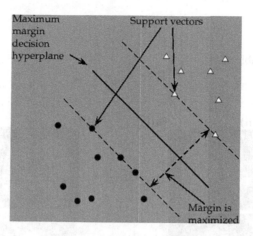

Fig. 5 Support vectors with margin classifiers for machine training

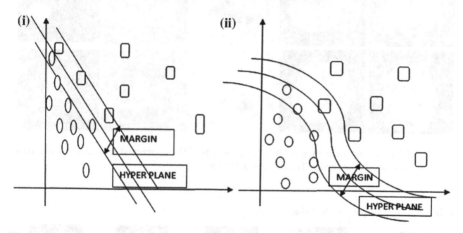

Fig. 6 Support vector Machine: **a** 2-D space linear support vector machine and **b** 2-D space non-linear support vector machine

ALGORITHM FOR MULTICLASS SVM

MULTI CLASS SVM PROCEDURE

INPUT Parameters $\{(\overline{x}_1, y_1), \ldots, (\overline{x}_m, y_m)\}$.

Initialize $\overline{\tau}_1 = 0, \ldots, \overline{\tau}_m = \overline{0}$.

Loop:

1. Choose an example 'p'.

2. Calculate the constants for the reduced problem:

- $Ap = K(\overline{X}_P, \overline{X}_P) \ldots (KERNEL\ FUNCTION)$

- $B_P = \sum_{I \neq P} K(X_I, X_P)\tau_I - \beta l$

3. Set $\overline{\tau}_p$ to be the solution of the reduced problem

Output: $H(X) = argmax_{r=1}^{k} \left\{ \sum_{l} \tau_{i,r} K(x, x_i) \right\}$

3 Results

See Fig. 7 and Table 2.

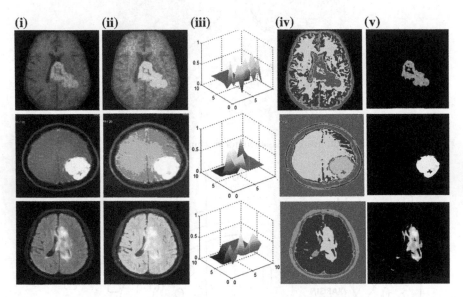

Fig. 7 MRI cross sectional images: **a** input image, **b** SOM segmented image, **c** Energy band graph, **d** K-means cluster and **e** Tumour extracted image

Table 2 Different parametric values

Fused image				
RMS	0.013559	0.095143	0.046042	0.012235
Entropy	6.8372	5.9159	7.0816	6.4666
Co-coeff	0.99947	0.99928	0.99895	0.99971
PSNR	85.4861	68.5632	74.8676	86.3784
STD DEV	0.2148	0.2204	0.3092	0.2789
Spatial frequency	0.0372	0.0306	0.1253	0.0371
Median frequency	0.1478	0.0175	0.0984	0.0578

4 Conclusion

Segmentation using SOM-K means cluster identifies the tumour part along with separating the tumour part from the image and the elevation of tumour part is shown in the results section, 3d view of energy elements is also showcased in the results. This helps most effectively and easily identifying the tumour in MRI images, tumour detection from the images and the area of spreading tumour in the MRI images or CT images, this will also provide at what stage the tumour is in and classification process helps whether the image is tumour or non-tumour image. This technology helps in identifying tumour and reduces the time for classification and detection of tumour spread area. Various texture features studied here are Energy, Contrast and area of the tumour that helps in the classification mode. The cluster model helps in detecting the area of the tumour. This provides the best accurate results of 96 % and this position of NEURAL NETWORKS with Multi class SVM classification is a novel idea to deal with.

References

1. Carlos, C.: Image segmentation with kohonen neural networks self organising maps. Comput. Sci. J. **20**, 24 (2002)
2. Chi, D.: Self organising Map-based color Image segmentation with K-means clustering and saliency map. Hindawi publications, vol. 2011 (2014)
3. Oritz, A., Gorriz, J.M., Ramirez, J., Salas-Gonzalenz D.: MR Brain image segmentation by growing hierarchical SOM and probability clustering. Electron. Lett. **47**, 585–586 (2011)
4. Wayne state university, Radiologic anatomy, http://www.med.wayne.edu/diagradiology/anatomy_modules/brain/brain.html
5. Shirsagar, V.K., Panchal, J.: Segmentation of brain tumour and its area calculation. Int. J. Adv. Res. Comput. Sci. Softw. Eng. (2014)
6. Selvakumar, J., Lakshmi, A., Arivoli, T.: Brain tumour segmentation and its ae calculation in brain MR images using K-mean clustering and fuzzy C-mean algorithm. In: IEEE conference (2012)
7. Pham, D.L., Hopkins, J.: Unsupervised tissue classification in medical images using edge-adaptive clustering. IEEE Trans. **1**, 634–637 (2003)
8. Kavitha, C.: An efficient approach for brain tumour detection based on modified region growing and neural network in MRI images, vol. 1. pp. 1087–1095 (2012)
9. http://nlp.stanford.edu/IR-book/html/htmledition/support-vector-machines-the-linearly-separable-case-1.html
10. Semenov, S., Seiser, B., Stoegmann, E., Auff, E.: Electromagnetic tomography for brain imaging: from virtual to human brain. In: IEEE Conference on Antenna Measurements and Applications (CAMA) (2014)
11. Abdulbaqi, H.S., Mat, M.Z., Omar, A.F., Bin Mustafa, I.S., Abood, L.K.: Detecting brain tumor in magnetic resonance images using hidden markov random fields and threshold techniques. In: IEEE Student Conference on Research and Development (SCOReD) (2014)

Non-Linear Classification using Higher Order Pi-Sigma Neural Network and Improved Particle Swarm Optimization: An Experimental Analysis

D.P. Kanungo, Janmenjoy Nayak, Bighnaraj Naik and H.S. Behera

Abstract In this paper, a higher order neural network called Pi-Sigma neural network with an improved Particle swarm optimization has been proposed for data classification. The proposed method is compared with some of the other classifiers like PSO-PSNN, GA-PSNN and only PSNN. Simulation results reveal that, the proposed IPSO-PSNN outperforms others and has better classification accuracy. The result of the proposed method is tested with the ANOVA statistical tool, which proves that the method is statistically valid.

Keywords Higher order neural network · Classification · PSO · Pi-Sigma neural network

1 Introduction

Introduced towards the end of 1980s, higher order neural network or popularly known as HONN, is the answer to the snowballing of complexity within the traditional Neural Networks. Outshining traditional Neural Networks, HONNs are found to be successfully performing a wide variety of Data Mining tasks such as classification, pattern recognition, prediction, optimization, etc. predominantly for

D.P. Kanungo (✉) · J. Nayak · B. Naik · H.S. Behera
Department of Computer Science Engineering and Information Technology,
Veer Surendra Sai University of Technology, Burla, Sambalpur
768018, Odisha, India
e-mail: dpk.vssut@gmail.com

J. Nayak
e-mail: mailforjnayak@gmail.com

B. Naik
e-mail: mailtobnaik@gmail.com

H.S. Behera
e-mail: mailtohsbehera@gmail.com

© Springer India 2016
H.S. Behera and D.P. Mohapatra (eds.), *Computational Intelligence in Data Mining—Volume 2*, Advances in Intelligent Systems and Computing 411, DOI 10.1007/978-81-322-2731-1_48

non-linear systems in a spectrum of application areas such as communication channel equalization, real time intelligent control and intrusion detection. HONNs are founded on correlators which are interconnected arrays with known relationships amongst them. They provide a higher edge to the Neural Networks by the introduction of interactions between the data as an input to the Neural Network model. HONNs are known to have built a niche for themselves owing to the fact that they have excellent capabilities not only in the terms of computation and storage but also in learning. The ability of customization of the order or structure of HONNs according to the order or structure of the problem is the reason behind this. Moreover, due to their property that the knowledge does not have to be learned, HONNs are effective and hence popular in solving problems which utilize this knowledge.

Particle Swarm Optimization or PSO is a very popular stochastic optimization technique developed by Kennedy and Eberhart in 1995 [1] which is based upon the social behaviour of flocking birds. Due to the fact that there are very few parameters to be set in PSO and also, it is faster and economical, PSO has found its application in a variety of scientific and research areas spanning across diversified fields. Merits notwithstanding, PSO is found to be lagging in a few scenarios. Inadequate in fine tuning solutions and a very slow searching around the global optimum are some of the limitations of PSO [2]. This led to the development of various refined versions of PSO like Improved PSO (IPSO), Modified PSO (MPSO), etc. In this paper, an improved version of PSO (IPSO) has been used to solve the classification problem in data mining. Simulation results reveal that classification accuracy in case of IPSO is found to be better than only PSO and GA.

The rest of the paper is divided as follows—Sect. 2 outlines some related literature survey. Some basic preliminaries like Pi-Sigma Neural Network, PSO and IPSO have been described in Sect. 3. Section 4 depicts the proposed methodology. Experimental setup and result analysis have been realized in Sect. 5. Section 6 constitutes the statistical analysis. Section 7 deals with the conclusion and future work.

2 Literature Survey

Wang and Cao [3] have investigated the exponential stability of stochastic reaction–diffusion Bi-directional Associative Memory (BAM) neural networks. The new formulated model is more general than the BAM neural networks investigated in previous works. Sermpinis et al. [4] have used HONNs for forecasting and trading the 21-day-ahead realized volatility of the FTSE 100 futures index. They have done this by benchmarking their results with those of the multi-layer perceptron (MLP) and the recurrent neural network (RNN), along with a traditional technique, Risk Metrics. Nayak et al. [5] have proposed a Firefly based higher order neural for data classification for maintaining fast learning and avoids the exponential increase of processing units. Also, the performance of the proposed method has been tested with various benchmark datasets from UCI machine learning repository and compared with the performance of other established models. Mehrabi et al. [6] have used an FCM-based neuro-fuzzy inference system and genetic algorithm-polynomial neural network as well as experimental data, to propose two models in order to forecast the thermal conductivity ratio of Alumina–water nano fluids. Jiang and Guo [7] have extended the permutation invariant polynomial-neural network (PIP-NN) method for constructing highly accurate potential energy surfaces (PESs) for gas phase molecules to molecule-surface interaction PESs. Oh et al. [8] have designed a polynomial-based radial basis function neural networks (P-RBF NNs) based on a fuzzy inference mechanism whose essential design parameters are optimized by means of PSO.

Panda et al. [9] have used Chou's pseudo amino acid composition along with amphiphillic correlation factor and the spectral characteristics of the protein has been used to represent protein data. They have used FLANN for the structural class prediction. With the use of hybrid gradient descent learning, Naik et al. [10] have used FLANN with hybrid PSO and GA for nonlinear data classification. They found that the proposed method works better as compared to their other considered methods. A honey bee based FLANN classifier has been proposed by Naik et al. [11] to classify real world data. They have made the comparison of the proposed method with some other classifier's performance like PSO, GA and only FLANN. Naik et al. [12] have proposed another hybrid FLANN with the use of the optimization algorithm harmony search for classification of real world data. Nayak et al. [13] have proposed Pi-Sigma neural network with the hybridization of a number of evolutionary algorithm like GA, PSO for effective data classification. They have also proposed a hybrid method [14] of both GA and PSO combined with Pi-Sigma neural network for classification of real world data. An improved version of Pi-Sigma neural network called Jordan Pi-Sigma neural network [15] have been combined with the most popularly used evolutionary algorithm such as GA to solve the classification problem of data mining.

Inspired from the above cited literature, in this paper an attempt have been made to use a combined approach of P-sigma neural network with an improved particle swarm optimization for efficient data classification.

3 Preliminaries

3.1 Particle Swarm Optimization (PSO)

Drawing inspiration from flocking of birds and schooling of fishes, Kennedy and Eberhart developed Particle Swarm Optimization—one of the most popular meta-heuristic used till date. The birds in a flock travel in search of food to a great distance without colliding with each other by communicating necessary information with each other. Few of the advantages of PSO over other optimization algorithms are easy to use, faster results and less parameter adjustment. Due to these merits, PSO has found its usage in application areas spreading across research, scientific and engineering domains. Each particle in the swarm is assigned a random initial velocity and position. The PSO consists of particles which make movements in the search space using the earlier experience of their best position as well as the best position of the entire swarm. The strategy is to update the position of each particle based on its velocity, the best known global position in the problem space and the best position known to a particle. Both the velocity and position are updated by using the Eqs. (1) and (2).

$$V_i^{(t+1)} = V_i^{(t)} + c_1 * rand(1) * (l_{best_i}^{(t)} - X_i^{(t)}) + c_2 * rand(1) * (g_{best}^{(t)} - X_i^{(t)}) \quad (1)$$

$$X_i^{(t+1)} = X_i^{(t)} + V_i^{(t+1)} \quad (2)$$

The cognition and social behavior of particles are controlled by Eq. 1 whereas the next position of the particles is updated using Eq. 2. V_i (t) and V_i(t+1) are the velocity of ith particle at time t and t+1 in the swarm respectively, c_1 and c_2 are acceleration coefficient usually set between 0 to 2 (may be same), Xi(t) is the position of ith particle and lbest$_i$(t) and gbest(t) represent the local best particle of ith particle and global best particle respectively among local bests at time t, rand(1) generates a random value between 0 and 1.

3.2   Improved Particle Swarm Optimization (IPSO)

Although popular, PSO has some limitations like accuracy heavily depends upon the selection of parameters, the parameters are constant irrespective of the model or problem [16], inadequacy in fine tuning solutions, slow searching capability in search space, etc. Hence, the need of an improvised version of PSO raised, leading to the development of Improved Particle Swarm Optimization (IPSO). In IPSO, an inertia vector (λ) is introduced to control the global and local search behavior of the particle in a swarm. More inertia ensures a more effective global search of particles; lesser inertia weight means a more efficient local search. The value of λ has to be decreased at a faster rate initially and may be reduced gradually during subsequent iterations. The updated velocity and position can be obtained by using Eqs. (3) and (4).

$$V_i^{(t+1)} = \lambda * V_i^{(t)} + c_1 * rand(1) * (l_{best_i}^{(t)} - X_i^{(t)}) + c_2 * rand(1) * (g_{best}^{(t)} - X_i^{(t)})$$
$$(3)$$

$$X_i^{(t+1)} = X_i^{(t)} + V_i^{(t+1)} \quad (4)$$

3.3 Pi-Sigma Neural Network

Shin and Ghosh [17] developed the Pi-sigma Neural Network in the year 1992. In contrast to traditional neural networks, PSNN has multiple neurons in the output layer and has a simple structure with faster convergence speed. Here, the weights from the hidden layer to output layer are set to unity. As a result of this, the time required for training is reduced drastically [18, 19]. The hidden layer consists of summation units and the output layer consists of the product units. Each unit in hidden layer is connected to only one unit in the output layer. However, we have to assume a linear activation for the hidden units. PSNN is widely applied in various domains of research due to its simple structure and good convergence speed.

Fig. 1 shows the architecture of Pi-Sigma neural network. where $x_1, x_2 \ldots x_n$ denotes the input vectors. The output at the hidden layer hj and output layer O can be computed by Eqs. (5) and (6) respectively.

$$h_j = B_j + \sum w_{ji} x_i \tag{5}$$

Here B is the bias, the weight components are w_{ij0}, w_{ij1}, w_{ij2}, $\ldots w_{ijn}$, where i=1 ... k and k is the order of the network.

$$O = f\left(\prod_{j=1}^{k} h_j\right) \tag{6}$$

where f(•) is an suitable activation function. Here, it should be noted that the weight from the hidden layer to output layer is fixed to 1.

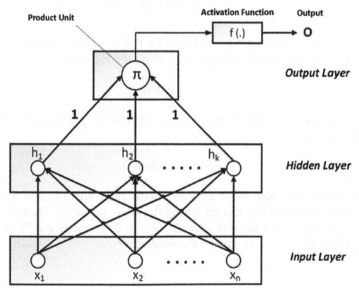

Fig. 1 Pi-sigma neural network architecture

4 Proposed Method

In this section, the proposed IPSO-PSNN has been explained with its learning strategy. The algorithm starts with the population of weight sets. The fitness of each of the individual weight sets is calculated and a local best solution is computed from the population by comparing the current and the previous best solutions. The fitness of each weight set is computed based upon the RMSE (Eq. 7) values.

$$RMSE = \sqrt{\frac{\sum_{i=1}^{n}(O_i - \hat{O}_i)^2}{n}} \tag{7}$$

In the Pi-Sigma network, a set of input is given and the aim is to compute the output in terms of a target vector. While calculating the output, the target output is compared with the expected output and the error term is computed by using Eq. (8).

$$E_j(t) = d_j(t) - O_j(t) \tag{8}$$

Where $dj^{(t)}$ is the expected output at time $(t - 1)$ and $Oj^{(t)}$ is the calculated output. The proposed IPSO-PSNN algorithm to solve the classification problem is described in Algorithm 1. The standard back propagation gradient descent learning (BP-GDL) [20] has been used to train the PSNN. The change in weight-set has been illustrated in Eq. (9).

$$\Delta w_j = \eta \left(\prod_{j \neq 1}^{m} h_{ji} \right) x_k \tag{9}$$

where h_{ji} the output of summing is layer and η is the rate of learning.

The weight-set updation is computed by using the Eq. (10).

$$w_i = w_i + \Delta w_i \tag{10}$$

To accelerate convergence of error, the momentum term α is added which can be realized in Eq. (11).

$$w_i = w_i + \alpha \Delta w_i \tag{11}$$

In Algorithm 1, the input to the algorithm is the initial population of weight sets 'P' with the bias 'B' and the data set having the target vector 't'. The output is the weight sets of the neural network 'w'. Algorithm 2 describes the Fitness From Training (FFT) algorithm to compute the fitness.

Algorithm – 1: PSO- PSNN for Classification

IPSO-PSNN (x, P, t, B)

Iter = 0;

WHILE (1)

      In the population P, the local best is calculated depending upon the fitness of each individual weight-set.

      Ic = Ic + 1; where Ic is the iteration count.

IF (Ic == 1)

           lbest = P;

ELSE

      The fitness of all the weight-sets is computed by using FFT algorithm.

      Choose the local best weight-sets.

      With the comparison between fitness of current and previous weight-sets, the 'lbest' is

generated.

END

      Based on RMSE values, the fitness of each weight-set in population is computed.

      Choose the global best weight set from the population based on fitness of all weight-sets by using the fitness vector F.

      The new velocity V_{new} of all weight-set is calculated by using eq. (3).

      The positions of all weight-sets are updated by using eq. (4).

      $P = P + V_{new}$;

      IF

           The population has 95 % similar weight sets

      or

           if maximum no. of iteration is reached,

               THEN stop iteration.

      END

END

Algorithm – 2: FFT Algorithm

---

1. **FUNCTION** F= **FFT** (x, w, t, B)

2. **FOR** i = 1 to n, n is the length of the dataset

3.   The hidden layer's output is calculated by using eq. (5)

4.   The network's output is calculated by using eq.(6).

5.   The error term is calculated by using eq. (8) and compute fitness F(i)=1/RMSE.

6. **END FOR**

7. RMSE is calculated by using eq. (7).

8. The updation in weight by using the BP-GDL algorithm is calculated by using (9).

9. Update the weight by using eq. (10).

10. The weight value can be calculated after adding the momentum term by using (11).

11. **IF** the stopping criteria like training error or maximum no. of epochs are satisfied, then Stop.

ELSE repeat the step from 2 to 11.

12. **END**

---

5 Experimental Setup and Result Analysis

The experiment has been carried out on a MATLAB 9.0 system with an Intel Core Duo CPU T5800, 2GHz processor, 2 GB RAM and Microsoft Windows-2007 OS. Five no. of standard bench mark data sets (Table 1) has been considered and those are prepared using fivefolds cross validation (Table 2) out of which fourfolds are

Table 1 Data set information

Dataset	Number of pattern	Number of attributes	Number of classes
PIMA	768	09	02
BALANCE	625	04	03
HEART	256	14	02
VEHICLE	846	18	04
ECOLI	336	07	08

Table 2 Fivefold cross validated Pima dataset

Dataset	Data files	Number of pattern	Task	Number of pattern in class-1	Number of pattern in class-2	Number of pattern in class-3
Pima	Pima -5-5trn.dat	278	Training	174	104	–
	Pima -5-1tst.dat	154	Testing	100	54	–
	Pima -5-2trn.dat	277	Training	171	106	–
	Pima -5-2tst.dat	154	Testing	100	54	–
	Pima -5-3trn.dat	278	Training	171	107	–
	Pima -5-3tst.dat	154	Testing	100	54	–
	Pima -5-4trn.dat	278	Training	167	111	–
	Pima -5-4tst.dat	153	Testing	100	53	–
	Pima -5-5trn.dat	277	Training	169	108	–
	Pima -5-5tst.dat	154	Testing	100	54	–

used for training and onefold is used for testing. The confusion matrix for classification accuracy is calculated by using Eq. (12). If cm is the confusion matrix then accuracy of classification is computed as:

$$\text{Accuracy} = \frac{\sum_{i=1}^{n} \sum_{j=1,}^{m} cm_{i,j} \atop i == j}{\sum_{i=1}^{n} \sum_{j=1}^{m} cm_{i,j}} \times 100 \qquad (12)$$

For data normalization, Min-Max normalization technique is used, which can be realized in Eq. (13).

$$v' = \frac{v - min_A}{max_A - min_A}(new\_max_A - new\_min_A) + new\_min_A \qquad (13)$$

The proposed IPSO-PSNN method has been designed for performing the classification task on various benchmark datasets like PIMA, BALANCE, HEART, VEHICLE, and ECOLI from the University of California at Irvine (UCI) machine learning repository [21]. The classification accuracies in average for all the classifiers have been presented in Table 3.

Table 3 Comparison of average performance of all classifiers

Dataset	Classification accuracy in average (%)							
	IPSO-PSNN		PSO-PSNN		GA-PSNN		PSNN	
	Train	Test	Train	Test	Train	Test	Train	Test
Pima	91.832	91.766	91.273	91.315	90.244	89.382	88.214	87.508
Balance	96.003	95.861	95.206	95.094	94.129	93.795	89.131	88.905
Heart	91.184	91.096	90.231	91.148	89.203	90.128	87.295	88.288
Vehicle	91.882	91.900	91.607	91.552	90.333	90.398	88.006	87.100
Ecoli	92.004	91.846	91.003	91.018	90.365	90.333	88.001	88.101

In both training and testing case, the proposed IPSO-PSNN performs better than the other classifiers like PSO-PSNN, GA-PSNN and PSNN.

The results of average value for each dataset after performing the fivefold cross validation [22] of the proposed IPSO-PSNN model clearly indicates that the classification accuracy of the proposed model is quite better than others in both training and testing.

6 Statistical Analysis

This section deals with a statistical test called one-way ANOVA test [23] which signifies for the obtained results statistically. ANOVA test divides the sum variability into the classifier's variability and the data sets as well as the residual (error) variability. The test being carried out using one-way ANOVA in Duncan multiple test range, with 95 % confidence interval, 0.05 significant level and linear polynomial contrast. The results are shown in Figs. 2, 3 and 4.

Descriptives

Result

	N	Mean	Std. Deviation	Std. Error	95% Confidence Interval for Mean		Minimum	Maximum
					Lower Bound	Upper Bound		
PSNN	10	88.0549	.64602	.20429	87.5928	88.5170	87.10	89.13
GA-PSNN	10	90.8310	1.70462	.53905	89.6116	92.0504	89.20	94.13
PSO-PSNN	10	91.9447	1.73164	.54759	90.7060	93.1834	90.23	95.21
IPSO-PSNN	10	92.5374	1.81528	.57404	91.2388	93.8360	91.10	96.00
Total	40	90.8420	2.29344	.36263	90.1085	91.5755	87.10	96.00

ANOVA

Result

			Sum of Squares	df	Mean Square	F	Sig.
Between Groups	(Combined)		118.584	3	39.528	16.441	.000
	Linear Term	Contrast	106.014	1	106.014	44.095	.000
		Deviation	12.569	2	6.285	2.614	.087
Within Groups			86.552	36	2.404		
Total			205.136	39			

Fig. 2 One way ANOVA statistical results

Multiple Comparisons

Dependent Variable:Result

	(I) Algo	(J) Algo	Mean Difference (I-J)	Std. Error	Sig.	95% Confidence Interval	
						Lower Bound	Upper Bound
Tukey HSD	PSNN	GA-PSNN	-2.77610*	.69343	.002	-4.6437	-.9085
		PSO-PSNN	-3.88980*	.69343	.000	-5.7574	-2.0222
		IPSO-PSNN	-4.48250*	.69343	.000	-6.3501	-2.6149
	GA-PSNN	PSNN	2.77610*	.69343	.002	.9085	4.6437
		PSO-PSNN	-1.11370	.69343	.388	-2.9813	.7539
		IPSO-PSNN	-1.70640	.69343	.084	-3.5740	.1612
	PSO-PSNN	PSNN	3.88980*	.69343	.000	2.0222	5.7574
		GA-PSNN	1.11370	.69343	.388	-.7539	2.9813
		IPSO-PSNN	-.59270	.69343	.828	-2.4603	1.2749
	IPSO-PSNN	PSNN	4.48250*	.69343	.000	2.6149	6.3501
		GA-PSNN	1.70640	.69343	.084	-.1612	3.5740
		PSO-PSNN	.59270	.69343	.828	-1.2749	2.4603
Dunnett t (2-sided)	PSNN	IPSO-PSNN	-4.48250*	.69343	.000	-6.1829	-2.7821
	GA-PSNN	IPSO-PSNN	-1.70640*	.69343	.049	-3.4068	-.0060
	PSO-PSNN	IPSO-PSNN	-.59270	.69343	.728	-2.2931	1.1077

*. The mean difference is significant at the 0.05 level.

a. Dunnett t-tests treat one group as a control, and compare all other groups against it.

Fig. 3 Tukey and Dunnett test

Result

	Algo	N	Subset for alpha = 0.05		
			1	2	3
Tukey HSD[a]	PSNN	10	88.0549		
	GA-PSNN	10		90.8310	
	PSO-PSNN	10		91.9447	
	IPSO-PSNN	10		92.5374	
	Sig.		1.000	.084	
Duncan[a]	PSNN	10	88.0549		
	GA-PSNN	10		90.8310	
	PSO-PSNN	10		91.9447	91.9447
	IPSO-PSNN	10			92.5374
	Sig.		1.000	.117	.398

Means for groups in homogeneous subsets are displayed.

a. Uses Harmonic Mean Sample Size = 10.000.

Fig. 4 Homogeneous test results

7 Conclusion and Future Work

Particle swarm optimization has been a frequent area of interest among all types of researchers due to its simple steps in execution of the algorithm. But, in case of some large data sets, PSO suffers from premature convergence because of rapid trailing of diversity. So in this paper, an attempt has been focused on the performance of a higher order neural network with optimization of its weights through an improved particle swarm optimization algorithm. The result of the proposed method has been compared with the results of some other evolutionary techniques like PSO, GA and also with only Pi-Sigma network. Simulation results reveal that the proposed IPSO-PSNN obtains better classification accuracy as compared to others. In case of all the considered five benchmark data sets, it performs better in terms of classification accuracy. Future work may comprise obtaining better accuracy. Also, a deep watch is required to reduce the time and computational complexity of the algorithm, for which the network may able to produce better results in an effective manner.

References

1. Kennedy, J., Eberhart, R.: Particle swarm optimization. Proc. 1995 IEEE Int. Conf. Neural Netw. **4**:1942–1948 (1995)
2. Dehuri, S., et al. An improved swarm optimized functional link artificial neural network (ISO-FLANN) for classification. J. Syst. Softw. **85**(6), 1333–1345, (2012)
3. Wang, Yangling, Cao, Jinde: Exponential stability of stochastic higher-order BAM neural networks with reaction–diffusion terms and mixed time-varying delays. Neurocomputing **119**, 192–200 (2013)
4. Sermpinis, Georgios, Laws, Jason, Dunis, Christian L.: Modelling and trading the realised volatility of the FTSE100 futures with higher order neural networks. Eur. J. Finance **19**(3), 165–179 (2013)
5. Nayak, J., Naik, B., Behera, H.S.: A novel nature inspired firefly algorithm with higher order neural network: performance analysis. In: Engineering Science and Technology, an International Journal (2015)
6. Mehrabi, M., Sharifpur, M., Meyer, J.P.: Application of the FCM-based neuro-fuzzy inference system and genetic algorithm-polynomial neural network approaches to modelling the thermal conductivity of alumina–water nanofluids. Int. Commun. Heat Mass Transf. **39**(7), 971–977 (2012)
7. Jiang, B., Guo, H.: Permutation invariant polynomial neural network approach to fitting potential energy surfaces. III. Molecule-surface interactions. J. Chem. Phys. **141**(3), 034109 (2014)
8. Oh, S.-K., et al.: Polynomial-based radial basis function neural networks (P-RBF NNs) realized with the aid of particle swarm optimization. Fuzzy Sets and Systems **163**(1), 54–77 (2011)
9. Panda, B., et al.: Prediction of protein structural class by functional link artificial neural network using hybrid feature extraction method. Swarm, Evolutionary, and Memetic Computing, pp. 298–307. Springer (2013)
10. Naik, B., Nayak, J., Behera, H.S.: A novel FLANN with a hybrid PSO and GA based gradient descent learning for classification. In: Proceedings of the 3rd International Conference on Frontiers of Intelligent Computing: Theory and Applications (FICTA) 2014. Springer (2015)
11. Naik, B., Nayak, J., Behera, H.S.: A honey bee mating optimization based gradient descent learning–FLANN (HBMO-GDL-FLANN) for classification. In: Emerging ICT for Bridging the Future-Proceedings of the 49th Annual Convention of the Computer Society of India CSI, vol. 2. Springer (2015)

12. Naik, B., et al.: A harmony search based gradient descent learning-FLANN (HS-GDL-FLANN) for classification. In: Computational Intelligence in Data Mining, vol. 2, pp. 525–539. Springer, India (2015)
13. Nayak, J., et al.: Particle swarm optimization based higher order neural network for classification. Computational Intelligence in Data Mining, vol. 1. Springer, India, 401–414 (2015)
14. Nayak, J., Naik, B., Behera, H.S.: A hybrid PSO-GA based Pi sigma neural network (PSNN) with standard back propagation gradient descent learning for classification. In: International Conference on Control, Instrumentation, Communication and Computational Technologies (ICCICCT), IEEE, 2014
15. Nayak, J., Kanungo, D.P. Naik, B. Behera, H.S.: A higher order evolutionary Jordan Pi-sigma neural network with gradient descent learning for classification. In: International Conference on High Performance Computing and Applications (ICHPCA). IEEE, 2014
16. Dai, Y., Niu, H.:. An improved PSO algorithm and its application in seismic wavelet extraction. International Journal of Intelligent Systems and Applications (IJISA) 3.5 (2011): 34
17. Shin, Y., Ghosh, J.: The pi-sigma networks: an efficient higher order neural network for pattern classification and function approximation. Proc. Int. Joint Conf. Neural Netw. Seattle, Washington 1, 13–18 (1991)
18. Nayak, J., Naik, B., Behera, H.S.: A novel nature inspired firefly algorithm with higher order neural network: Performance analysis. Engineering Science and Technology, an International Journal (2015)
19. Nayak, J., Naik, B., Behera, H.S.: A novel chemical reaction optimization based higher order neural network (CRO-HONN) for nonlinear classification Ain Shams Engineering Journal (2015)
20. Rumelhart, D.E., Hinton, G.E., Williams, R.J.: Learning representations by back-propagating errors, Nature, 323(9), 533–536
21. Bache, K., Lichman, M.: UCI machine learning repository (http://archive.ics.uci.edu/ml), Irvine, CA: University of California, School of Information and Computer Science (2013)
22. Alcalá-Fdez, J., Fernandez, A., Luengo, J., Derrac, J., García, S., Sánchez, L., Herrera, F.: KEEL data-mining software tool: data set repository, integration of algorithms and experimental analysis framework. J. Multiple-Valued Logic Soft Comput. 17(2–3), 255–287 (2011)
23. Fisher, R.A.: Statistical Methods and Scientific Inference, 2nd edn. Hafner Publishing Co., New York (1959)

Author Index

© Springer India 2016
H.S. Behera and D.P. Mohapatra (eds.), *Computational Intelligence
in Data Mining—Volume 2*, Advances in Intelligent Systems
and Computing 411, DOI 10.1007/978-81-322-2731-1

Printed in the United States
By Bookmasters